TWELFTH EDITION

AN INTRODUCTION TO Physical Science

Laboratory Guide

James T. Shipman
Ohio University

Clyde D. Baker
Ohio University

Preparation and Editing at Ohio University
Karen M. Baker
Jennifer L. Gwilym

HOUGHTON MIFFLIN COMPANY

Boston New York

Vice President and Publisher: Charles Hartford
Senior Development Editor: Rita Lombard
Senior Marketing Manager: Nicole Moore
Supplements Editor: Kathryn White
Editorial Associate: Henry Cheek
Editorial Assistant: Jill Clark
Marketing Assistant: Kris Bishop
New Title Project Manager: Susan Peltier

Please note that all website addresses included in this manual were accurate at the time of publication. Houghton Mifflin is not responsible for content changes of any websites referenced by an external link.

Printed in the U.S.A.

ISBN 13: 978-0-618-93579-6
ISBN 10: 0-618-93579-7

123456789-POO-11 10 09 08 07

CONTENTS

Preface

This laboratory manual is designed to be used at the college, university, or high school honor's level in conjunction with any good physical science textbook. The experiments are written so that not only can the entire experiment be completed in a two-hour laboratory period, but many parts of these experiments can also become the basis for demonstrations of the principles of physical science in the classroom. We, of course, recommend the use of *An Introduction to Physical Science*, by Shipman, Wilson, and Todd, twelfth edition, as the classroom textbook. These two resources have been prepared to cover the many facets of physical science in an interesting and comprehensive way and along with the *online instructor's resource guides* for these two publications, provide many ideas and information that will be useful in preparing lesson plans for many levels of study from elementary school through junior and even senior high school.

The experiments covered in this manual can be done in any order and there are more than required by a normal physical science lab sequence so many will not be done at all. The ones that are not used as laboratory experiments can, therefore, be used as demonstrations or enhancement material for the lecture or recitation portions of the course. We have attempted to choose not only interesting ways of teaching the basic principles of physical science, but have tried to use simple, inexpensive equipment that can be easily obtained for use in the laboratories themselves or for demonstrations designed to illustrate particular principles and ideas about the world around us.

Laboratory work, with its practical applications, helps familiarize students with nature's physical laws. Students often have difficulty visualizing the physical concepts discussed in lectures, and practice in the laboratory with apparatus and raw materials often clarifies the picture. The experience of solving physical problems by applying knowledge gained in the classroom to laboratory activity is highly rewarding. However, the benefits derived from laboratory work are in direct proportion to students' interest, curiosity, and keenness of observation. The laboratory simply provides the means for carrying out such experiments.

Students should approach the laboratory experiments with a desire for understanding and a willingness to learn. They will profit from critically evaluating the results of each experiment with an open mind and a commitment to discovering the truth.

The *Laboratory Guide* contains fifty-five experiments in the five major divisions of physical science: physics, chemistry, astronomy, geology, and meteorology. Each experiment includes an introduction, learning objectives, a list of apparatus, procedures for taking data, and questions. In addition, many experiments call for calculations and the plotting of graphs, and this guide provides space and graph paper for those purposes.

The *Laboratory Instructor's Resource Manual* has been revised and is available on the instructor's web site via the following link: college.hmco.com/PIC/shipman12e. This resource contains useful information and sample data for many of the experiments in this manual, and has worked out calculations and even typical answers for the exercises and questions. This material has been prepared to help both experienced and inexperienced laboratory instructors, and will be especially useful to laboratory assistants assigned to do the grading for these experiments when they are used in a formal laboratory setting, but anyone needing to prepare lecture or demonstration material for physical science classes at any level can benefit from this information.

In this edition we have attempted to allow some freedom with the use of units. A few experiments still tend to lend themselves to the use of centimeter-gram-second (cgs) units, but we are recommending the adaptation of System International (SI) units in most of the experiments presented in this manual. It cannot be overstressed that the use of units in all calculations and final answers is as necessary as the numbers themselves. We feel very strongly that it is always important for the student to be aware of the necessity of choosing and recording units for all measurements and in using all measuring equipment to its maximum accuracy, that is making all readings to one estimated place past the least count of the measuring device being used.

Student backgrounds often vary considerably, especially in their mathematical ability. We have tried to keep the math requirements for these labs as simple as possible, but science cannot be completely understood without the knowledge of how to take data and how to analyze it. Graphing is an important mathematical tool for visualizing and understanding experimental results. Experiment 1 on graphing can be done first or it can be assigned later in the course. It can, of course, be omitted entirely but we feel that this is an important aid in understanding the work done in several experiments, and this experiment should be reviewed by the student even if it is not given as a regular lab assignment.

ACKNOWLEDGMENTS

We wish to thank our colleagues and students for the many contributions made to this edition of the *Laboratory Guide*. A special thanks to Karen Baker and Jennifer Gwilym at Ohio University, to the staff of M&N Toscano, and to the staff at Houghton Mifflin Company.

James T. Shipman
Clyde D. Baker

Laboratory Safety

Safety in the laboratory is the responsibility of BOTH the instructor and EVERY STUDENT. The instructor should emphasize safety at all times, and the students should follow all instructions, both verbal and printed.

No one wants a pleasant laboratory learning experience to be marred by an accident or avoidable injury. Here are a few critical rules to follow that we hope will minimize the possibility of such an unpleasant occurrence.

▼ Do not play with or attempt to use the laboratory equipment until AFTER it has been presented and fully explained by the laboratory instructor.

▼ Read all safety information given in the laboratory manual and follow it.

▼ Listen carefully to the information provided by the instructor, especially about safety precautions that need to be followed.

▼ Remember that even the most basic equipment can be dangerous if you are not paying attention to what you and your lab partners are doing at all times. Do not use the equipment for any purpose other than that outlined in the experiment.

▼ If you have a problem, no matter how small, ask your instructor for help BEFORE it escalates into a dangerous situation.

We have instituted a system of providing safety information and tips on the laboratory experiment in this *Laboratory Guide.* A distinctive safety tip logo has been inserted at the beginning of most experiments to remind everyone of any special precautions that should be taken while performing that experiment. Instructors and students should pay close attention to this aspect of each experiment to avoid potential problems or injuries during the laboratory period.

A safety reminder like this one will be found at the beginning of each relevant experiment to remind students of any special precautions necessary when performing that experiment.

The experiments in this *Laboratory Guide* have been chosen in such a way that there are no serious dangers inherent in any of the procedures, but there are certain precautions that should be taken to ensure that even minor mishaps do not occur. Please follow the safety guidelines presented in this *Laboratory Guide* and also try to exercise good common sense in handling all laboratory equipment. In this way students will have an opportunity to learn basic physical science principles in a safe manner.

The laboratory should have adequate first aid equipment and personnel with the knowledge to use this equipment in case of injury. Fire extinguishers should be in plain view and in good working order, proper ventilation maintained, safe methods readily available for the disposal of waste chemicals and other materials, and all electrical equipment examined for shock hazards.

The experiments in this *Laboratory Guide* are designed to be done safely, but students should exercise caution at all times. Most accidents in the laboratory result from carelessness and failure to take proper precautions in carrying out experiments. *Do not perform any unauthorized experiments.*

Be careful in handling glass and sharp tools. Also be careful to avoid chemical spills and splashing hot liquids that can cause serious eye and skin injuries and damage clothing. Spills on the floor of the laboratory may produce a slippery surface, setting the stage for falls. When handling electrical equipment, be extra careful to avoid electric shock.

Before touching any laboratory experiment, wait until the instructor has explained safety precautions and the basic operating processes for the equipment and has given permission to start the experiment.

When performing experiments with chemicals, never use more of a chemical than is called for in the experiment. Never pour a liquid from a bottle into a test tube; instead, use a beaker or other convenient pouring container as a transfer device. Solid chemicals can be handled more easily by using a piece of glazed paper to transfer the solid into the test tube. As a precaution against contamination, never return unused chemicals to the supply bottle or container. Dispose of any excess chemicals carefully as directed by your laboratory instructor.

All electrical experiments using a source of electrical energy must be checked by the instructor before they are connected to the source of energy. This is both for the safety of the student and to protect the delicate electrical measuring instruments used in these experiments. Even though low-voltage batteries are recommended in the experiments on electric current that are performed in this *Laboratory Guide*, it is good procedure to handle all electrical equipment carefully and to stress the point that higher voltage electrical equipment, especially 100 V ac appliances or tools, must be treated with extreme care.

If you have any doubts about the safety precautions to be taken when performing an experiment, ask the instructor for assistance, and follow his or her instructions carefully. Physical science experiments can be fun to do, but not if commonsense safety procedures are ignored.

Experiment 1

Graphs

INTRODUCTION

A *graph* is a pictorial representation of ordered pairs of numbers from which the reader may quickly determine relationships between these quantities. In many of the experiments performed in this *Laboratory Guide,* quantities that change in value will be studied. A change in one quantity may cause another quantity to change. We say one of these quantities is a function of the other. For example, the area of a circle is a function of the radius. That is, the area depends on the length of the radius. When we increase the length of the radius, the area of the circle increases accordingly. In this example the radius is called the independent variable, and the area is called the dependent variable.

After the ordered pairs of numbers [for example, (x_1, y_1) or (r_1, A_1)] are plotted on graph paper, the plotted points are connected by a *smooth line*. The line that is drawn may be straight or it may be curved. The general name *curve* is used in reference to all graphs whether the line is actually curved or straight.

LEARNING OBJECTIVES

After completing this experiment, you should be able to do the following:

- Define and explain the term *graph.*
- Plot a graph of a linear relationship using a mathematical equation or formula to obtain ordered pairs of numbers.
- Plot a graph of a nonlinear relationship using a mathematical equation or formula to obtain ordered pairs of numbers.
- Distinguish between dependent and independent variables.
- Define *slope,* determine the slope of a straight line, and give a physical interpretation of the slope.
- State the equation for a straight line that passes through the origin in terms of the plotted variables.
- State the equation for a parabola.
- Replot a parabola as a straight line.

APPARATUS

Several sheets of linear graph paper, pen or sharp pencil, ruler. (A hand calculator is useful but not necessary.) Graphing can also be done using a graphing calculator or a computer. For one example of this automated graphing technique see the information below and Experiment 45.

GRAPHING TECHNIQUES

In recent years it has become possible to prepare graphs of ordered pairs of numbers as measured in physical science experiments by using graphing calculators and/or computer software. There are some definite advantages to using these methods. These include rapid data acquisition, high-speed data processing, and automatic plotting where the axis scales and actual drawing of the graphs is totally automated. This can greatly increase the amount of experimentation that can be done and reduce the time of data analysis, but in studying the use of such programs by beginning physics and physical science students we have found that these graphing methods also have some disadvantages.

Most students tend to accept the final results of such data acquisition and processing as true without questioning the accuracy or understanding the actual physical principles involved in the experiments under study. We have found that nearly every student attains a higher degree of understanding if they process the data by hand and plot at least one graph in the old manual fashion. This does not preclude the usefulness of these graphing methods in the study of physical science as demonstrations of actual physical processes or the easy and rapid processing of additional runs of data after the initial first hand graphing is completed. In other words, students can gain a great deal using these automated graphing methods after they have learned to plot and actually have constructed at least one basic hand-drawn graph of the physical phenomenon under study. For this reason we recommend that even if calculator and computer graphing capability is available, that the student do at least one graph of the data by hand and then proceed to use the automated graphing techniques for additional data processing.

Since there are so many different systems available for automated graphing, we will defer to the manufacturer's instructions for the use of these specific products. Any one of them can make the student's laboratory experience more fun and productive after they have learned the basic graphing techniques, and we believe that this experiment is an excellent way for them to get these basic skills. Any combination of manual and automatic graphing can be used following the procedures laid out in this experiment, from total manual processing to fully automatic graphing. We hope that whatever combination you and your instructor choose, that it will give you some genuine insight into the power and presentation capabilities of graphs in analyzing physical data such as that outlined in the procedures for this experiment.

EXPERIMENT 1 NAME (print) ______________________________ DATE ______________
LAST FIRST

LABORATORY SECTION ______ PARTNER(S) ______________________________

PROCEDURE 1

A graph is a picture of ordered pairs of numbers. Thus, to draw a graph, we must use graph paper and a set of ordered pairs of numbers. Graph paper is provided in the back of this *Laboratory Guide.* You must calculate or experimentally determine the ordered pairs of numbers. In this experiment, calculate the ordered pairs using the mathematical equation or formula for the circumference of a circle as a function of the diameter of the circle (Eq. 1.1); then record the values in Data Table 1.1. Values for the circumference should be rounded off to two decimal places. The circumference of a circle equals the value pi (π) times the diameter of the circle. This can be written in symbol notation as

$$C = \pi d \qquad \text{Eq. (1.1)}$$

where C = the circumference
d = the diameter
$\pi = 3.14$

DATA TABLE 1.1 (Record units with all data values.)

Diameter (*d*) Independent variable	Circumference (*C*) Dependent variable	Ratio *C*/*d*
1.00 m		
2.00 m		
3.00 m		
4.00 m		
5.00 m		
6.00 m		
7.00 m		
	Average value of *C*/*d*	

Plot a graph (Graph number 1) of the circumference of a circle as a function of the diameter using the information in Data Table 1.1. The following guidelines should be used when plotting graphs:

1. Make the proper choice of graph paper.
2. Use a pen or a sharp pencil.
3. Write neatly and legibly.
4. Choose scales so that the major portion of the graph paper is used.
5. Choose scales for the *x* and *y* axes that are easy to read and plot.

6. Plot the independent variable on the horizontal or x axis and the dependent variable on the vertical or y axis.
The two quantities (circumference and diameter) plotted for this experiment are called *variables.* For the data given, the diameter of the circle is the independent variable and the circumference the dependent variable.
7. Plot each ordered pair of numbers as a single point (d_1, C_1), (d_2, C_2), etc.
8. Plot each point clearly using a dot surrounded by a small circle ⦿ .
9. Label the x and y axes with the quantity plotted. Examples: distance, time, area, volume, etc.
10. State the units of each quantity plotted. Examples: centimeters, seconds, cm^2, cm^3, etc.
11. Draw a smooth line connecting the plotted points. *Smooth* suggests that the line does not have to pass exactly through each point but connects the general areas of significance.
12. Give the graph a title, and place the title on the graph (usually upper center of graph). The name of the graph is taken from the labels of the x and y axes plus reference to the object to which the data refer. Examples: volume versus radius for a sphere; period versus length for a simple pendulum.
13. Place your name and the date on the graph (usually lower right side of graph).
14. Determine the slope of the curve. The *slope* of a curve (for this graph the curve is a straight line) is defined as the change in the y or vertical values divided by the change in the x or horizontal values. This can be stated in symbol notation as $\Delta y/\Delta x$. Read this as "delta y over delta x" (delta means "change in"). Using, s as the symbol for the slope we can write

$$\text{Slope} = s = \frac{\Delta y}{\Delta x} = \frac{y_2 - y_1}{x_2 - x_1} \quad \text{(General equation)}$$

In the special case that if the straight line passes through the origin (ordered pair, 0,0), then we can write

$$\text{Slope} = s = \frac{y_2 - y_1}{x_2 - x_1} = \frac{y_2 - 0}{x_2 - 0} = \frac{y_2}{x_2} \quad \text{(Special case)}$$

$$s = \frac{y}{x}$$

$$y = sx$$

This is the general equation for a straight line that passes through the origin. Remember, the slope (s) of a straight line has a constant value. Therefore, the values (x_1, y_1) and (x_2, y_2) can be any ordered pair of points on the plotted graph.

The point where the curve crosses the y axis is known as the *y intercept.* In the equation above, when x is zero, y is zero. Thus the curve crosses the y axis at the origin. Later we shall have curves that do not pass through the origin, and the y intercept will have a definite numerical value. In the case where the curve does not go through the origin the equation for a straight line becomes:

$$y = sx + y_0$$

Here y_0 is the y intercept of this value can be either plus, minus or in some cases zero.

See Fig. 1.1 for an example of a graph plotted according to the above instructions.

EXPERIMENT 1 NAME (print) ______________________________ DATE __________
LAST FIRST

LABORATORY SECTION ______ PARTNER(S) ______________________________

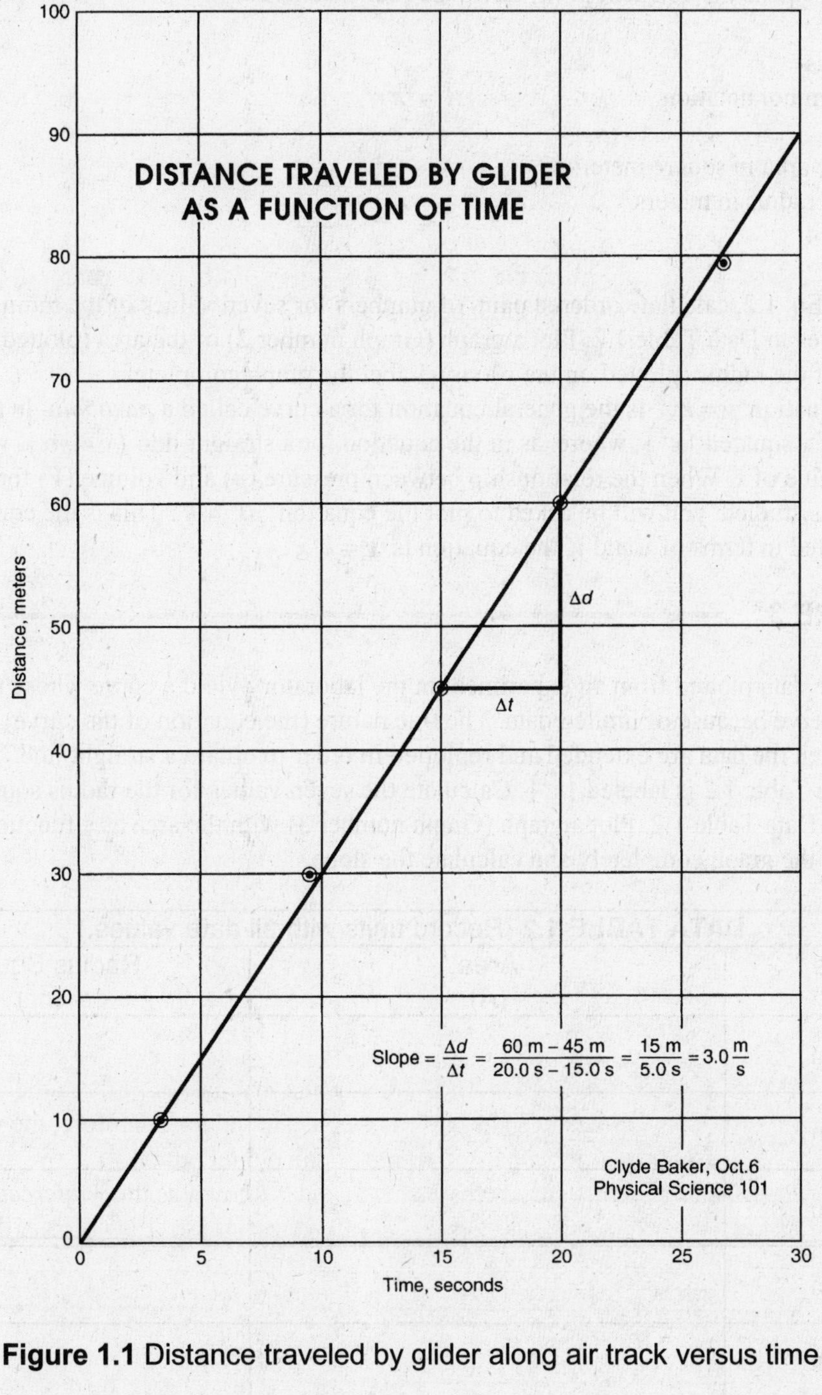

Figure 1.1 Distance traveled by glider along air track versus time

Note: Sample graph (your variables, units, and data values may differ). In this case the *y* intercept is again zero.

PROCEDURE 2

The relationship between the area of a circle and the radius of the circle can be stated as follows: The area of a circle equals the constant pi (π) times the radius squared,

$$\text{Area} = \pi(\text{radius})^2$$

or written in symbol notation,

$$A = \pi r^2 \qquad \text{Eq. (1.2)}$$

where A = the area in square meters,
r = the radius in meters
$\pi = 3.14$

Using Eq. 1.2, calculate ordered pairs of numbers for seven values of the radius. Record the calculated values in Data Table 1.2. Plot a graph (Graph number 2) of the area (plotted on the y axis) as a function of the radius (plotted on the x axis). Label the graph completely.

The equation $y = kx^2$ is the general equation for a curve called a *parabola.* In this equation y is a function of x squared (x^2), where, as in the equation for a straight line $(y = mx)$, y is a function of the linear value of x. When the relationship between pressure (p) and volume (V) for an ideal gas (Boyle's law) is studied, you will be asked to plot the equation $pV = k$. This is the equation for a *hyperbola.* Stated in terms of x and y, the equation is $y = k/x$.

PROCEDURE 3

Many times the data plotted from an experiment in the laboratory yield a curve whose true nature is difficult to observe because of limited data. The true nature (the equation of the curve) can be determined when the data are extended and replotted in order to obtain a straight line. The third column in Data Table 1.2 is labeled (r^2). Calculate the seven values for the radius squared and record them in Data Table 1.2. Plot a graph (Graph number 3) with the area as a function of the radius squared. Label the graph completely and calculate the slope.

DATA TABLE 1.2 (Record units with all data values.)

Radius (r)	Area (A)	Radius Squared (r^2)
0.200 m		
0.400 m		
0.600 m		
0.800 m		
1.000 m		
1.200 m		
1.400 m		

EXPERIMENT 1 NAME (print) ______________________________ DATE ______________
LAST FIRST

LABORATORY SECTION ______ PARTNER(S) ______________________________

QUESTIONS

1. Explain how you determined the magnitude of the divisions for both the x and y axes in order to plot a graph that utilizes the major portion of the graph paper.

2. Distinguish between dependent and independent variables.

3. Name the dependent and independent variables in Procedure 2.

4. What must be included in the title used on a graph?

5. How are the units for the slope of a graph determined?

6. Examine your graph for Procedure 1 and determine if the relationship between the circumference and the diameter of a circle is a linear relationship? Justify your answer.

7. When the value for *x* increases faster than the values for *y*, which way (toward the *y* axis or away from the *y* axis) will the graph curve?

8. State in your own words the relationship between the area of a circle and its radius squared [r^2] for a series of circles that have increasingly larger radii. Use the graph plotted in Procedure 3 to help you formulate this answer.

Experiment 2

Measurement

INTRODUCTION

All experiments to be performed in this laboratory require one or more measurements. A *measurement* is defined as the ratio of the magnitude (how much) of any quantity to a standard value. The magnitude is first determined by using some device, such as a ruler, to compare the unknown quantity to a standard and then expressing the comparison as a number. The number representing the magnitude is not sufficient to express the measurement, however; a unit also must be assigned to the number. To record a person's height as 162 is meaningless. We must record it as 162 cm or 1.62 m (this is about 5 ft 4 inches tall). All measurements, therefore, must be recorded by using both numbers and units.

Many devices (rulers, balances, clocks, speedometers, thermometers, voltmeters, oscilloscopes, spectrometers, etc.) are used to make measurements. The information obtained from these instruments must be recorded and evaluated in order to obtain a true value of the properties of a physical quantity. Such terms as *least count, significant digits, precision, accuracy,* and *percent error* must be learned and applied to experimental data to get the most useful information from it.

Laboratory experiments require the taking and recording of data. Values read from the measuring instrument must be expressed with numbers known as significant figures. By definition, a *significant figure* is a number that contains all known digits plus one doubtful digit. That is, the significant figure includes all digits as read from the instrument plus one estimated digit. This estimated digit represents some fractional portion of the smallest scale division on the instrument. For each measurement in all experiments, the student must record as many digits as possible when measuring the variables under study, in other words you must record as many *significant digits* for each measurement as possible.

Once the measurements have been made and recorded, it may be necessary to perform arithmetical operations using the significant figures. In the multiplying or dividing of two or more measurements, the number of significant digits in the final answer can be no greater than the number of significant digits in the measurement with the least number of significant digits. When adding or subtracting, the last digit retained in the sum or difference should correspond to the first doubtful decimal place.*

Example:

$$\begin{array}{r l} 16.3 & \text{cm} \\ 203.4 & \text{cm} \\ \underline{+\,4.07} & \underline{\text{cm}} \\ 223.8 & \text{cm} \end{array}$$

* See Appendix I for more information.

Since the tenths position is present in all 3 numbers but the hundredths position is not, we must round back as shown.

The following procedure is usually used to round off significant figures to fewer digits. If the last significant digit on the right is less than 5, drop it and leave the preceding digit unchanged. It is often necessary to insert one or more zeros to maintain the proper decimal location in the reading. If the last significant digit is 5 or greater, drop it and increase the preceding digit by 1.

EXAMPLES

Round off the following numbers to two significant digits.

247 Since the last digit on the right is greater than 5, drop it and increase the preceding digit by 1, for the result 250.

243 Since the last digit on the right is less than 5, drop it and insert zero instead, for the result 240.

275 Since the last digit on the right is 5, drop it and increase the preceding digit by 1, for the result 280.

A number representing a measured quantity also may be expressed by means of the powers-of-10 notation. To express a number in this way, place the decimal point after the first significant digit, then use a power of 10 to locate the true position of the decimal point. For example, the average distance from the Sun to Earth is 93,000,000 mi. Using powers-of-10 notation, this can be written as 9.3×10^7 mi. Likewise, a very small number can be expressed using the powers-of-10 notation. For example, if the thickness of a piece of paper is 0.00011 m, this can be written as 1.1×10^{-4} m.

Just recording a numerical value for the measured quantity is not sufficient to express a physical quantity. A unit also must be indicated. For example, a measurement is taken for the length of the laboratory table and recorded as 183. This number has no real meaning unless expressed with a unit. A correct recording might be 183 centimeters (cm).

The process of taking any measurement always involves some uncertainty. Such uncertainty is usually called *experimental error.* Two methods may be used to calculate the amount of error:

1. When an accepted or standard value of the physical quantity is known, the *percentage error* is calculated in a comparison of the experimental measurement with a standard:

$$\text{Percentage error} = \frac{\text{absolute difference}}{\text{accepted value}} \times 100$$

$$= \frac{|E_v - A_v|}{A_v} \times 100$$

where E_v is the experimental value and A_v is the accepted value (also known as the *standard value*). Absolute difference means that the smaller value of the two is always subtracted from the larger value.

2. When no accepted value exists, a *percentage difference* is calculated to compare two or more experimental measurements. Begin by taking the average of all your experimental values:

$$\text{Average} = \frac{E_1 + E_2 + E_3 + \ldots E_n}{n}$$

where E_1, E_2, and E_3 represent the experimental values measured in the experiment, and n represents the number of experimental values being averaged. The percentage difference can then be calculated:

$$\text{Percentage difference} = \frac{(E_L - E_s)}{\text{Average}} \times 100$$

where E_L = largest experimental value and E_s = smallest experimental value.

The metric system is used throughout this *Laboratory Guide.* Some of the more common units of length and mass are listed in the accompanying tables.

Length Units Based on the Meter (m)

Unit	Abbreviation	Expression in Meters
millimeter	mm	0.001
centimeter	cm (cgs unit)	0.01
decimeter	dm	0.1
meter	m (SI standard)	1.0
dekameter	dam	10.0
hectometer	hm	100.0
kilometer	km	1000.0

Mass Units Based on the Gram (g)

Unit	Abbreviation	Expression in Grams
milligram	mg	0.001
centigram	cg	0.01
decigram	dg	0.1
gram	g (cgs unit)	1.0
dekagram	dag	10.0
hectogram	hg	100.0
kilogram	kg (SI standard)	1000.0

Note: In SI units, length is measured in meters, mass in kilograms, and time in seconds. In the cgs system, length is measured in centimeters, mass in grams, and time in seconds.

LEARNING OBJECTIVES

After completing this experiment, you should be able to do the following:

- ▼ Define the terms *accuracy, measurement, least count, precision,* and *significant digits.*
- ▼ Make measurements using a meter stick, beam balance, and stopwatch or electric timer.
- ▼ Calculate the density of a substance when the mass and volume are known or can be determined.
- ▼ Determine experimentally and theoretically the period of a simple pendulum.
- ▼ Differentiate between percent error and percent difference.

APPARATUS

Meter stick, balance, stopwatch or electric timer, wooden block, simple pendulum.

EXPERIMENT 2 NAME (print) ______________________ LAST ______________ FIRST ______________ DATE ______________

LABORATORY SECTION ______ PARTNER(S) ______________________

PROCEDURE

1. (a) The *least count* is the smallest subdivision marked on a measuring instrument; that is, it is the smallest reading that can be made with the instrument without estimating. Determine the least count of the three measuring instruments listed in Data Table 2.1. Record the numerical value of the least count and the unit of measurement. Example: Take a good look at the meter stick you will be using in this experiment. Most meter sticks are divided and numbered into 100 equal divisions. Each of these numbered divisions is called 1 cm. One centimeter means 1/100 m. Each centimeter is further divided with markings into 10 equal divisions. This is the smallest subdivision on most meter sticks. The least count can be expressed as 1/1000 (numerical value) of a meter (unit of measurement) or as 1 (numerical value) millimeter (unit of measurement). However the least count is expressed, it must also contain both the correct units of measurement and the numerical value.

DATA TABLE 2.1 Least Count of Three Measuring Instruments
(Record units with all data values.)

Measuring Instrument	Numerical Value	Unit of Measurement
Meter stick		
Balance		
Stopwatch or electric timer		

(b) Measure the length (*L*) and width (*W*) of the top of your laboratory table and determine the surface area (*A*) in square centimeters.

$A = L \times W =$ __________ cm × __________ cm = ____________ cm^2

Convert the number of square centimeters to square meters. Show your work. See Appendix II for the conversion procedure and the back inside cover for conversion factors.

Surface area (A) = ____________ $meters^2$

2. Determine the volume of the wooden block in cubic centimeters. Take three measurements of each dimension of the wooden block, and record all significant digits in Data Table 2.2. *Significant digits* are digits read from the measuring instrument *plus one digit estimated by the observer*. The estimate will be a fractional part of the least count of the instrument.

The more precise the measurement, the more exact the information obtained concerning the physical properties of the object that is being measured. *Precision* indicates the maximum amount a measurement varies from other measurements taken of the same quantity. A measurement to 0.01 cm is more precise than a measurement to 0.1 cm. The maximum error may be expressed as a plus or minus value. For example, the length of a wood block may be 15.4 cm ± 0.1 cm. The maximum error given as ± 0.1 cm means we know that the length must be between 15.3 cm and 15.5 cm.

When using the meter stick, place it on the object to be measured as shown in Fig. 2.1. Never use the end of the meter stick if it can be avoided because it may have been damaged.

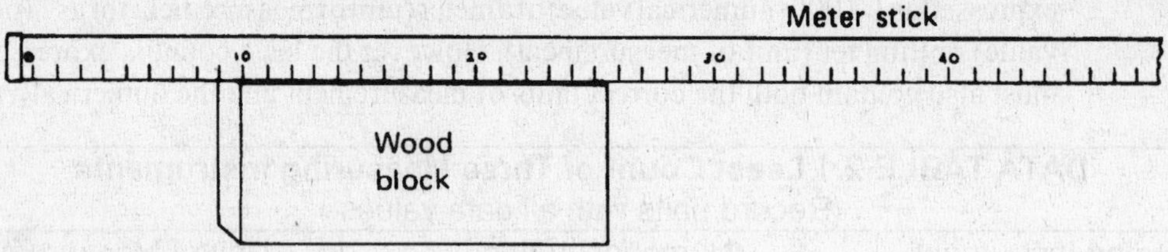

Figure 2.1

DATA TABLE 2.2 (Record units with all measurements.)			
Trial	Length	Width	Height
1			
2			
3			
Average			

Calculate the volume using the average values from Data Table 2.2.

Volume = length × width × height = ______________________ *

3. Determine the mass of the wood block in grams. This will require the use of a balance, like the triple beam balance shown in Fig. 2.2 or an accurate laboratory grade electronic balance. Either of these must be leveled and properly adjusted before an accurate measurement can be made. Ask the instructor for help if needed.

Mass of wood block = ______________________________ *

* Don't forget to record the units as part of your answer.

EXPERIMENT 2 NAME (print) ______________________ DATE ______________
LAST FIRST

LABORATORY SECTION ______ PARTNER(S) ______________________

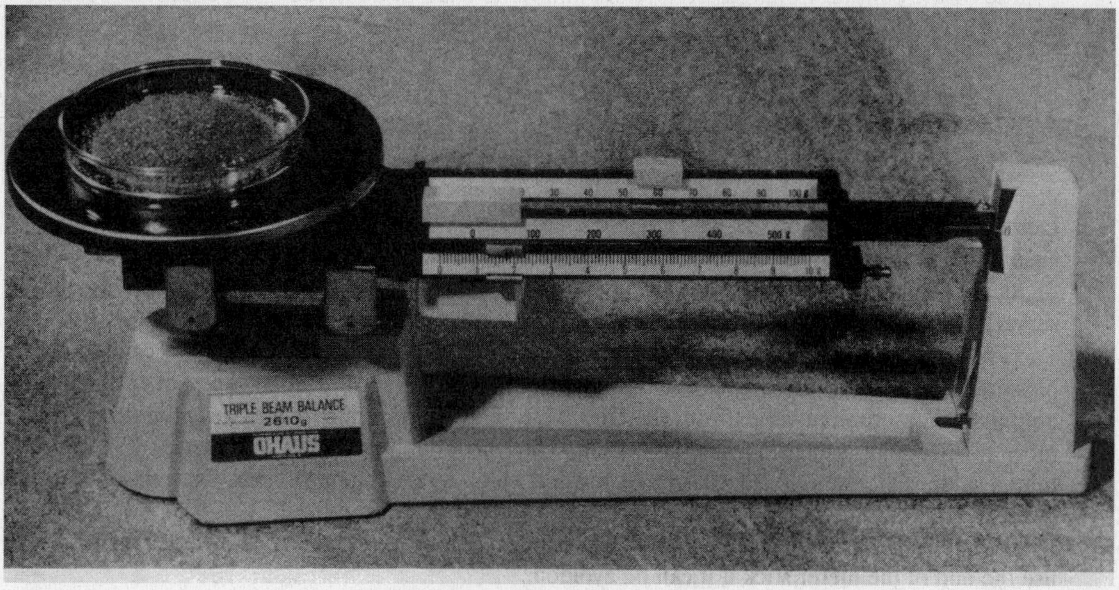

Figure 2.2 Beam balance

4. The density of a substance is defined as the ratio of the mass to the volume. Determine the density of the wood block and record below.

$$\text{Density} = \frac{\text{mass}}{\text{volume}} = \underline{\hspace{6cm}}$$

5. The *second* is defined as a fractional part of a mean solar day. Express one second as a fraction of a mean solar day based on your knowledge that there are 60 s in 1 min, 60 min in 1 h, and 24 h in one mean solar day. Show your work.

6. With the use of the stopwatch or electric timer, determine your pulse rate. Record the data in Data Table 2.3 and compute the average value.

DATA TABLE 2.3

Trial	Pulses per Minute
1	
2	
3	
Average	

7. A simple pendulum consists of a small heavy mass attached to a light string and suspended from a rigid support. The pendulum is set swinging by displacing the mass (called a *bob*) slightly from its equilibrium position. Use a length-of-arc-to-pendulum-length ratio (see Fig. 2.3) of no greater than 1:10. That is, displace the pendulum bob less than one-tenth the value of the pendulum length. The word *simple* is used to describe this pendulum because most of the mass is concentrated in the bob. (An *ideal simple pendulum* would be one in which the entire mass is concentrated at a point. This is impossible to obtain because a bob of any size will have a distribution of mass; also the lightest of strings has mass.) Although the system we are using is not actually an ideal simple pendulum, a very good approximation of the period can be obtained from the equation

$$T = 2\pi\sqrt{\frac{L}{g}},$$

where the period T is defined as the time of one complete swing of the bob. Squaring both sides of this equation, we obtain

$$T^2 = 4\pi^2 \frac{L}{g} \qquad \text{Eq. (2.1)}$$

where T = the period in seconds
L = the length of the pendulum measured from the point of support to the center of the bob
g = the acceleration due to gravity ($g = 9.80\ \text{m/s}^2$ in SI units or $980\ \text{cm/s}^2$ in cgs units)

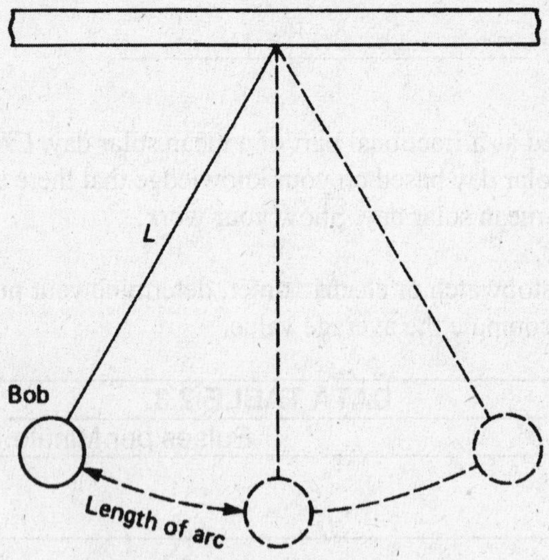

Figure 2.3 Simple pendulum

EXPERIMENT 2 NAME (print) ______________________ ______________________ DATE ______________________
LAST FIRST

LABORATORY SECTION ______ PARTNER(S) ____

(a) Determine, experimentally, the period of a simple pendulum with a length of 0.49 m. Construct your own data table in the space below and record the data taken. Perform five determinations and record the average experimental value of the period in the space provided below.

Period (experimental value) = _____________

(b) Calculate the theoretical value of the period of a simple pendulum when $L = 0.49$ m. Use Eq. 2.1.

Period (theoretical value) = _______________

How accurate is the preceding experimental measurement? In theory, the *accuracy* of a measurement refers to how well the experimental value agrees with the true or accepted value. However, the perfectly true value can never be obtained, so the experimental value is compared with an accepted value known as a *standard.* How well the experimental value agrees with the standard is usually expressed as a percentage.

$$\text{Percentage error} = \frac{\text{Absolute difference}^{*}}{\text{accepted value}} \times 100 = \frac{|E_v - A_v|}{A_v} \times 100$$

where E_v is the experimental value and A_v is the accepted value (also know as the standard value).

*Absolute difference means the smaller value is subtracted from the larger value in all cases so the percentage error can never come out to be negative.

EXPERIMENT 2 NAME (print) ________________________ DATE ____________
LAST FIRST

LABORATORY SECTION ______ PARTNER(S) ____________________________

(c) Calculate the percentage error using the theoretical value of the period as the standard value. Show your work.

Percentage error in period determination = ____________

(d) If no standard value is obtainable, then a percentage difference can be determined as follows:

$$\text{Percentage difference} = \frac{\text{largest experimental value} - \text{smallest experimental value}}{\text{Average of the experimental values}} \times 100$$

Calculate the percentage difference between your largest and smallest experimental value of time. Show your work.

Percentage difference in the time values measured = ____________

QUESTIONS

1. Define *least count*, and then determine the least count of your (or someone else's) wristwatch.

2. Define *measurement*, and give an example.

3. Differentiate between *accuracy* and *precision*.

4. Why should the experimenter avoid using the end of a meter stick when making a measurement?

5. Calculate the length in meters of a simple pendulum that has a period of 1.00 second?

6. Explain under what conditions you would use percentage error and when you would use percentage difference. Why would you sometimes use both?

7. How can accuracy of a measurement be increased?

8. How can precision of a measuring instrument be increased?

Experiment 3

The Simple Pendulum

INTRODUCTION

In the experiment on measurement, the simple pendulum was used to take data on the length and time to acquire knowledge about the concepts of percent error and percent difference. This experiment will be concerned with how the period of a simple pendulum varies with the length of the pendulum, the mass of the bob, and the magnitude of the bob's displacement from the equilibrium position.

Pendulums are classified as compound or simple. The *compound* pendulum is one in which the mass of the pendulum is distributed throughout its length. Examples are the pendulum in a grandfather clock and a meter stick held at one end and allowed to swing back and forth. A *simple* pendulum is one in which the mass of the pendulum is concentrated at a point. It is obvious that the mass of a simple pendulum cannot really be located at a point, since a point is defined as position without dimensions. However, when the length of the string or wire supporting the mass (called the *bob*) is large in comparison with the diameter of the bob, the pendulum can be treated as a simple pendulum. Figure 3.1 illustrates the parameters of such a simple pendulum.

The equation showing the relationship between the parameters of a simple pendulum is

$$T = 2\pi\sqrt{\frac{L}{g}} \qquad \text{Eq. (3.1)}$$

where T = the period
L = the length
g = the acceleration due to gravity $= 9.8\ \text{m/s}^2 = 980\ \text{cm/s}^2 = 32\ \text{ft/s}^2$

When both sides of the equation are squared, we obtain

$$T^2 = 4\pi^2 \frac{L}{g}$$

Note: The mass of the simple pendulum does not appear in this equation.

LEARNING OBJECTIVES

After completing this experiment, you should be able to do the following:

- ▼ State the difference between a simple and a compound pendulum.
- ▼ Name the parameters that determine the period of a simple pendulum.
- ▼ Determine experimentally the parameter values for a simple pendulum.
- ▼ State how the period of a simple pendulum varies with the (a) length of the pendulum, (b) mass of the bob, and (c) magnitude of the displacement.
- ▼ Calculate the length or period for a simple pendulum when one or the other is given.

APPARATUS

Two different masses for pendulum bobs (hooked weights can be used if regular bobs are not available), string, a 2-m stick (a 1-m stick may be used here), stopwatch or electric timer, balance, and support holder.

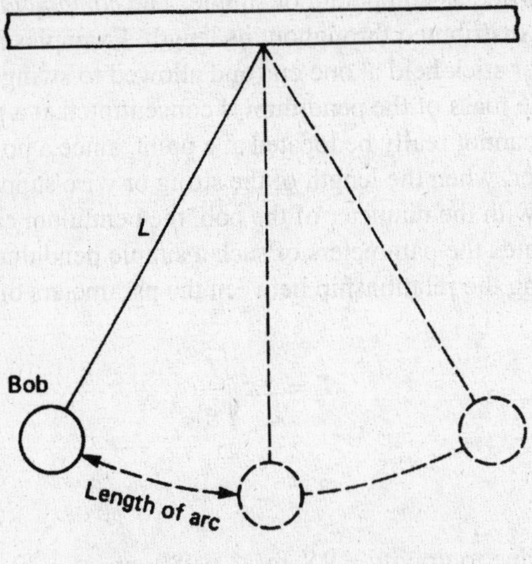

Figure 3.1 Simple pendulum

EXPERIMENT 3 NAME (print) ______________________________ DATE ______________
LAST FIRST

LABORATORY SECTION ______ PARTNER(S) ______________________________

PROCEDURE 1

This procedure is for determining the period of a simple pendulum for various lengths.

Step 1 Construct a simple pendulum as shown in Fig. 3.1. Use either a given mass (call this mass number one) and make the length of the pendulum for this first step fairly large (160 cm) or whatever is reasonably possible with the means of support you have for the pendulum. The length is the distance from the point of support to the center of the bob.

Displace the pendulum bob approximately 16 cm from its equilibrium position. If you use a length other than 160 cm, use a length-of-arc-to-pendulum-length ratio that does not exceed 1:10. That is, displace the pendulum bob no more than one-tenth the value of the pendulum length. Determine the time it takes for the pendulum to swing through 10 complete cycles. Before you start your timer, allow the bob to swing through a few cycles. **Caution:** Do not count the first cycle until one complete cycle has occurred. When the timer is turned on, say “start”; then when the bob comes back to this same position, say “one,” after the next complete swing say “two,” etc. up to “ten.” Stop the timer when you say “ten.” The best place to start is at one of the end points of the swing where the bob is stopped momentarily. Make three trials and record the data in Data Table 3.1. From this data calculate the period of the pendulum and record its value in the data table. The period is equal to the average time for the three trials divided by ten.

Step 2 Adjust the length of the pendulum to 80 cm or one-half the length used in Step 1. Complete three trials to determine the time of 10 cycles, and record the data in Data Table 3.1. From the data calculate the period of the pendulum and record this in the data table. Remember the length of arc is to be approximately one-tenth the length of the pendulum.

DATA TABLE 3.1 (for Procedure 1)

Mass of first pendulum bob: ______________ grams

	Time in Seconds for 10 Cycles					
Length of Pendulum (cm)	Time for Trial 1 (*s*)	Time for Trial 2 (*s*)	Time for Trial 3 (*s*)	Average Time for Three Trials (*s*)	Time of One Cycle Period (*s*)	Period^2 $(s)^2$
*160 cm or ________ cm						
* 80 cm or ________ cm						
* 40 cm or ________ cm						
* 20 cm or ________ cm						

* Cross out this value and insert correct value, if different.

Step 3 Adjust the length of the pendulum to 40 cm or one-fourth the length used in Step 1. Displace the pendulum bob 4 cm or one-tenth the value of the pendulum length you used. Complete three trials to determine the time of 10 cycles and record the data in Data Table 3.1. **Caution:** Allow the pendulum to swing through enough cycles so that a minimum of 10 seconds are required. If the time for 10 complete cycles is less than 10 seconds, then allow the pendulum to swing through 20 complete cycles for the time measurement. Now the period is equal to the average time divided by n, where n is whatever number of swings you counted for this length of a simple pendulum.

Step 4 Repeat Step 3 with a pendulum length of 20 cm or one-eighth of your original length.

Step 5 Construct a graph with the period of the pendulum in seconds (*y* axis) versus the length of the pendulum in centimeters (*x* axis). Refer to Experiment 1 for the correct procedure for constructing a graph.

Step 6 Square all calculated periods and record them in Data Table 3.1. Construct a graph with the period squared $(\text{seconds})^2$ on the *y* axis and the length (centimeters) of the pendulum on the *x* axis. Determine the slope of the curve. This plot should be a straight line. Show your work plus the numerical value of the slope with units on the graph. Refer to Experiment 1 for an example of such a graph.

PROCEDURE 2

This procedure is for determining the period of a simple pendulum for different masses. Repeat Procedure 1 using mass number two. Make sure the length of the pendulum is measured from the point of support to the center of the bob. Record the data in Data Table 3.2, and compare the periods with those in Data Table 3.1.

DATA TABLE 3.2 (for Procedure 2)

Mass of second pendulum bob: ______ grams

Length of Pendulum (cm)	Time in Seconds for 10 Cycles				Time of One Cycle Period (*s*)	Period^2 $(s)^2$
	Time for Trial 1 (*s*)	Time for Trial 2 (*s*)	Time for Trial 3 (*s*)	Average Time for Three Trials (*s*)		
* 160 cm or ______ cm						
* 80 cm or ______ cm						
* 40 cm or ______ cm						
* 20 cm or ______ cm						

* Cross out this value and insert correct value, if different.

EXPERIMENT 3 NAME (print) ______________________________ DATE ______________
LAST FIRST

LABORATORY SECTION ______ PARTNER(S) ______________________________

PROCEDURE 3

This procedure is for determining the period of a simple pendulum for various displacements of the pendulum bob.

Step 1 Construct a simple pendulum as shown in Fig. 3.1. Use mass number two and make the length of the pendulum 160 cm or the same length you used in Procedure 1, Step 1. Displace the pendulum bob 16 cm or one-tenth the pendulum length you used. Make three trials and record the data in Data Table 3.3. From the data calculate the period of the pendulum and record in the data table. *Note:* For this length of pendulum, the period should be the same as you recorded in Data Tables 3.1 and 3.2 if you have set the length up correctly in each case.

Step 2 Using the same length and mass as in Step 1, displace the pendulum bob 32 cm from its equilibrium position or one-fifth the pendulum length you used. Make three trials and record the data in Data Table 3.3. From the data calculate the period of the pendulum and record in the data table.

Step 3 Repeat Step 2 displacing the pendulum bob 48 cm from its equilibrium position or three-tenths the pendulum length you used. Record all data in Data Table 3.3.

Step 4 Repeat Step 2 displacing the pendulum bob 64 cm or four-tenths the pendulum length you used. Record all data in Data Table 3.3.

DATA TABLE 3.3 (for Procedure 3)

Length of pendulum: (L) ______________ cm; mass of bob ______________ grams

	Time in Seconds for 10 Cycles				
Initial Displaced Length of Arc (cm)	Time for Trial 1 (*s*)	Time for Trial 2 (*s*)	Time for Trial 3 (*s*)	Average Time for Three Trials (*s*)	Period (Time of One Cycle) (*s*)
*16 cm or ________ cm					
* 32 cm or ________ cm					
* 48 cm or ________ cm					
* 64 cm or ________ cm					

* Cross out this value and insert correct value, if different.

QUESTIONS

1. Why do you think you were asked to complete more cycles in Procedure 1, Step 3, if the total time was less than 10 seconds for 10 cycles?

2. What type of graph (straight line, parabola, hyperbola) was obtained in Procedure 1, Step 5?

3. Which variable (period or length) does the graph show to be increasing the fastest? (*Note:* If they were increasing at the same rate, the graph would be a straight line.)

4. What type of graph was obtained in Procedure 1, Step 6?

5. State how the period of the simple pendulum varied experimentally with the (a) length of the pendulum, (b) length of arc (displacement of the bob), and (c) mass of the bob.

6. A simple pendulum has a length L and a period T. If the length is increased to $4L$, how will be the new period be affected? Refer to data tables for the answer.

7. Using Equation 3.1, calculate the period of your pendulum for a length of 160 cm or for the maximum length you used. How does this compare with your experimental value? Calculate the percent error. Use the calculated value of the period as the standard value.

8. Assuming the length remains the same, calculate the period of the same simple pendulum on the surface of the Moon? [*Note:* The acceleration (g) due to gravity on the Moon is one-sixth of that on Earth.]

Experiment 4

Uniform and Accelerated Motion

INTRODUCTION

An object that is undergoing a change in position is said to be moving. The motion may be either simple straight-line motion, where the change in distance is the same for each period of time, or it may be very complex, as in a tumbling satellite orbiting Earth. Our study in this experiment deals with simple straight-line motion with uniform change in the distance and straight-line motion with uniform change in the velocity. A good example of straight-line motion is the movement of a glider along a linear air track. A close observation of the glider will reveal that its position changes at a uniform rate; that is, the distance traveled is the same for equal periods of time. We can say it another way—the distance traveled is directly proportional to the time. This can be written as

$$\text{distance } (d) \propto \text{time } (t)$$

or

$$d = \bar{v}t \qquad \text{Eq. (4.1)}$$

where $\bar{v}$ (pronounced "v bar") is a proportionality constant. Solving for $\bar{v}$, we obtain

$$\bar{v} = \frac{d}{t} \quad \text{(defining equation for average velocity)}$$

The quantity $\bar{v}$ is known as the *average velocity* of the glider.

An object whose velocity is changing is said to be *accelerating. The velocity may be increasing, decreasing, or changing direction. Any of these causes a change in the velocity.* In this experiment we will be dealing with a change in velocity due to an increase. In observing the glider traveling down the inclined plane, you will discover that the velocity change will be the same for equal periods of time. This means that the velocity change is directly proportional to the change in time. This can be written as

$$\text{Change in velocity } (\Delta v) \propto \text{change in time } (\Delta t)$$

or

$$\Delta v = a\Delta t$$

where a (acceleration) is a proportionality constant. Solving the equation for a, we obtain

$$a = \frac{\Delta v}{\Delta t} \quad \text{(defining equation for simple acceleration)}$$

Since $\Delta v = v_f - v_0$,

$$a = \frac{v_f - v_0}{\Delta t}$$

where a = acceleration of the glider
v_0 = original velocity of the glider (v_0 is often written v_i for initial velocity)
v_f = final velocity of the glider
Δt = change in time

LEARNING OBJECTIVES

After completing this experiment, you should be able to do the following:

- Differentiate between *velocity* and *acceleration.*
- For a moving object, measure the distance traveled and the time required to travel that distance.
- Calculate the velocity and acceleration of a moving object from measurements of distance and time.
- State the relationship between the distance traveled by an object and the time it takes to travel the distance in uniform *linear* motion.
- State the relationship between the distance traveled by an object and the time it takes to travel the distance in uniform *accelerated* motion.

APPARATUS

Linear air track, glider, meter stick, stopwatch or electric timer.

EXPERIMENT 4 NAME (print) ________________________________ DATE ______________
LAST FIRST

LABORATORY SECTION ______ PARTNER(S) ________________________________

PROCEDURE 1

The air track will be assembled and leveled by the laboratory technician. Please do not attempt any adjustments of the air track. If you have trouble, ask the instructor for assistance.

Place the glider in motion by applying a very small force on the glider in a direction parallel to the air track. Do not attempt to move the glider if the air supply is not turned on. Determine the time required for the glider to travel the distances listed in Data Table 4.1. Calculate the velocity v as called for in the data table. Several students should work together, each having a timer, and take simultaneous time readings of each pass of the glider along the track.

DATA TABLE 4.1 (Record units with all data values.)

Distance (d)	d_1 100 cm	d_2 150 cm	d_3 200 cm	d_4 250 cm	d_5 300 cm
Time (t)					
Average velocity ($\bar{v}$)					

(*Graph number 1*) Plot the experimental results with time t on the x axis and the distance d on the y axis. Refer to Experiment 1 for the correct procedure for plotting graphs. Determine the slope of the curve on the graph itself and record the final slope value here.

$$\text{Slope} = \frac{\Delta d}{\Delta t} = \underline{\qquad\qquad}$$

PROCEDURE 2

When Procedure 1 has been completed, ask the instructor to adjust the air track at an angle with the table top. This can be done by placing a small riser block under one end. In this position the glider will have an acceleration along the air track.

Place the glider on the air track and determine the time required to travel the distances given in Data Table 4.2. Calculate the values of v_f and a, as called for in the table. Make the calculations with the glider starting from rest. The quantity v_f is the instantaneous velocity at the distance indicated, that is, the velocity at the end of 100 cm, or 150 cm, or 200 cm, and so on. The average velocity can be determined as follows:

$$\bar{v} = \frac{v_f + v_0}{2}$$

where $\bar{v}$ = average velocity
v_f = instantaneous velocity at distance indicated
v_0 = velocity at rest

Since the glider is started from a condition of rest, $v_0 = 0$ and $\bar{v} = \frac{v_f}{2}$, so $v_f = 2\bar{v}$. Substituting d/t for $\bar{v}$ yields

$$v_f = \frac{2d}{t}$$

The acceleration is constant; therefore, its value can be calculated from a definition of acceleration, which can be written

$$a = \frac{v_f - v_0}{t} \qquad \text{Eq. (4.2)}$$

Using the equation for displacement,

$$d = v_0 t + \frac{1}{2}at^2$$

Since the glider started from rest, $v_0 = 0$, and $d = 0 + \frac{1}{2}at^2$

Solving this equation for a gives $a = \frac{2d}{t^2}$ Eq. (4.3)

We will use both Eqs. 4.2 and 4.3 to find the acceleration in the following table.

DATA TABLE 4.2 (Record units with all data values.)

Distance (d)	d_1 100 cm	d_2 150 cm	d_3 200 cm	d_4 250 cm	d_5 300 cm
Time (t)					
Final velocity (v_f) $v_f = \frac{2d}{t}$					
Experimental acceleration $a = \frac{v_f - v_0}{t}$					
Theoretical acceleration $a = \frac{2d}{t^2}$					

(*Graph number 2*) Plot the experimental data from Data Table 4.2 with time t on the x axis and the velocity v_f on the y axis. Determine the slope of the curve on the graph itself and record the final value of its slope here.

$$\text{Slope} = \frac{\Delta v_f}{\Delta t} = \_\_\_\_\_\_\_\_\_\_$$

(*Graph number 3*) Plot the experimental data from Data Table 4.2 with time t on the x axis and the distance d on the y axis. This graph will not be a straight line. Explain why it is not.

EXPERIMENT 4 NAME (print) ______________________________ DATE ______________________
LAST FIRST

LABORATORY SECTION ______ PARTNER(S) ____

QUESTIONS

1. Define *velocity,* and give an example.

2. Define *acceleration,* and give an example.

3. How does the distance traveled vary with time in uniform linear motion? (*Hint:* Examine Eq. 4.1.)

4. How does the distance traveled vary with time in uniform accelerated motion? (*Hint:* Solve Eq. 4.3 for *d.*)

5. In Procedure 1, what is the relationship between the velocity and the slope of the curve? (*Hint:* What are the units for the slope on the graph that you plotted?)

6. In Procedure 2, what is the relationship between the acceleration and the slope of the curve? (*Hint:* How does Eq. 4.2 for acceleration compare with the equation given for slope?)

7. What kinematic quantity is indicated by the speedometer of an automobile?

8. Give an example of an object having a velocity but no acceleration.

Experiment 5

Determining g, the Acceleration of Gravity

INTRODUCTION

A glider sliding down an inclined air track or a steel ball rolling down a V-grooved 2 in × 6 in wooden plank has the force of gravity pulling vertically downward on it. The force acting is equal to the weight, w, of the object. A second force is also acting on the object. The second force is exerted by the supporting air track or the wood plank, which acts perpendicular to the surface of the supporting air track or wood plank and is called the normal force. Figure 5.1 shows the two forces and the resultant of the two forces that act parallel to the incline to produce an acceleration of the object down the incline. An analysis of the two forces shows that the resultant force F down the incline is equal to $mg\sin\theta$. (See Fig. 5.2.) The sine of the angle θ is equal by definition to the side of a right triangle opposite the angle divided by the hypotenuse. Thus the height of the incline divided by the length of the incline equals $\sin\theta$, or $\sin\theta = h/L$. This force of $mg\sin\theta$ or mgh/L is the unbalanced force acting to accelerate the mass down the incline. Newton's second law of motion states the relationship between the unbalanced force, mass, and acceleration of the object. This can be written as

$$a = \frac{F}{m}$$

Since $F = mgh/L$,

$$a = \frac{mgh/L}{m} = \frac{gh}{L}$$

or solving for g:

$$g = \frac{aL}{h} \qquad \text{Eq. (5.1)}$$

To determine the value of g, we need to measure the values of h and L and determine the value of the acceleration a. The acceleration a can be determined by the following relationship:

$$a = \frac{v_f - v_0}{t} \quad \text{by definition}$$

If $v_0 = 0$, then $a = v_f/t$. Since $v = (v_f + v_0)/2$, if $v_0 = 0$, then $v = v_f/2$ or $v_f = 2\bar{v}$. But $\bar{v} = d/t$ by definition. Therefore,

$$a = \frac{2\bar{v}}{t} = \frac{2d/t}{t} = \frac{2d}{t^2} \qquad \text{Eq. (5.2)}$$

where d is the distance traveled and t is the time for the object to travel the distance d.

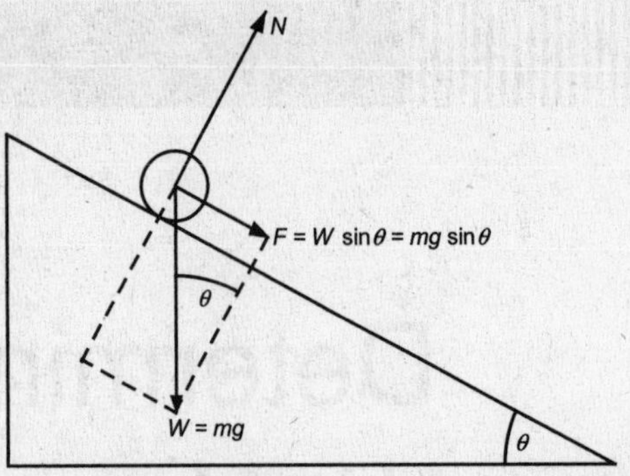

Figure 5.1 Diagram illustrating the forces acting on an object on an inclined plane. Note: Surface friction and air resistance are both neglected. *W* represents the weight of the object, *N* represents the normal or perpendicular force on the object, and *F* is the component of *W* parallel to the incline.

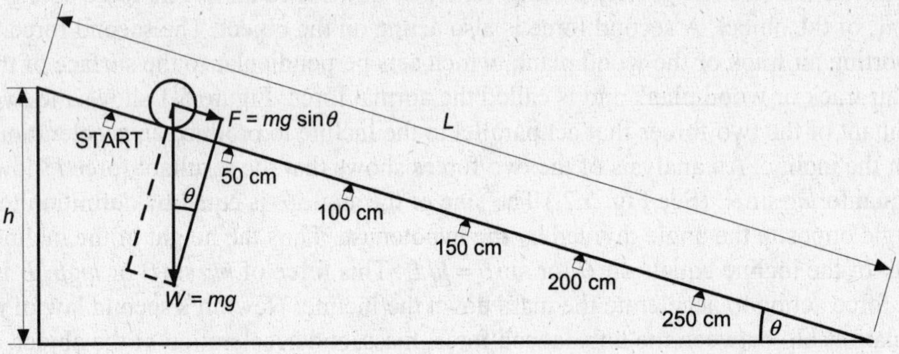

Figure 5.2 Steel ball rolling down an incline due to the pull of gravity.

LEARNING OBJECTIVES

After completing this experiment, you should be able to do the following:

- ▼ Determine experimentally the acceleration of gravity (g) at the surface of Earth.
- ▼ Calculate the percent error of your measurement with the accepted value of g at your surface location on the surface of Earth.

APPARATUS

Linear air track, glider, meter stick, five timing devices (stopwatches or electric timers), and masking tape. If the air track is not available, then a 2 in × 6 in wood plank 10 ft long plus a 3/4-in-diameter steel ball can be used. The wood plank must have a groove milled into one edge of the plank. A V-grooved trough is preferred.

EXPERIMENT 5 NAME (print) ______________________________ DATE ______________
LAST FIRST

LABORATORY SECTION ______ PARTNER(S) ______________________________

PROCEDURE

1. If the air track is to be used, ask the instructor to adjust the air track at an angle with the table top. This can be done by placing a small block (about 2 cm high) under one end of the air track. If the 2 in × 6 in plank is used, ask the instructor to elevate the plank at an angle of about 5 degrees. Use small pointed pieces of masking tape and mark off five equal distances (50 cm) along the incline. Place the glider or the steel ball on the incline at the piece of tape farthest up the incline. This is the starting point. Release the object and start all timers simultaneously. Determine the time for the moving object to travel the distance from rest at the starting point to each of the other markers. Make three trials and record the values obtained for each distance in Data Table 5.1. Then calculate the average time for each distance, square this average time and record all values in Data Table 5.1.

2. Measure the height h and the length L of the incline and also record in the Data Table 5.1.

DATA TABLE 5.1

Distance traveled (d)	50 cm	100 cm	150 cm	200 cm	250 cm
TRIAL 1, time of travel in (s)					
TRIAL 2, time of travel in (s)					
TRIAL 3, time of travel in (s)					
Average time, t in (s)					
Average time squared, t^2 squared in (s^2)					

Height of the incline (h) __________ cm*

Length of the incline (L) __________ cm

* Note: If an airtrack is used, h will be the height of the riser block inserted under one of the leveling screws and L will be the distance between the leveling screws. L is *not* the total length of the track.

CALCULATIONS

1. Plot a graph with distance, *d*, on the vertical axis and time squared, t^2, on the horizontal axis. Make sure the graph contains all the information concerning graphs as explained in Experiment 1.

2. Calculate the acceleration of the object down the incline. Use Eq. 5.2. Show your work for each of the distances measured from the data recorded in Data Table 5.1 above. Now calculate the average acceleration for the various distance-time combinations recorded in this table. Show your work below.

3. Calculate your experimental value for the acceleration of gravity. Use the average value of *a* from Calculation 2 using the relationship: $g = \frac{aL}{h}$

4. Calculate the percent error using 980 cm/s^2 as the accepted value of *g*. Show your work.

EXPERIMENT 5 NAME (print) ______________________________ DATE ________________
LAST FIRST

LABORATORY SECTION ______ PARTNER(S) ________________________________

QUESTIONS

1. What do you think is the greatest source of error in this experiment? Justify your answer.

2. What kinematic concept (distance, velocity, acceleration, etc.) of the moving object does the slope of the graph represent?

3. Calculate the slope of the "distance vs time squared" graph on the graph page itself. How does the value compare to the value for the acceleration found in calculation 2?

4. If the initial velocity was not zero, how would your experimental value of g compare with the value you obtained by assuming the initial velocity was zero? For instance, suppose you accidentally gave the object a slight shove down the incline at the starting point. Would your calculated value for g be larger or smaller?

5. What will be the value of the acceleration when θ (theta) in Fig. 5.2 is increased to 90°? (Draw a small sketch to help you visualize this.)

6. Is the acceleration due to gravity always pointing vertically downward even for an object whose velocity is vertically upward? Explain your answer.

Experiment 6

Newton's Second Law of Motion

INTRODUCTION

A mass whose velocity is changing at a constant rate is said to be undergoing *uniform acceleration.* The velocity may be increasing; if so, the acceleration is considered to be positive. Or the velocity may be decreasing; then the acceleration has a negative value. In either case the change in velocity must be the same for each period of time for uniform acceleration.

Newton's second law of motion gives the relationship between mass, acceleration, and the unbalanced force causing the mass to accelerate:

$$F = ma \quad \text{or} \quad a = \frac{F}{m}$$

where a = acceleration of total mass in direction of the unbalanced force (m/s^2 or cm/s^2)
F = the unbalanced force acting to produce acceleration (Newtons or dynes)
m = total mass accelerated (kg or g)

Remember to change mass units to force units, multiply the mass by *g*, the acceleration due to gravity.

To study the relationships among these quantities, we shall use a single pulley that has very little friction and masses supported as shown in Fig. 6.1. The mass of the pulley and the mass of the string will be neglected in performing the experiment.

LEARNING OBJECTIVES

After completing this experiment, you should be able to do the following:

- State Newton's second law of motion both in words and in symbol notation.
- Determine experimentally the relationship between the concepts mass, acceleration, and the unbalanced force causing the mass to accelerate.

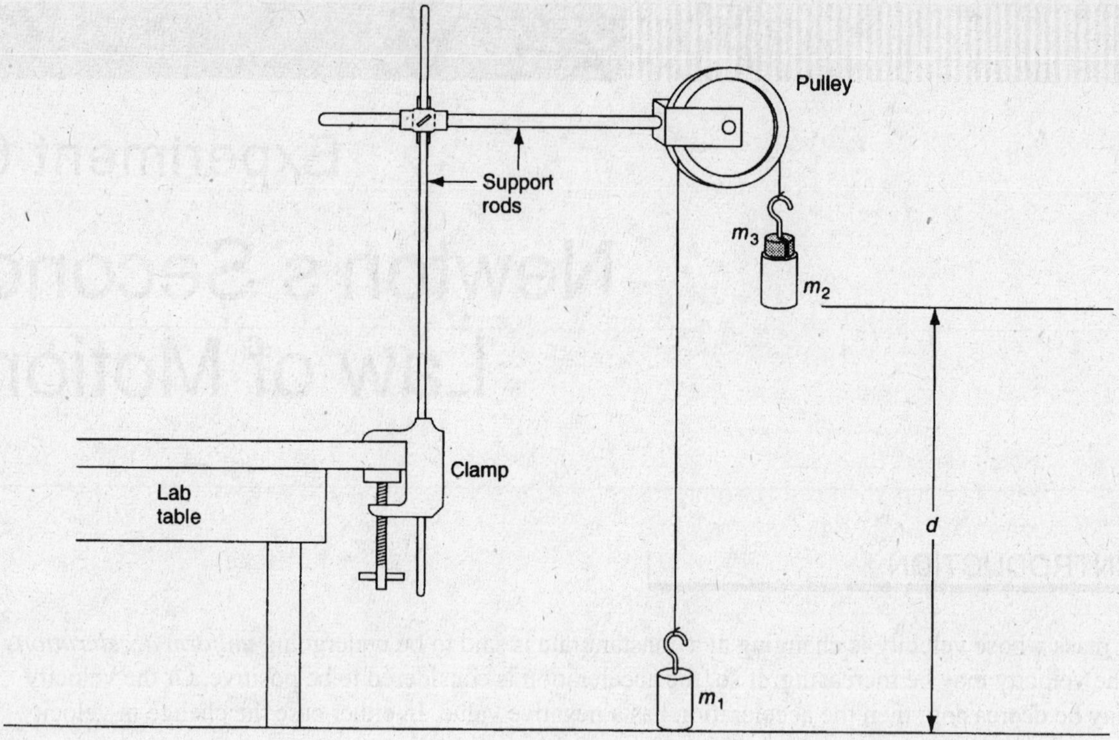

Figure 6.1

APPARATUS

Precision ball-bearing pulley, masses, cord, stopwatch or electric timer, support rods, clamps, and 2-m meter stick.

String may slip off pulley, especially if masses are moving too fast. Do not get hands or feet below weights. It is advisable, and safer, to place a small carpet square or rubber pad, (about 1 foot on a side), directly under the descending mass to cushion its landing.

EXPERIMENT 6 NAME (print) ______________________ LAST FIRST DATE ______________

LABORATORY SECTION ______ PARTNER(S) ______________________

PROCEDURE 1

1. Assemble the apparatus as shown in Fig. 6.1 if this has not already been done by a previous laboratory group. It is important that the pulley be positioned to turn freely in a vertical plane. Otherwise, there will be an opposing force that will give poor results for the experiment. *Keep the unbalanced force constant*, and determine experimentally how the acceleration varies as the total mass is varied. Perform two trials using two different sets of matched masses, (Example: $m_1 = m_2 = 500$ g and $m_1 = m_2 = 1000$ g). Enter data in Data Table 6.1.

   To obtain the data, proceed as follows: With one mass resting on the floor and a suitable rider or riders (m_3) on the top mass, simultaneously release the system and start the timer. The rider or riders should not be more than 6 percent of the total mass. When the bottom of the upper mass reaches the floor, stop the timer. Make six independent determinations of this time, and take the average. Distance d is measured from the bottom of the upper mass to the floor. See Fig. 6.1.

DATA TABLE 6.1

			Trial 1 $m_1 = m_2 =$ ______ g	Trial 2 $m_1 = m_2 =$ ______ g
Distance (d) the masses travel				
Unbalanced force F, $F = m_3g$ (constant for all trials)				
Total mass moved $M = m_1 + m_2 + m_3$* (different for each trial) (kg)				
Time to travel distance d, (s)	Three runs with rider on left	1		
		2		
		3		
	Three runs with rider on right	1		
		2		
		3		
	Average for 6 runs			

* m_3 = rider or riders = ________

m_3 is the same for both trials.

CALCULATIONS FOR PROCEDURE 1

Compute the acceleration for each trial, first, from the measurements taken of the distance traveled and the time. Then calculate the acceleration by using Newton's second law. Show your work. Then enter your two calculated values for acceleration in Data Table 6.2 and calculate and record the percent difference for these acceleration values in each of the two trials.

1. $a = \frac{2d}{t^2}$

2. $a = \frac{F}{M} = \frac{[(m_2 + m_3) - m_1]g}{M}$ where M is the total mass moved.

DATA TABLE 6.2

	Trial 1	Trial 2
Acceleration from measurements of distance and time		
Acceleration from Newton's second law		
Percent difference		

PROCEDURE 2

2. *Keep the total mass constant*, and determine how the acceleration varies with the variation of the unbalanced force. *Note:* To keep the total mass constant, the experimenter should change some of the riders from one side of the pulley to the other. Perform two trials using two different rider values. Enter data in Data Table 6.3. In this procedure the m in Newton's Second Law ($F = mg$) is the difference between the total mass on one side of the pulley and the total mass on the other side of the pulley.

EXPERIMENT 6 NAME (print) ______________________________ DATE ____________
LAST FIRST

LABORATORY SECTION ______ PARTNER(S) ________________________________

DATA TABLE 6.3

			Trial 1 $m_1 = m_2 =$ ________ g	Trial 2 $m_1 = m_2 =$ ________ g
Distance (*d*) the masses travel (m)				
Unbalanced force *F*, $F = m_3g$* (different for each trial) (N)				
Total mass moved (constant for all trials) (kg)				
Time to travel distance *d*, (*s*)	Three runs with rider on left	1		
		2		
		3		
	Three runs with rider on right	1		
		2		
		3		
	Average for 6 runs			

* *m* = the difference between total mass on one side of the pulley and total mass on the other side.

CALCULATIONS FOR PROCEDURE 2

Compute the acceleration (1) from the measurements taken of the distance traveled and the time and (2) by using Newton's second law. Show your work on a separate sheet of paper, and record the answers in the space provided in Data Table 6.4. Then find the percent difference for each trial.

DATA TABLE 6.4

	Trial 1	Trial 2
Acceleration from measurements of distance and time		
Acceleration from Newton's second law		
Percent difference		

QUESTIONS

1. What have the data and calculations shown concerning the acceleration, *first*, as a function of the mass when the unbalanced mass is held constant, and *second*, as a function of the unbalanced force when the total mass is held constant? Answer this by completing the following two statements:

   (a) When the total mass that is accelerating increases (the unbalanced force is held constant), the acceleration will ______________________________.

   (b) When the unbalanced force increases (the total mass is held constant), the acceleration will ______________________________.

2. Distinguish between mass and weight, i.e., define both and explain the difference.

3. Should the mass of the string be added to the total mass being moved by the unbalanced force? Why or why not?

4. Since it was the addition of m_3 to the system that produces the unbalanced force on the system in Procedure 1, why is the unbalanced force in Procedure 2 equal to $g\left[(m_2 + m_3) - m_1\right]$ instead of simply $m_3 g$?

5. Explain where at least three (3) possible sources of error could have occurred in this experiment.

6. Which of the sources of error that you cited in question 5 do you think most affected *your* experiment? Why?

Experiment 7

Hooke's Law for a Vibrating Spring

INTRODUCTION

A common type of vibrating system is a spring with a heavy mass (m) suspended from one end. If the mass is set in an up-and-down (vertical) motion, the repeated stretching and compression of the spring follows certain rules of periodicity that demonstrate some of the characteristics of wave motion.

The period of the vibration (time for one complete up and down oscillation) may be defined as

$$T = 2\pi\sqrt{\frac{M}{k}}$$

where T = the period in seconds
$\pi = 3.14$
M = the sum of the attached mass (m) plus one-third of the mass of the spring itself
$[M = m + 1/3 \text{ (spring mass)}]$
k = the spring constant

The same spring constant (k) is also found in Hooke's law. This law applies to the stretching of a spring when a mass is hung on the end of it and gravity pulls on the mass to produce a stretching force. The proportionality between force (F) and stretch (X) is known as *Hooke's law.* This relationship can be written in equation form as

$$F = kX$$

where F = the applied force mg
X = the distance the spring is stretched
k = the same spring constant as used in the equation for period.

LEARNING OBJECTIVES

After completing this experiment, you should be able to do the following:

- State and explain Hooke's law.
- Write down and explain the equation for the period of an oscillating mass on a spring.
- Determine the spring constant for any spring using a set of weights and a meter stick.
- Calculate the period of oscillation for a mass on a spring both theoretically and experimentally.
- Access the accuracy of your period calculations using a percentage error calculation.

APPARATUS

Spring, set of calibrated masses, triple-beam balance or spring balance or electronic balance to weigh the spring, support stand, clamp, stopwatch or electric timer.

EXPERIMENT 7 NAME (print) ______________________________ DATE ______________

LAST FIRST

LABORATORY SECTION ______ PARTNER(S) ______________________________

PROCEDURE

First we must measure the spring constant of the particular spring you are going to use in this experiment. Attach the spring to the support clamp and *carefully* hang enough mass on the spring to stretch it between 2.0 and 3.0 cm. This mass will be referred to as the *incremental mass* throughout the rest of the experiment. Record the value of the incremental mass in the space provided below. Remove the incremental mass from the spring.

Incremental mass = ______________

Now adjust the support clamp so that the bottom end of the spring is about 30 cm above the top of the table. Carefully measure the distance from the tabletop to some identifiable point near the lower end of the spring. This distance is h_0 (See Fig. 7.1.) Record this value in the space provided at the top of Data Table 7.1.

Again hang the same incremental mass on the spring that you used in your earlier calibration of the spring. The spring should stretch between 2.0 and 3.0 cm, as it did before. Carefully measure the distance from your reference point on the spring to the top of the table. This is the distance h for Trial 1. Record this value in Data Table 7.1 along with the value of the mass hanging on the spring. This should be the same as the incremental mass that you recorded in a previous step.

Repeat the above procedure for values of mass equal to 2, 3, 4, 5, and 6 times the incremental mass. The spring should stretch an additional 2.0 to 3.0 cm for each trial. Carefully measure the height h above the table of your reference point on the spring for each trial. Record the height, h, and the mass, m, for each trial in Data Table 7.1. You may use either SI or cgs units.

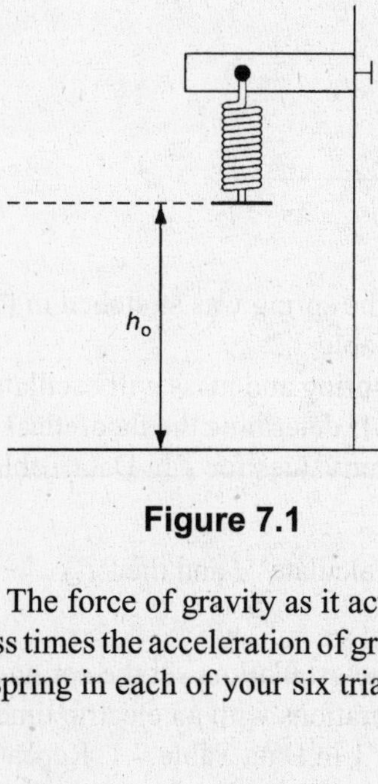

Figure 7.1

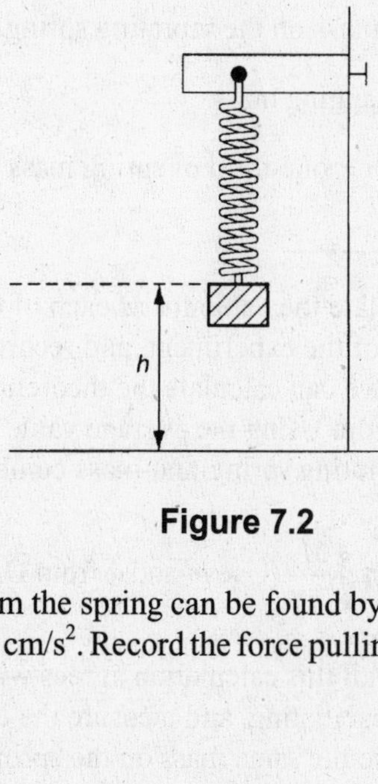

Figure 7.2

The force of gravity as it acts on the mass hung from the spring can be found by multiplying this mass times the acceleration of gravity g, 9.80 m/s^2 or 980 cm/s^2. Record the force pulling downward on the spring in each of your six trials in Data Table 7.1.

DATA TABLE 7.1 (Record units with all data values.) Starting height h_0 = ___________					
Trial	Total Mass on Spring m	Height Above the Table h	$X = h_0 - h$	Force $F = mg$	$k = \frac{F}{X}$
1					
2					
3					
4					
5					
6					
				Average value of *k* for all six trials	

Calculate the spring constant *k* using Hooke's law for each of the six trials and record them in the last column in Data Table 7.1, also find the average of these values and record it as well.

Weigh the entire spring to find the spring mass.

Spring mass = _____________

The effective mass on the vibrating spring will now be:

m = hanging mass

$M = m$ + one-third of spring mass = (__________) + 1/3 (__________)

M = __________

Calculate the value for *M* each of the six trials for which the spring was stretched in the previous part of the experiment, and record these values in Data Table 7.2.

Now we can calculate the theoretical period at which the spring and mass will oscillate if they are set in motion. Using the average value of *k* from Data Table 7.1, determine the theoretical period for the six vibrating spring-and-mass combinations, and record your values for *T* in Data Table 7.2.

$T = 2\pi\sqrt{\frac{M}{k}}$ (Use *m* and *k* from Data Table 7.1 above to calculate *M* and then *T*.)

To see how well this calculation agrees with the actual time for one oscillation, set the spring-and-mass system oscillating, and measure the time for 20 complete vibrations with an electric timer or stopwatch. Use the same mass on the spring that you used in Trial 1 in Data Table 7.1. Repeat this procedure for the other five trial masses in Data Table 7.1, and record the times for 20 oscillations for each mass value in Data Table 7.2.

EXPERIMENT 7 NAME (print) ______________________________ DATE ______________
LAST FIRST

LABORATORY SECTION ______ PARTNER(S) ______________________________

Calculate the experimental period for each trial by dividing your measured time (t) for each trial by the number of oscillations (20) and filling in these data in Data Table 7.2.

$$T = \frac{t}{20} \qquad \text{(Experimental period)}$$

How does your experimental value for T compare with the theoretical value found by calculation? To determine this, do a percentage error calculation for each trial using the theoretical period as the accepted value.

DATA TABLE 7.2 (Record *units* with all data values.)

Trial Number	Effective Mass (M) of Spring and Hanging Mass	Theoretical Period (T) $T = 2\pi\sqrt{\frac{M}{k}}$	Measured Time (t) for 20 Oscillations	Experimental Period, Time for 1 Vibration $T = \frac{t}{20}$	Percent Error in Experimental Compared to Theoretical Periods
1					
2					
3					
4					
5					
6					

QUESTIONS

1. Why must you use the same reference point (h_0) on the spring for each subsequent height measurement?

2. Why did you use the average value for k from Data Table 7.1 rather than one of the individual trial values?

3. Do a calculation of the theoretical value of the period for the oscillating mass for Trial 4 *without* correcting the effective mass by adding one-third of the mass of the spring. Again determine the percentage error between the experimental and theoretical values of the period of oscillation. Do you agree that the effective mass must be corrected in this way to obtain good results?

Experiment 8

Centripetal Acceleration and Force

INTRODUCTION

A mass moving in a circle with constant speed is said to be in *uniform circular motion*. Although the speed of the mass is constant, its velocity is *not* constant because the direction of motion is continually changing. Because the velocity is continually changing, the mass is undergoing continual acceleration. However, only the direction of travel of the mass is changing, therefore, the acceleration must be directed at right angles to the direction in which the mass is moving (otherwise its speed would increase or decrease): that is, the acceleration must be directed toward the center of the circle. The force causing this acceleration is called the *centripetal force* (F_c). In this experiment, a spring will supply the centripetal force. (See Fig. 8.1.) The resulting acceleration is, therefore, referred to as the *centripetal acceleration*.

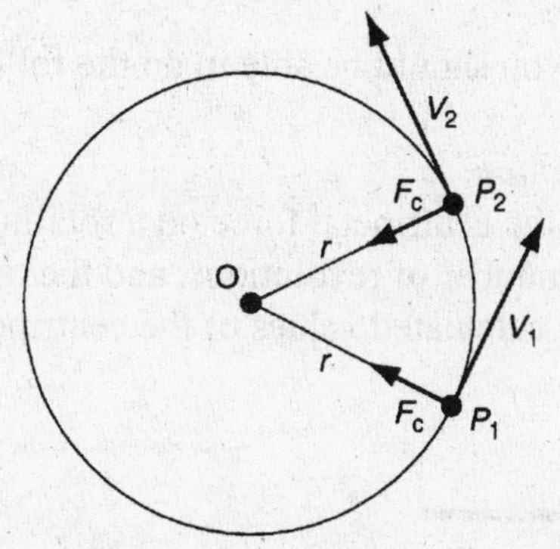

Figure 8.1 Velocity of a mass moving in uniform circular motion showing the centripetal force, F_c, at two separate locations.

The centripetal acceleration of a mass moving in uniform circular motion is given by

$$a_c = \frac{v^2}{r} \qquad \text{Eq. (8.1)}$$

where r is the radius of the circle in which the mass is traveling and v is the linear velocity of the mass along the circumference of the circle. According to Newton's second law, the centripetal force F_c producing this acceleration is

$$F_c = ma_c = \frac{mv^2}{r} \qquad \text{Eq. (8.2)}$$

It is convenient to express v in terms of other quantities. The circumference of the circle c that the mass follows as it travels is simply

$$c = 2\pi r$$

where r is the radius of the circle. If the mass makes n revolutions in a time t, the total distance d traveled by the mass in this time period is

$$d = nc = n2\pi r$$

and the velocity of the particle is simply

$$v = \frac{d}{t} = \frac{n2\pi r}{t} \qquad \text{Eq. (8.3)}$$

By use of Eqs. 8.2 and 8.3, it is now possible to determine the centripetal force on a rotating body by measuring the mass and calculating the linear velocity in terms of the radius of rotation, the number of revolutions, and the total time for these revolutions.

LEARNING OBJECTIVES

After completing this experiment, you should be able to do the following:

- Define *uniform circular motion.*
- Determine experimentally the centripetal force on a rotating object by measuring the mass, the radius of rotation, the number of revolutions, and the time for the revolutions.
- Compare the observed and calculated values of the centripetal force.

APPARATUS

Centripetal force apparatus, stopwatch or electronic timer, weight hanger, assorted slotted masses, rods and clamps.

Caution Safety Tips

Rod and crossbar rotate during experiment as does the mass on the string. Be careful not to reach in or get in the path of these moving parts as they are turning.

EXPERIMENT 8 NAME (print) ______ LAST ______ FIRST ______ DATE ______

LABORATORY SECTION ______ PARTNER(S) ______

PROCEDURE

1. This experiment makes use of the apparatus shown in Fig. 8.2. The apparatus is made up of a rotating vertical rod with a crossbar attached to a support rod and horizontal brace. The rotating rod and crossbar assembly can be turned easily by pulling a string wound around the vertical rotating rod.

   Hang the known mass (m) from the rotating crossbar so that it is a few centimeters from the end of the unstretched spring and its tip is just above the plastic ruler. Note: The hanging mass is *not* attached to the spring in this step. Record the measurement to the point just below the tip of the mass. Place an identifying mark on the ruler at this point using a triangular piece of masking tape as shown. This measurement is made from the center of the vertical rotating rod. It is r in Eq. 8.3.

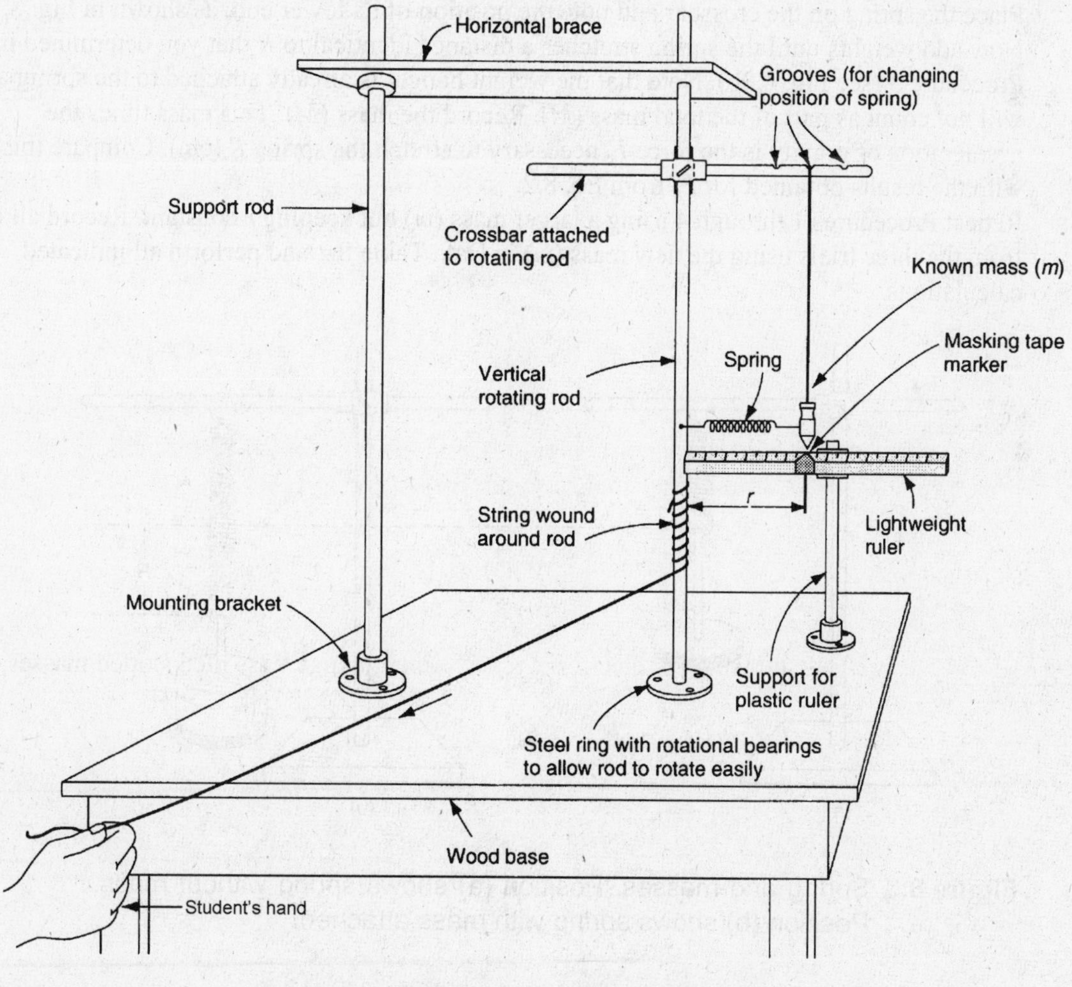

Figure 8.2 Centripetal force apparatus.

2. Attach the mass (m) to the spring (it will be drawn inward, toward the rotating rod). Wind the string from the student's hand around the vertical rod and pull smoothly until the mass moves with uniform velocity and consistently passes, vertically, over the tape mark on the plastic ruler. Practice this two or three times. It is important that the string be pulled uniformly and that the mass passes directly over the tape mark on the plastic ruler on each rotation. Count the number of revolutions (n) and observe the time interval (t) from the instant when the mass passes over the tape mark in the first revolution to the time when the mass passes over the tape mark in the last revolution. The latter must occur just before the string is completely unwound. Make three trials for n and t and record in Data Table 8.1. Compute the linear velocity (v) using Eq. 8.3, and then determine the centripetal force (F_c) using Eq. 8.2 for all trials.
3. While the mass is still attached to the spring, pull the mass out until it is over the tape mark on the plastic ruler. Measure the distance from the center of the rotating rod to some identifiable mark on the mass. Call this R_1. Now allow the mass to be pulled in by the spring as far as it will go toward the rotating rod. Again, measure the distance to the same point on the mass. Call this distance R_2. The difference $R = R_1 - R_2$ is the amount the spring was stretched. Record this in Data Table 8.1.
4. Place the spring on the crossbar and note the position of its lower end, as shown in Fig. 8.3a. Now add weights until the spring stretches a distance identical to R that you determined in Procedure 3 (see Fig. 8.3b). Note that the weight hanger is already attached to the spring and will *not* count as part of the total mass (M). Record the mass (M). This mass times the acceleration of gravity is the force F_s necessary to stretch the spring R (cm). Compare this with the results obtained for F_c from Eq. 8.2.
5. Repeat Procedures 1 through 4 using a larger mass (m) but keeping r constant. Record all data from the three trials using the new mass (m) in Data Table 8.2 and perform all indicated calculations.

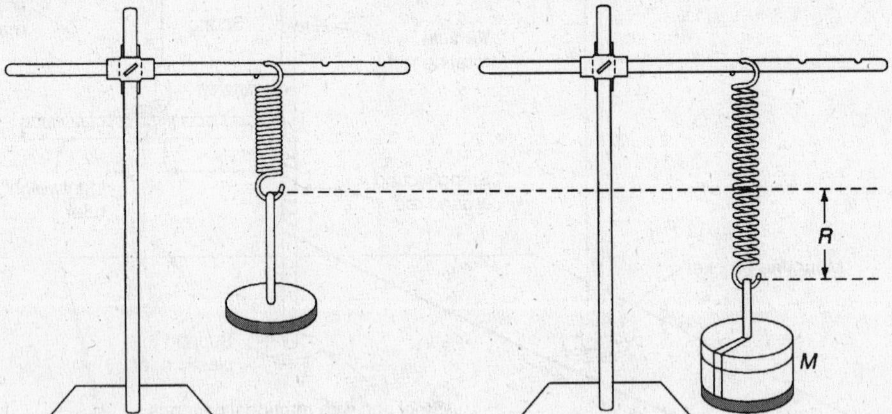

Figure 8.3 Spring and masses. Position (a) shows spring without mass. Position (b) shows spring with mass attached.

EXPERIMENT 8 NAME (print) ______________________ DATE ______________
LAST FIRST

LABORATORY SECTION ______ PARTNER(S) ______________________

DATA TABLE 8.1 For first suspended mass, m_1

Trial	n	t	r	$v = \frac{n2\pi r}{t}$	$F_c = m_1\frac{v^2}{r}$	Total Mass (M)	$R = R_1 - R_2$	$F_s = Mg$	Percent Difference Between F_s and F_c
1									
2									
3									

DATA TABLE 8.2 For second suspended mass, m_2

Trial	n	t	r	$v = \frac{n2\pi r}{t}$	$F_c = m_2\frac{v^2}{r}$	Total Mass (M)	$R = R_1 - R_2$	$F_s = Mg$	Percent Difference Between F_s and F_c
1									
2									
3									

QUESTIONS

1. Assume that the string breaks while the mass (m) was in uniform circular motion. If the spring is also suddenly unhooked from the mass at the same instant that the string breaks, what will happen to the mass?

2. How did F_c change when you redid the experiment using the larger mass? Did you expect this? Why or why not?

3. Why must the string held in the student's hand be pulled at a uniform rate?

4. Calculate the centripetal acceleration for each of the two trials.

Experiment 9

Laws of Equilibrium

INTRODUCTION

An object is in equilibrium when the summation of all forces and torques acting on the object equals zero.

$$\text{First law of equilibrium: } \sum F = 0$$

$$\text{Second law of equilibrium: } \sum \tau = 0$$

where Σ means summation
F stands for force
τ stands for torque

A state of equilibrium does not mean that the object is at rest. The object may be at rest or it may be in uniform motion at a constant velocity, but if it in equilibrium, the summation of all forces and the summation of all torques must each be equal to zero and it can not be accelerating.

LEARNING OBJECTIVES

After completing this experiment, you should be able to do the following:

- Define the terms *torque* and *center of gravity.*
- State in words and symbols the two laws of equilibrium.
- Determine experimentally that $\sum F = 0$ and $\sum \tau = 0$ when a system is in equilibrium.

APPARATUS

Moment-of-force apparatus (uniform meter stick, knife-edge clamp, support for meter stick, set of hooked weights), one unknown weight, table supports and crossbar, two spring balances (~50 – 200 g), string, second meter stick.

Caution Safety Tips

When placing weights on the meter sticks, be careful not to unbalance the apparatus or drop the masses. Do not leave loaded meter sticks unattended. Do not get hands or feet below weights.

EXPERIMENT 9 NAME (print) ______________________________ DATE ______________
LAST FIRST

LABORATORY SECTION ______ PARTNER(S) ______________________________

PROCEDURE 1

In this part of the experiment we will study the conditions for equilibrium of a horizontal meter stick under the action of forces that tend to cause rotation.

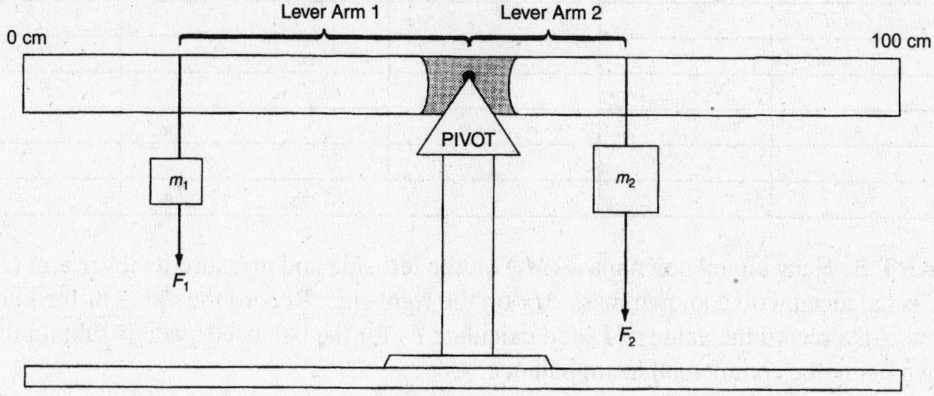

Figure 9.1

Equilibrium condition: $\sum \tau = 0$ where Σ is the summation sign, counter-clockwise torques are conventionally positive, and clockwise torques are considered negative. Thus when we transfer the negative clockwise torques to the other side of the equal sign the equation becomes:

$$\sum \tau_{ccw} = \sum \tau_{cw} \qquad \text{or} \qquad L_1F_1 = L_2F_2$$

We will be using masses calibrated in grams to hang on the meter stick. The force of gravity on these masses will be expressed in Newtons by first converting the mass from grams to kilograms and then multiplying by 9.8 m/s^2. Thus forces in this experiment actually will be the weights of the various masses involved.

PART A: Clamp the knife edge on the meter stick near its center, at the point where the meter stick balances. Make sure that the knife edge is placed on the bar so that the edge is near the top of the bar and the calibrations on the bar are erect with the zero on the left side (see Fig. 9.1). This means that the center of gravity of the meter stick will be below the point of suspension of the fulcrum. When the meter stick is balanced, one end will be just as far above the table top as the other end. *Note*: All meter sticks are not perfectly uniform so the location of the knife edge may not be exactly at the 50 cm point.

Weigh the meter stick, recording the mass and the point of balance in Data Table 9.1. Place a single mass on one side of the meter stick, and balance it with a different mass on the opposite side. Use a loop of string with "negligible mass" to hang the large masses from the meter stick. Record the magnitude of each force (in Newtons) and its lever arm (in meters) under Part 1 in Data Table 9.1.

Repeat the previous procedure with the masses at different positions. Record the data under Part 2 in Data Table 9.1. Now hang two masses on one side of the fulcrum and balance with one mass on the opposite side. Record the data under Part 3 in Data Table 9.1.

Do all calculations as required in Data Table 9.1. Some spaces in the torque columns will not be filled because a force cannot produce a clockwise and a counterclockwise torque at the same time. The percent difference in torques is calculated between the *total clockwise torques* and the total *counterclockwise torques.* In Parts 1 and 2 there will be only one clockwise and one counterclockwise torque to deal with, but in Part 3 there will be two torques in one direction, so these must be added together to get the total torque for this case.

DATA TABLE 9.1 Mass of the bar ________________ kg Point of balance of the bar ___________ m (Should be near the 0.50 m point)					
Part	Force (N)	Lever Arm (m)	Clockwise Torque (N m)	Counter-clockwise Torque (N m)	Percent Difference in Torques
1					
2					
3					

PART B: Hang an unknown mass (M_1) on the left side and measure its lever arm (L_1) when the system is balanced with a known mass M_2 on the right side. Record the value of the known mass (M_2) below. Also record the value of L_2 and calculate F_2 for the balanced system. Obtain the mass of the unknown using the commercial beam balance.

Mass of unknown (M_1) using balance = __________ kg Known mass, M_2 = __________ kg

L_1= __________ m L_2 = __________ m $F_2 = M_2 g$ = __________ N

Set counterclockwise torque equal to the clockwise torque, and solve for F_1:

$$L_1 F_1 = L_2 F_2 \quad \text{or} \quad F_1 = \frac{L_2 F_2}{L_1}$$

Calculate the mass M_1 and compare it with the value of M_1 found using the commercial balance. Then determine the percent difference between the two values of M_1.

$M_1(\text{calculated}) = \frac{F_1}{g} =$ ______________ kg

% difference = ____________ %

PROCEDURE 2

We will now use the two laws of equilibrium to study *nonconcurrent parallel forces*.

PART A: Hang two spring balances from the hooks on the horizontal support rod above the table. Check the "zero" reading of the balances with no forces acting on them. If the reading is not zero on a balance, each observation hereafter on the balance will have to be corrected for this "zero" reading. Your meter stick should have its center of gravity at or very near 0.50 m. First obtain the mass of this meter stick (M_m) using the beam balance; then suspend the stick from the spring balances with the scales so that the calibrations are erect and the zero end is at the left. Now suspend a 0.5 kg and a 1.0 kg mass from the meter stick as shown in Fig. 9.2. *Note*: There are three forces acting downward; $F_1 = -M_1 g$, $F_2 = -M_2 g$, and $F_m = -M_m g$. F_m acts at the center of gravity of the meter stick. There are also two upward forces, F_A and F_B, that can be read directly from the spring balances.

EXPERIMENT 9 NAME (print) ______________ LAST ______________ FIRST DATE ______________

LABORATORY SECTION ________ PARTNER(S) ______________

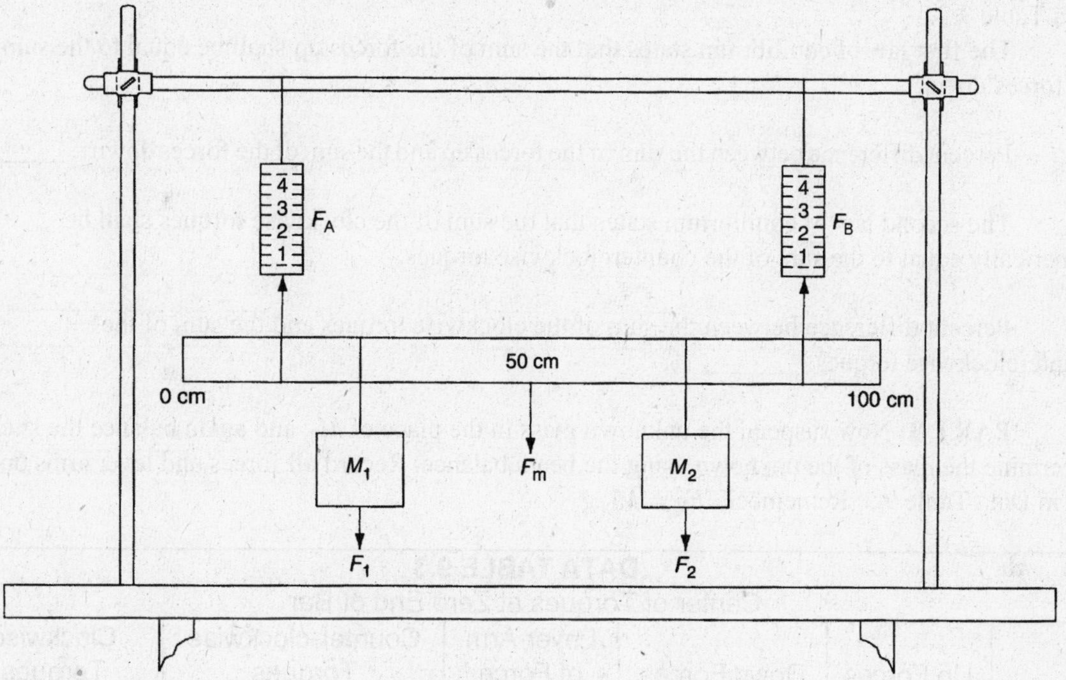

Figure 9.2

Note: Since each force can only be either up or down and each torque can only be either clockwise or counter-clockwise, not all spaces in the following two tables will have numbers in them.

DATA TABLE 9.2
Center of Torques at Zero End of Bar

Symbol	Up Forces (N)	Down Forces (N)	Lever Arm of Force (m)	Counter-clockwise Torques (N m)	Clockwise Torques (N m)
F_A					
F_B					
F_1					
F_2					
F_m					
Sum					

Note: Measure all lever arms from the *zero end of the meter* stick.

Adjust the positions of the balances and masses along the meter stick until each end of the stick is at the same distance from the table top. The balances must be in a vertical position. The lengths of the cords attached to the meter stick may have to be adjusted to obtain a balance. Read F_A and F_B and correct for any error in the "zero" of the balances. Record all forces and lever arms in Data Table 9.2.

The first law of equilibrium states that the sum of the forces up shall be equal to the sum of the forces down.

Percent difference between the sum of the forces up and the sum of the forces down = ______ %.

The second law of equilibrium states that the sum of the clockwise torques shall be numerically equal to the sum of the counterclockwise torques.

Percent difference between the sum of the clockwise torques and the sum of the counterclockwise torques = ______ %.

PART B: Now suspend the unknown mass in the place of M_2, and again balance the stick. Determine the mass of the unknown using the beam balance. Record all forces and lever arms on the bar in Data Table 9.3. Remember: $F_X = M_X g$

DATA TABLE 9.3
Center of Torques at Zero End of Bar

Symbol	Up Forces (N)	Down Forces (N)	Lever Arm of Force (m)	Counter-clockwise Torques (N m)	Clockwise Torques (N m)
F_A					
F_B					
F_1					
F_X		F_X*			
F_m			0.50		

* *Note:* Treat F_X, as an unknown to be used in the equations for the summation of forces (Eq. 9.1) and the summation of torques (Eq. 9.2) below. Record L_X as a measured quantity in this table; it will be used in the summation of torques equation when you solve the second time for F_X.

From the first law of equilibrium, we have Eq. 9.1:

Forces up = Forces down

$$F_A + F_B = F_1 + F_X + F_m \quad \text{so} \quad F_X = F_A + F_B - F_1 - F_m \qquad \text{Eq. (9.1)}$$

Solution F_X = ____________ N.

EXPERIMENT 9 NAME (print) ________________________________ DATE ____________________
LAST FIRST

LABORATORY SECTION ______ PARTNER(S) ________________________________

From the second law of equilibrium, we have Eq. 9.2:

$$\text{Sum of clockwise torques} = \text{sum of counterclockwise torques} \qquad \text{Eq. (9.2)}$$

$$\text{Sum of CW torques} = \text{Sum of CWW torques}$$
$$F_1 L_1 + F_X L_X + F_m L_m = F_A L_A + F_B L_B$$

so:

$$F_X = \frac{F_A L_A + F_B L_B - F_1 L_1 - F_m L_m}{L_X}$$

F_X = ____________________ N

Measured:

Mass of M_X from beam balance = ______________ kg $F_X = M_X g$ = ______________ N

Percent of difference F_X from your beam balance measurement and F_X as calculated using Eq. 9.1.

% difference = ____________ %

Percent of difference F_X from your beam balance measurement and F_X as calculated using Eq. 9.2.

% difference = ____________ %

QUESTIONS

1. Explain the difference between *force* and *torque*.

2. State the two laws of equilibrium.

3. How does the fact that the center of gravity of the meter stick, that is balanced, may not be exactly at the 50-cm point affect your measurements?

Experiment 10

Principle of Work Using an Inclined Plane and Pulleys

INTRODUCTION

Work can be calculated by taking the applied force times the parallel distance through which that force moves. This experiment deals with two applications of this concept: (1) the force needed to move a load up an inclined plane and (2) the force needed to raise a load using a pulley system. In addition, the actual mechanical advantage and theoretical mechanical advantage for these simple machines can be calculated, and their efficiency can be determined.

LEARNING OBJECTIVES

After completing this experiment, you should be able to do the following:

- ▼ Analyze the application of force to simple machines and learn how much work is done in the process.
- ▼ Define *theoretical mechanical advantage* (TMA), *actual mechanical advantage* (AMA), and *efficiency*.
- ▼ Determine experimentally the TMA, AMA, and efficiency of some simple machines.

APPARATUS

Hooked weights, inclined plane, Hall's carriage, two spring balances (one about 2.5 N (250 gm) and one about 20 N (2000 gm)), weight hanger, set of slotted weights, one unknown weight, two single pulleys, two five-string pulleys, cord, and table supports.

PROCEDURE 1

Set the inclined plane at an angle of 25°, and adjust the pulley and car as shown in Fig. 10.1 so that part of the cord below the pulley will be pulling parallel to the plane. Make sure the cord is not dragging on the pulley holder. With no load in the car, pull the spring balance up *vertically* so that the car moves *slowly* and *uniformly* (without acceleration) up the plane. Observe the force registered on the spring balance during the movement. The spring balances may be calibrated in grams, and the loads on the plane will be given or measured in grams. We will be dealing primarily with forces in this experiment, so we must convert the masses to the force applied to these masses by gravity in units of Newtons by multiplying mass (kg) by 9.80 m/s^2. Remember that you must first convert grams to kilograms. If the spring balance is calibrated in Newtons, the force can be read directly from the attached scale. Record your data in Data Table 10.1. The car should be pulled up several times to make sure you have the best value for the force. Determine the mass of the car with the beam balance. This will be the load (W) for the first trial.

Locate the car near the bottom of the inclined plane, and carefully measure the height (H_1) of the rear axle of the car above the table top. Now arrange the meter stick so that you can move the spring balance a distance (D) of about 20 cm. The car also will move up the track. Holding the spring balance very still, measure the height (H_2) of the rear axle of the car in the new position. The difference between H_2 and H_1 will be the vertical height (H) that the car has moved against the force of gravity. Record both D and H in the appropriate places in Data Table 10.1.

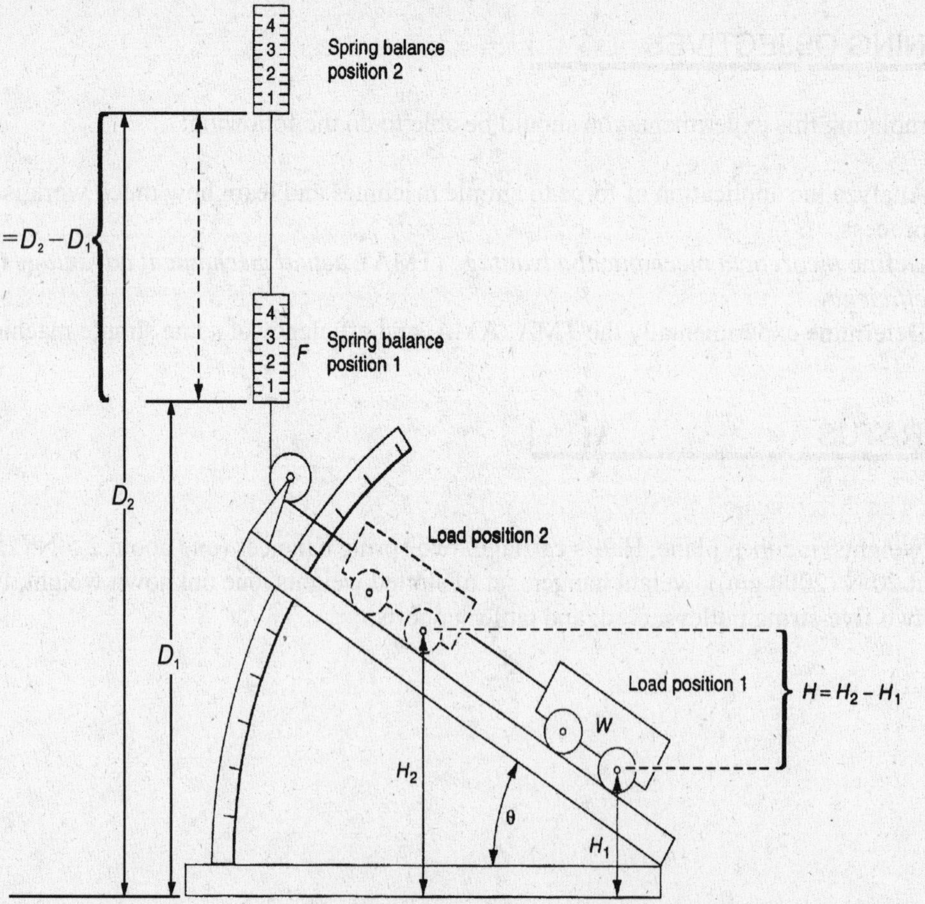

Figure 10.1

EXPERIMENT 10 NAME (print) ______________________ DATE ______________
LAST FIRST

LABORATORY SECTION ______ PARTNER(S) ______________________

Repeat the preceding procedure at the same angle but with some additional load placed in the car (say 400 to 500 g).

Adjust the plane and repeat the procedure with two different loads on the plane for an angle of 45°. Record the data in the Data Table 10.1.

DATA TABLE 10.1 *Note:* Weight = mass × *g*

Angle of Plane	*W* *Weight of* Car and Added load (if any) (N)	*F* (from spring balance) Applied Force *F* (N)	Actual M.A. $= \frac{W}{F}$	*D* $(D_2 - D_1)$ (m)	*H* $(H_2 - H_1)$ (m)	Theoretical M.A. $= \frac{D}{H}$	Percent Difference Between Actual and Theoretical M.A.
25°							
25°							
45°							
45°							

PROCEDURE 2

Assemble the pulley system as shown in Fig. 10.2 (In the actual pulley system used, the pulleys are all the same size, but for purposes of visualization, Fig. 10.2 is drawn with different size pulleys.). With a load of 1000 g at *L*, lift the spring balance slowly and uniformly (without acceleration) and observe the force required. Observe that there are seven strands of cord supporting the lower pulley.

Hang the pulley system on the crossbar assembly, and then measure the height (H_1) of the bottom of the load (*L*) suspended from the pulley system to the table top. Now arrange the meter stick so that you can move the spring balance a distance (*D*) of 30 cm. The load (*L*) also will move upward. Holding the spring balance very still, measure the height (H_2) from the bottom of the load (*L*) to the table top. The difference between H_2 and H_1 will be the vertical distance that the load has moved. Record both *D* and *H* in the appropriate places in Data Table 10.2.

DATA TABLE 10.2

	Load *L* (N)	Force *F* (N)	Actual M.A. $= \frac{L}{F}$	*D* $(D_2 - D_1)$ (m)	*H* $(H_2 - H_1)$ (m)	Theoretical M.A. $= \frac{D}{H}$	Percent Difference Between Actual and Theoretical M.A.
7 strands							
5 strands							

Repeat these observations with five strands supporting to the bottom pulley. Record the data in Data Table 10.2.

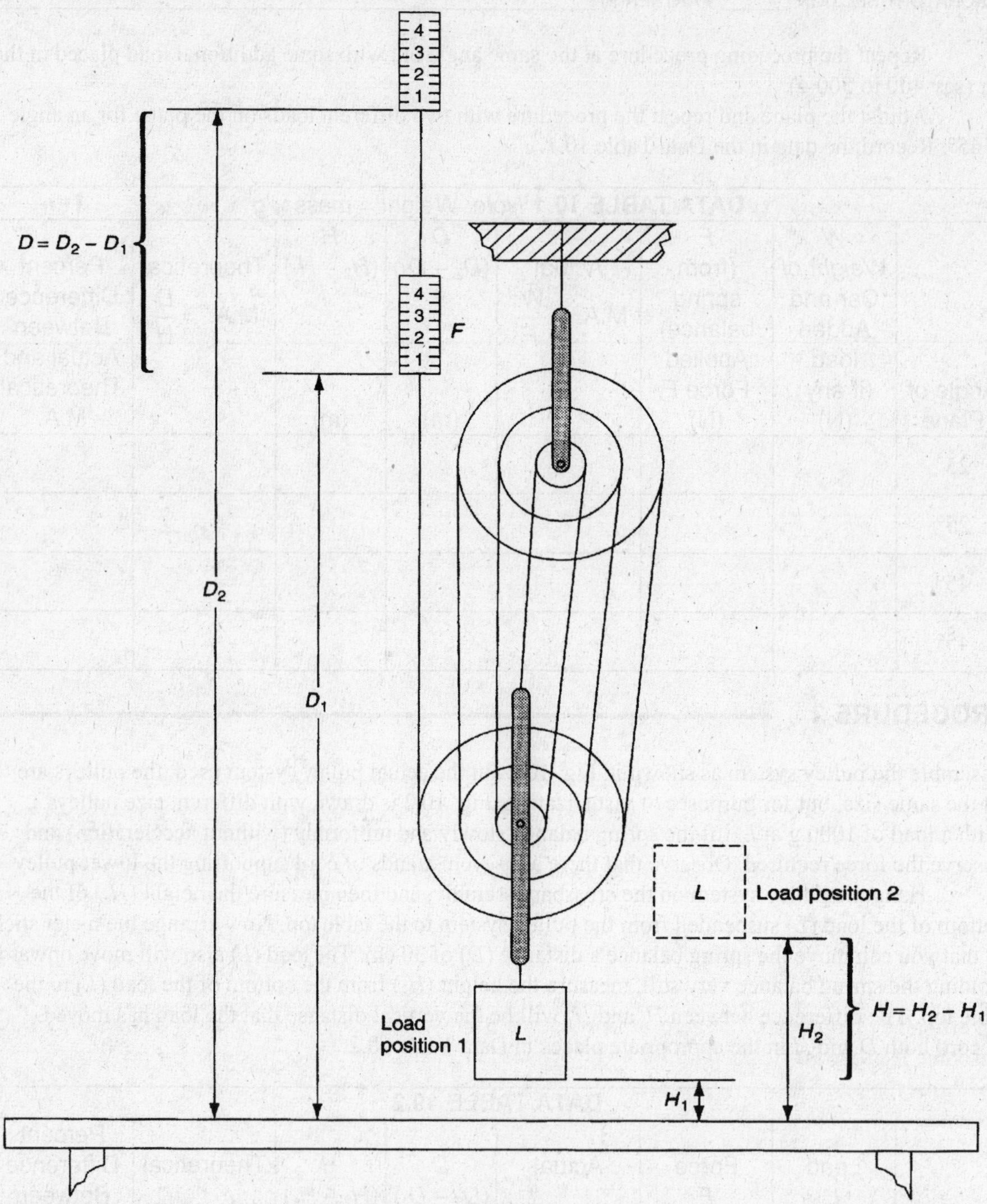

Figure 10.2

Note: Load position 2 has been moved to the right in the figure to show its vertical position after the spring balance has been raised a distance *D*. The lower pulley will also move upward during this process but that is not shown in the diagram. In fact it will be directly above Load position 1 when lifted.

EXPERIMENT 10 NAME (print) ______________________________ DATE ______________________
LAST FIRST

LABORATORY SECTION ______ PARTNER(S) ____

CALCULATIONS

The *actual mechanical advantage* of a machine is defined as the ratio of the force exerted on the load by the machine to the force applied to the machine. Treating the inclined plane as a machine, the force exerted on the load by the machine is the lifting of load W and the force applied to the machine is F. Therefore,

$$\text{Actual M.A.} = \frac{W}{F}$$

The *theoretical mechanical advantage* (that which would be obtained if the machine were 100 percent efficient) is the ratio of the distance moved by the applied force to the distance moved by the load. Referring to Fig. 10.1, the distance moved by the applied force is D, and the corresponding vertical height that the weight (W) raises is H.

$$\text{Theoretical M.A.} = \frac{D}{H}$$

In the case of the pulley system, the ratio of the distance (D) moved by the force to the height (H) moved by the load is just equal to the number of strands of cord supporting the lower pulley. Therefore, the theoretical mechanical advantage is equal to *7* and 5 in the two cases here. We also can calculate the theoretical mechanical advantage using the equation involving the ratio of the distances H and D.

Calculate the actual and theoretical mechanical advantages of both the inclined plane and pulley systems as required in Data Tables 10.1 and 10.2. Also calculate the percent differences between the AMA and the TMA and record all values in the data tables.

The *efficiency* of a machine is defined as the work done by the machine divided by the work put into the machine in order to perform this work output. It turns out that by applying the physical definition of work to simple machines, the efficiency can be calculated easily as the ratio of the actual mechanical advantage divided by the theoretical mechanical advantage. As a percentage, this can be written

$$\%\ \text{Efficiency} = \frac{\text{actual mechanical advantage}}{\text{theoretical mechanical advantage}} \times 100$$

By multiplying by 100, we can express the efficiency as a percentage. No machine can have an efficiency larger than 100 percent, and because of losses due to friction in the operation of the machine itself, the efficiency of most machines is much less than 100 percent. The more complex the machine, the more losses can be expected due to friction and the less efficient the machine will be. Lubricants often are applied to reduce the friction and thus increase the efficiency of a machine.

Calculate the efficiency of the various machines studied in this experiment.

Inclined plane 25° (Trial 1) % Efficiency = ____________
Inclined plane 25° (Trial 2) % Efficiency = ____________
Inclined plane 45° (Trial 1) % Efficiency = ____________
Inclined plane 45° (Trial 2) % Efficiency = ____________
7-String pulley system % Efficiency = ____________
5-String pulley system % Efficiency = ____________

Show your calculations on a separate sheet of paper

QUESTIONS

1. Define *work* in terms of applied force and distance.

2. Why must the spring balance be moving slowly at a constant speed when the applied force is measured?

3. Distinguish between TMA and AMA.

4. Is the TMA always greater than the AMA for a machine? Explain.

5. Define *efficiency*. Give an example.

Experiment 11

Waves

INTRODUCTION

Waves are a means of propagating energy by particle disturbances. Most students are familiar with water waves, sound waves, light waves, and waves in stretched strings like piano or guitar strings. Particle disturbance means that the particle has been displaced from its equilibrium position by some agent. Once displaced, the particle tends to return to its original position. In the process, a wave will be propagated outward from the disturbance, transferring energy. The propagation of energy from the disturbance is called *wave motion.*

Waves are classified as transverse or longitudinal. *Transverse waves* have the particle displacement perpendicular to the motion of the wave. *Longitudinal waves* are those in which the particle displacement is in the same direction, that is parallel to the direction of the wave motion.

All waves have certain fundamental properties. A few of these properties are wave velocity, wavelength, period, frequency, and amplitude. These are illustrated in Fig. 11.1 on the next page.

The *wave velocity* of the wave is the distance the wave travels per unit of time, and in this experiment it is measured in centimeters per second. The *wavelength* is the distance between two consecutive wave crests or wave troughs and is measured in centimeters or some other unit of length. The *period* of a wave is the time required for one wavelength to pass any point along the direction of travel. The *frequency* is the number of vibrations or cycles per unit time the particle disturbance is taking place. Frequency is usually measured in cycles per second, or Hertz. The period is the reciprocal of the frequency. You will recall that the reciprocal of a number is one divided by the number. The relation here, therefore, would be written as

$$T = \frac{1}{f}$$

where T is the period in seconds and f is the frequency in cycles per second.

The relation among velocity, frequency, and wavelength is shown by the equation

$$v = \lambda f$$

where v = velocity
λ = wavelength
f = frequency

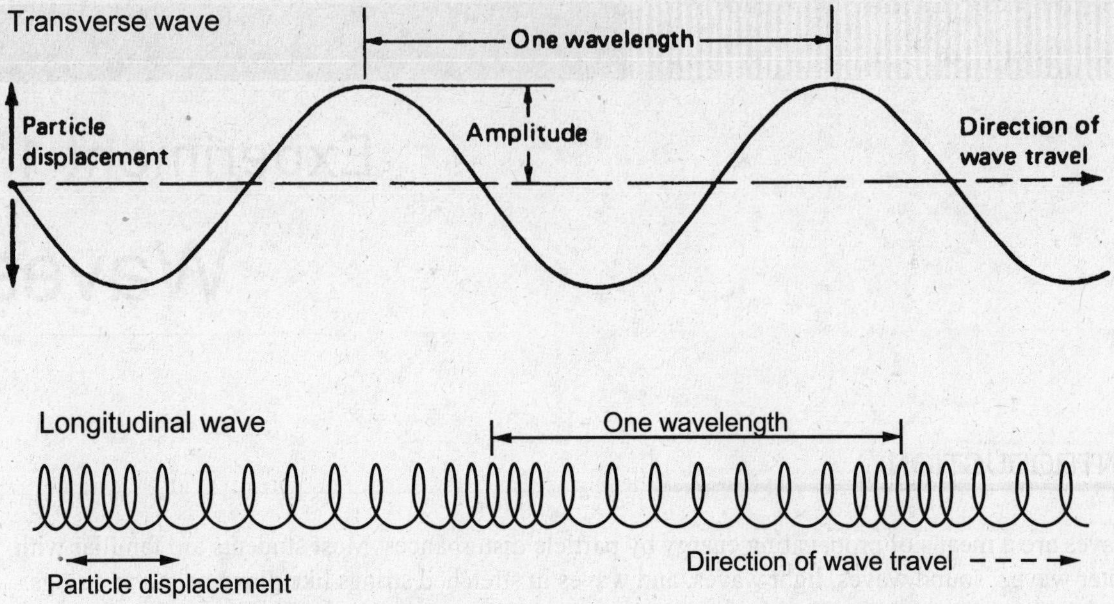

Figure 11.1 Transverse and Longitudinal Wave Properties.

LEARNING OBJECTIVES

After completing this experiment, you should be able to do the following:

- ▼ Define the terms *wave velocity, wavelength, frequency, period,* and *amplitude* for a wave.
- ▼ Differentiate between longitudinal and transverse waves.
- ▼ Demonstrate the transfer of energy by wave motion.
- ▼ Measure the wave velocity in a stretched rubber rope or hose.
- ▼ Measure the wavelength of a sound wave.

APPARATUS

Metal can with plastic lid (large-sized coffee can with about 1 inch hole cut into the flat metal end), candle, 2-m meter stick, rubber cord, timer, coiled spring, tuning forks (use frequencies high enough to give two or three maximum and minimum points for air column in closed pipe), a 1,024 cps tuning fork, rubber hammer, resonance apparatus, large water tray, cork stopper.

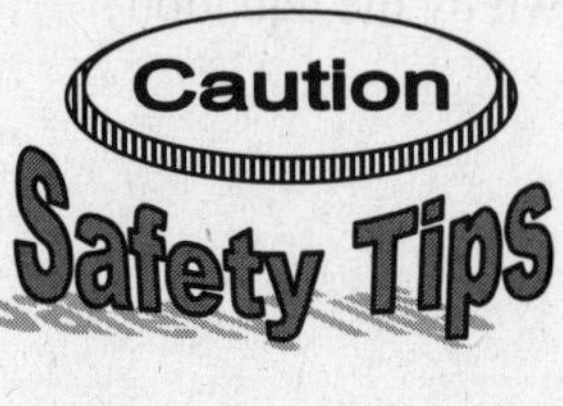

Open flame from a candle can be dangerous. Do not leave lighted candles unattended or reach across flame.

EXPERIMENT 11 NAME (print) ____________________ DATE ____________
LAST FIRST

LABORATORY SECTION ______ PARTNER(S) ____________________

PROCEDURE

1. Assemble the metal can, meter stick, and candle as shown in Fig. 11.2. Determine if you are able to extinguish the candle flame one meter (or less) from the can by lightly thumping the plastic lid with a rubber hammer. Explain your results.

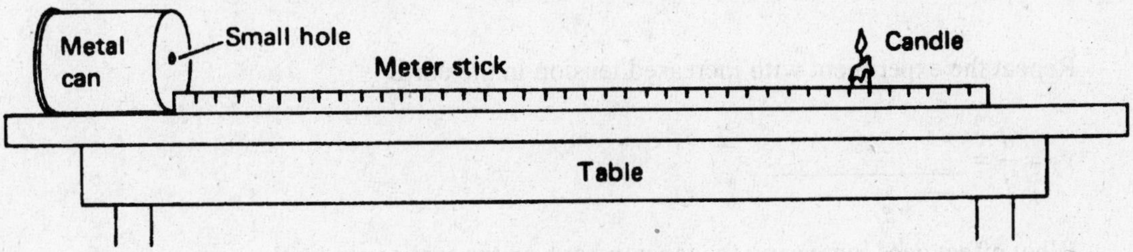

Figure 11.2

2. The velocity of a wave in a stretched string or rubber cord is given by the following:

$$v = \sqrt{\frac{F}{m/L}}$$

where v = velocity
F = tension in cord measured in units of force
m/L = mass per unit length

Attach the rubber cord to one wall of the laboratory, then apply tension to stretch the cord across the room. Displace the cord as shown in Fig. 11.3. Observe the wave travel the length of the cord, be reflected at the wall, and return to your hand. Note: Do not stretch the hose too tight or the wave pulse will travel too fast for you to time accurately.

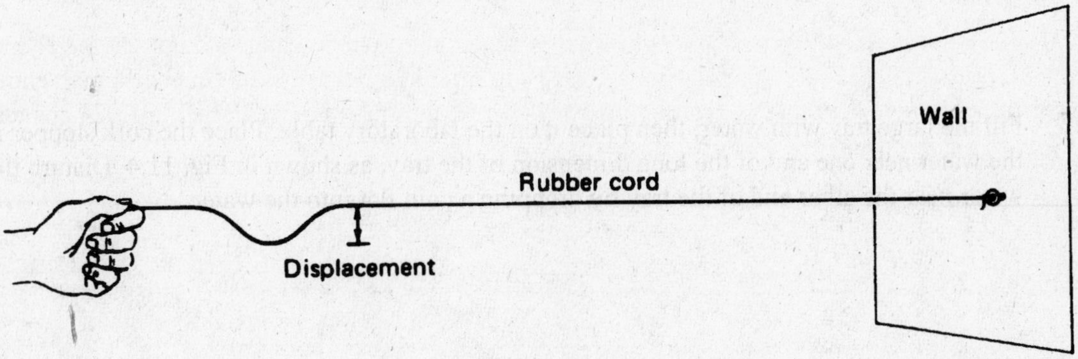

Figure 11.3

With the timer determine the time (three trials) it takes a wave to travel the total distance (d). Measure the distance traveled, and calculate the velocity of the wave using the average value from your three time trials. Since the wave moves quite rapidly you may wish to measure the time the wave pulse to travel back and forth between your hand and the wall several times and then divide the *total* distance by the *total* time to find the wave velocity.

Remember the definition of velocity:

$$v_1 = \frac{d}{t} = \_\_\_\_\_\_\_\_\_\_\_\_$$

Repeat the experiment with increased tension in the cord.

$$v_2 = \frac{d}{t} = \_\_\_\_\_\_\_\_\_\_\_\_$$

What effect does increasing the tension have on the wave velocity?

What type of wave is traveling in the cord? Explain your answer.

3. Stretch the coiled spring along the floor. Displace the spring at one end by compressing a few coils and releasing. Observe the wave motion along the spring. Explain the type of wave observed.

4. Fill the large tray with water; then place it on the laboratory table. Place the cork stopper in the water near one end of the long dimension of the tray, as shown in Fig. 11.4. Disturb the water near the other end of the tray by dropping a coin flat into the water.

EXPERIMENT 11 NAME (print) ______________________ DATE ______________
LAST FIRST

LABORATORY SECTION ______ PARTNER(S) ______________________

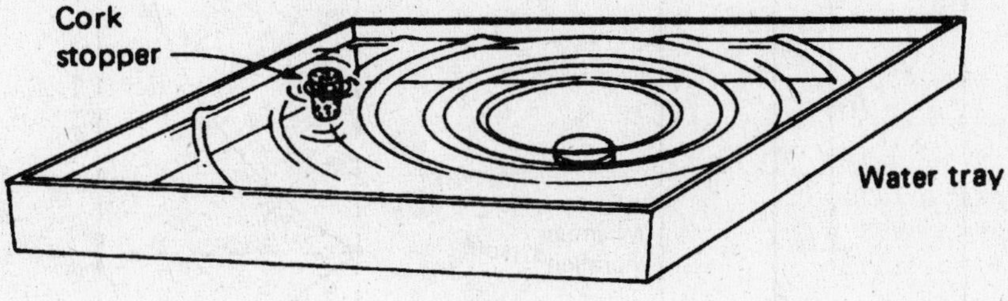

Figure 11.4

Observe the wave motion on the water and the movement of the cork. What type of wave are you observing? Explain your answer.

5. In the introduction, wavelength was defined and illustrated. Fig. 11.5 illustrates the component sections of one complete wavelength.

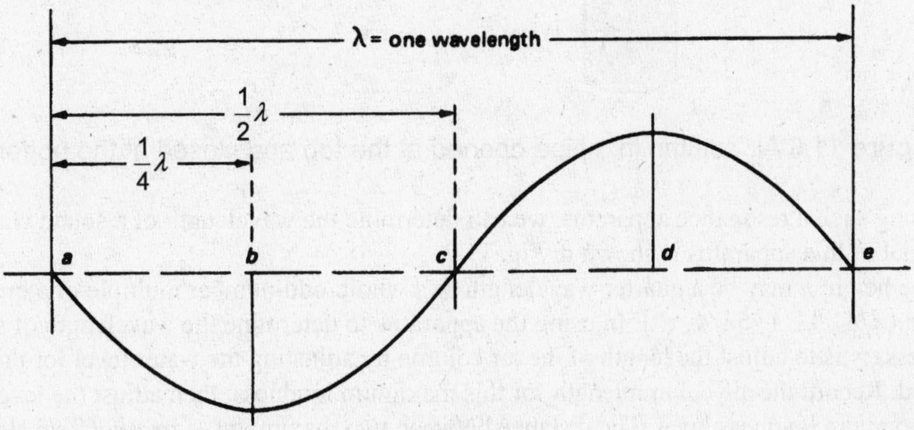

Figure 11.5

Note that minimum particle displacement occurs at positions *a, c,* and *e*. These points are called *nodes*. Maximum displacement occurs at *b* and *d* which are referred to as *antinodes*. The distance *ab* is one-fourth wavelength. Likewise the distances *bc, cd,* and *de* are one-fourth wavelengths. Observe that one maximum and one minimum value occur in each quarter-wavelength. Figure 11.6 illustrates the particle displacement along a quarter-wave tube containing a vibrating column of air produced by a tuning fork. The air at the open end of the tube will vibrate at the maximum value because at the open end the air is free to move. At the lower end of the tube, minimum particle displacement will take place because the tube is closed and particle vibration is restricted.

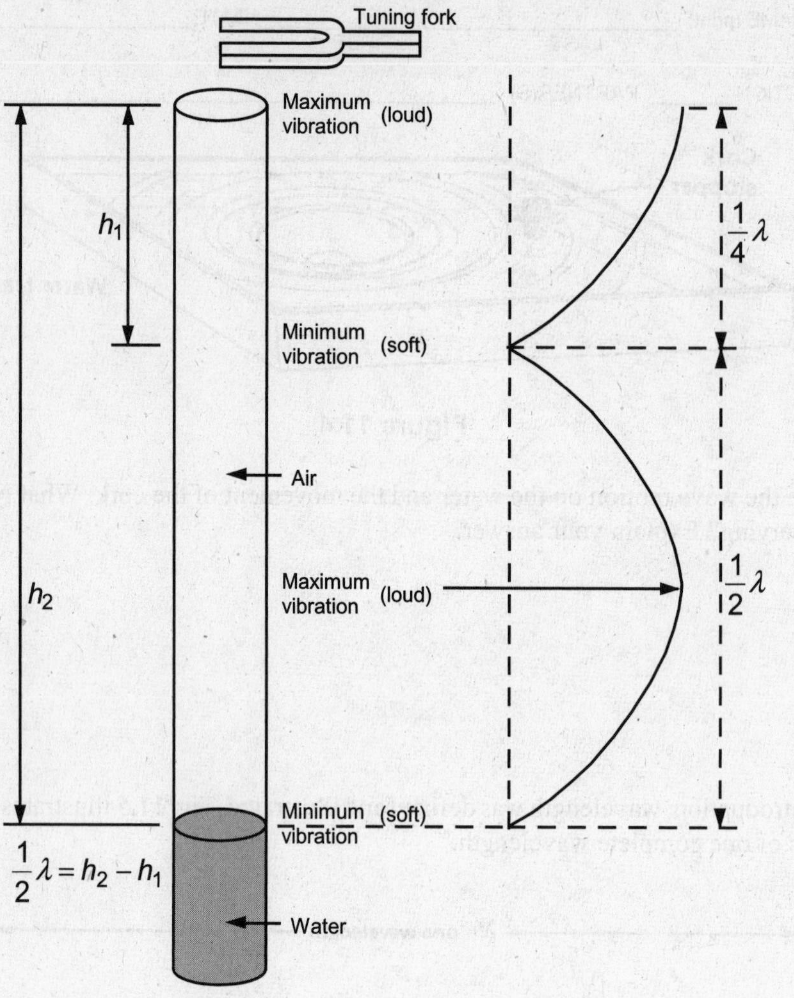

Figure 11.6 Air column in a pipe opened at the top and closed at the bottom.

Using such a resonance apparatus, we can determine the wavelength of a sound wave. A simple model of this apparatus is shown in Fig. 11.7.

The height *h* may be a quarter-wavelength or a whole-odd-number multiple of a quarter-wavelength $(\lambda/4, 3\lambda/4, 5\lambda/4, \ldots)$. In using the apparatus to determine the wavelength of sound, all that is necessary is to adjust the length of the air column by adjusting the water level for the loudest sound heard. Record the air-column length for this maximum loudness; then adjust the level again for the next maximum loudness level. The distance between two maximums is *one-half wavelength.*

The velocity of sound in air in meters per second can be calculated from the following equation:

$$v = 332 \text{ m/s} + (0.60 \text{ m/s})T$$

where v = velocity of sound in air
332 m/s = velocity of sound in air at 0° Celsius
T = temperature of air in degrees Celsius

EXPERIMENT 11 NAME (print) ______________________________ DATE ______________
LAST FIRST

LABORATORY SECTION ______ PARTNER(S) ______________________________

Using this information and the equation relating velocity, wavelength, and frequency, $\lambda = v/f$ calculate the theoretical value of one wavelength for the waves produced in the air column by a 1024-cycles-per-second tuning fork.

Theoretical value of wavelength $\lambda =$ __________

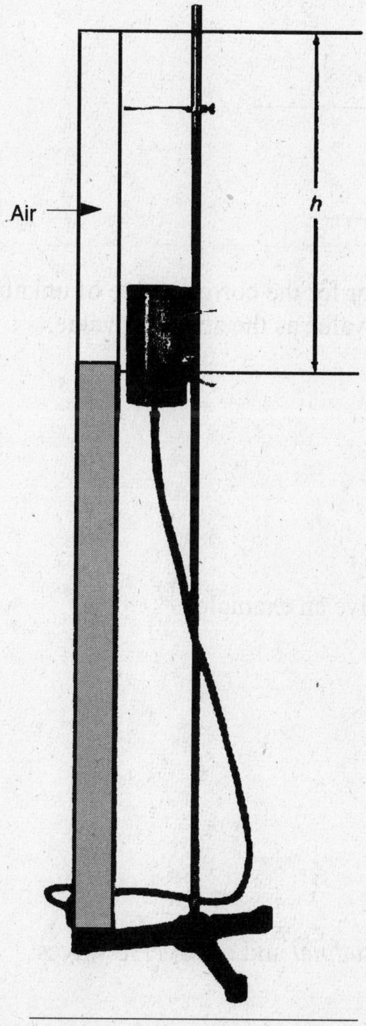

Figure 11.7 Resonance apparatus.

Experimental Determination of Wavelength:

A: Strike the 1024-cps tuning fork with the rubber hammer. Place the vibrating fork at the top of the air column and adjust the water level until the air column is vibrating at its maximum loudness. You will be able to ascertain the maximum loudness by listening. Record the length of the air column. Readjust the water level to the next maximum loudness position. Record the length of the air column and calculate the value of a half-wavelength by subtracting these lengths.

Experimental value of half-wavelength $\frac{\lambda}{2} =$ ____________

Wavelength $\lambda = 2\left(\frac{\lambda}{2}\right) =$ ____________

Percent difference between measured and theoretical wavelength % difference = __________

B: Determine the wavelength and frequency of an unknown tuning fork using the resonance apparatus.

Wavelength $\lambda = 2\left(\frac{\lambda}{2}\right) =$ ____________

Frequency $f = \frac{v}{\lambda} =$ ____________

Ask the laboratory instructor for the correct value of unknown frequency, and determine the percent error using the instructor's value as the accepted value.
f (given by instructor) = ____________

Percent error = ______________

QUESTIONS

1. Define *wave motion,* and give an example.

2. Distinguish between *longitudinal* and *transverse* waves.

3. Define the *velocity, wavelength,* and *frequency* of a wave and give an example of each. Use symbol notation in the definitions.

4. State the relationship between velocity, wavelength, and frequency of a wave.

5. Define the *amplitude* of a wave, and give an example.

6. State the relationship between the frequency and the period of a wave.

7. How do your measurements in Procedure 2 predict the way that the velocity of a wave in a rubber cord varies with the tension in the cord?

8. Why is there a maximum vibration of the air molecules at the open end of a resonance tube?

9. How does the velocity of sound in air vary with the air temperature?

Experiment 12

Interference of Light Waves

INTRODUCTION

When matter is disturbed, energy radiates from the disturbance; this radiation of energy is called *wave motion.* Wave motion has physical properties such as velocity, frequency, period, wavelength, and amplitude as defined and explained in Experiment 11. Another property of a wave is phase. The *phase* of the particle undergoing the disturbance is a measure of its position in respect to its equilibrium position. Two or more waves from a disturbance may reinforce one another when they are changing in the same direction or tend to cancel one another when varying in the opposite directions. When two or more waves have their displacement at all times in the same direction, they are said to be *in phase* with one another. See Fig. 12.1.

When interference occurs between waves, the two waves must remain exactly in phase for constructive interference (waves reinforce each other) and 180° out of phase for destructive interference (waves completely cancel each other). For complete destructive interference, the two waves must have the same amplitude. They also must have the same wavelength and be traveling in the same plane.

The preceding conditions can be obtained with the use of a single source of light. The waves spreading out from the source as shown in Fig. 12.2 fall on a double slit S_1 and S_2. Thus the waves spreading out from sources S_1 and S_2 have the same wavelength, the same amplitude, and in this case are traveling in the same direction. The interference of these waves produces alternate bright and dark areas as shown in Fig. 12.2. The bright areas are regions where constructive interference is occurring, and the dark areas are regions where destructive interference is occurring.

The relationship among the wavelength (λ), the distance (d) between S_1 and S_2 (called the *slit separation*), the order number between two wave paths showing where wave reinforcement takes place (n), and the ratio of a value of (y) to a corresponding value of (x) (see Fig. 12.3) is given by the following equation:

$$n\lambda = d\left(\frac{y}{x}\right)$$

or

$$\lambda = \frac{d}{n}\left(\frac{y}{x}\right) \qquad \text{where } n = 1, 2, 3, 4, \ldots \qquad \text{Eq. (12.1)}$$

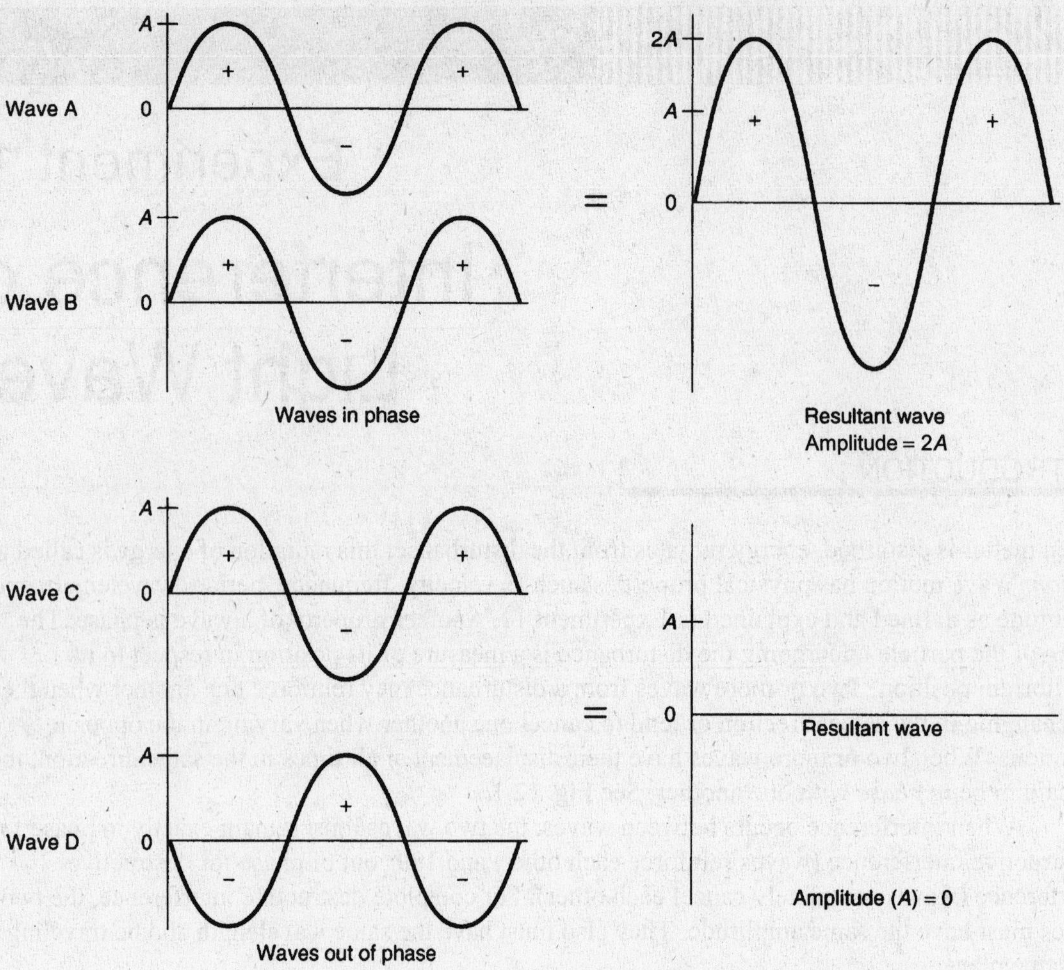

Figure 12.1 Diagram illustrating the phase relationship between two waves that are in phase with one another, and two waves that are 180° out of phase. The resultant wave is also shown. Waves A and B are in phase. That is, their displacement at all times is in the same direction. Waves C and D are out of phase. That is, their displacement at all times is in the opposite direction. The waves shown have the same amplitude and wavelength.

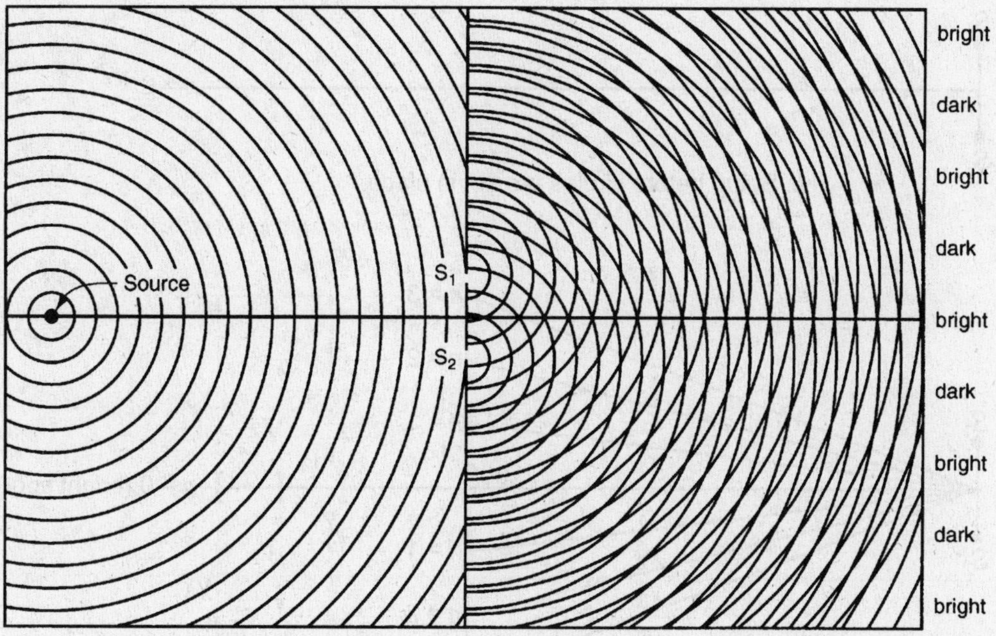

Figure 12.2 Diagram illustrating the interference pattern of bright and dark regions produced by waves interfering with one another. Since the waves from S_1 and S_2 are from the same source, they have the same wavelength, the same amplitude, and are traveling in the same direction.

LEARNING OBJECTIVES

After completing this experiment, you should be able to do the following:

- Define and explain *phase* as it relates to wave motion.
- Distinguish between constructive and destructive interference.
- Calculate the wavelength of wave motion from data obtained in the experiment.

APPARATUS

Millimeter rule, two Keuffel and Esser wave-pattern transparencies, sheet of white paper, hand calculator.

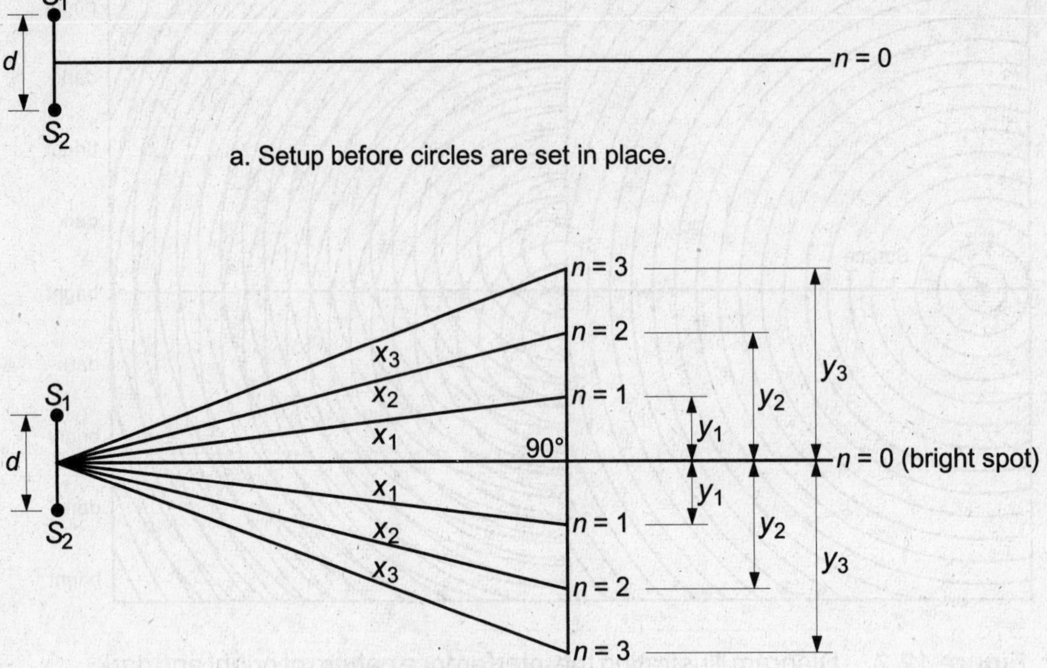

a. Setup before circles are set in place.

b. Completed drawing.

Figure 12.3

PROCEDURE 1

1. Take a sheet of white paper and place it on your laboratory table with the long side parallel to the edge of the table next to you. Draw a line 2 cm long near and parallel to the left edge of the paper.
2. Draw a dot at each end of the 2-cm line. These dots represent two sources of waves similar to S_1 and S_2 as shown in Fig. 12.2.
3. From the midpoint on the line between the two dots, draw a perpendicular line (call it the $n = 0$ line) to the right across the paper to a point near the opposite edge of the paper. See Fig. 12.3a. Label this $n = 0$.
4. Position the two plastic transparencies of concentric circles so that the center of the circles coincides with the two dots. This will produce an observable interference pattern.
5. Slightly rotate the two transparencies as a unit, if necessary, until the bright region near the perpendicular line $(n = 0)$ is spaced evenly on each side of the perpendicular line you drew.
6. With a pen, place a mark on the white paper at the right edge of the transparencies to coincide with the centers of the first, second, and third bright regions above and below the central bright region located on the perpendicular line $(n = 0)$. See Fig. 12.3b.
7. Remove the transparencies and draw lines x_1, x_2, and x_3 from the midpoint of the 2-cm line to the marks made in Step 6. Also draw lines y_1, y_2, and y_3.
8. Label all x lines and all y lines. See Fig. 12.3b.
9. Measure with the millimeter rule the length of all x and y lines and record their values in Data Table 12.1.
10. Calculate the wavelength (λ_c) using Eq. 12.1. Record the calculated wavelength values in Data Table 12.1. Then calculate the average value, $\bar{\lambda}_c$, for your six experimental values.

EXPERIMENT 12 NAME (print) ______________________ DATE ______________
LAST FIRST

LABORATORY SECTION ______ PARTNER(S) ______________________

PROCEDURE 2

Repeat Procedure 1 with the distance between S_1 and S_2 equal to 3 cm. Record the data in Data Table 12.1 for the 1st, 2nd, and 3rd order bright spots.

DATA TABLE 12.1

d (cm)	n	y (cm)	x (cm)	λ_c Calculated (cm)
2.00	1	$y_1 =$	$x_1 =$	
2.00	2	$y_2 =$	$x_2 =$	
2.00	3	$y_3 =$	$x_3 =$	
3.00	1	$y_1 =$	$x_1 =$	
3.00	2	$y_2 =$	$x_2 =$	
3.00	3	$y_3 =$	$x_3 =$	
			Average value for λ_c	

PROCEDURE 3

The wavelength of the concentric wave pattern on the transparencies can be measured directly with the millimeter rule. The accuracy of the measurement can be increased by measuring several wavelengths, rather than just one, and then dividing by the number of wavelengths measured.

1. Measure the distance for 20 wavelengths and record the value below.
   Measured value for 20 wavelengths $d =$ ______________

2. Find the value for 1 wavelength.
   $\lambda_m = \frac{d}{20} =$ ______________

3. Determine the percent difference between the average calculated wavelength (λ_c) from Data Table 12.1 and the measured wavelength (λ_m) found above.

QUESTIONS

1. Define and explain *phase* as it relates to wave motion.

2. What would you observe when two light waves meet in phase with one another?

3. When referring to wave motion, what is the meaning of the expression *out of phase?*

4. What three conditions must be met for complete destructive interference between two waves?

5. What happens to the spacing between the 1st, 2nd, and 3rd bright spots (y_1, y_2, and y_3) in the interference pattern when the slit separation (d) is increased?

Experiment 13

Plane Mirrors and Index of Refraction of Light

INTRODUCTION

Light rays change direction when they intercept a new medium. If the ray bounces back into the same medium, we say that it is *reflected.* If it passes on into the new medium, it is often *deflected* (its direction of travel is changed). This is called *refraction.* The law dealing with reflection states that the incident angle of the approaching ray will always be exactly equal to the angle of reflection. These angles are measured with respect to a perpendicular line drawn to the reflecting surface, called the *normal line.* In equation form,

$$\theta_i = \theta_r$$

For light that passes into the new medium, the law of refraction, also called *Snell's law,* will apply. Again, measured with respect to a normal line, the angle of incidence and the angle of refraction are related to each other in the following way:

$$n_1 \sin\theta_i = n_2 \sin\theta_r$$

where n_1 is the index of refraction for the initial medium the ray is traveling through, θ_i is the angle of incidence of the ray, n_2 is the index of refraction of the medium into which the light passes, and θ_r is the angle of refraction. This experiment studies both laws and allows for the experimental verification of them using simple ray tracing methods.

LEARNING OBJECTIVES

After completing this experiment, you should be able to do the following:

- ▼ Observe experimentally the principles involved in the reflection of light from a plane mirror.
- ▼ Trace the path of a ray of light through a rectangular plate of glass.
- ▼ Determine the index of refraction of glass.

APPARATUS

Plane mirror with holder, glass plate, several pieces of 8 1/2 in. × 11 in. white paper, pins, soft wood or Celotex board, protractor, ruler.

PROCEDURE 1

Place an 8 1/2 in. × 11 in. sheet of paper on the horizontal soft wood or Celotex board. Place the mirror near one end of the paper as shown in Fig. 13.1. Place two pins at *C* and *D*. Start with angle θ_r about 30°; then sight along the reflected line and place two pins at *E* and *F* so that pins *C, D, E,* and *F* appear to be in a straight line when pins *F* and *E* are viewed directly and pins *D* and *C* are seen as reflections in the mirror. You will have to position your eye very close to the pin position *F* and just above table level to "see" all four pins in a straight line. This may be most easily accomplished if the "board" is placed near one edge of your lab table so that you may look over the edge (while stooping or kneeling) to see the pins and mirror. Draw a line along the front and back of the mirror surfaces. Then remove the mirror and the pins. Draw lines through the holes left by the pins so that they intersect at or near the mirror's surface.

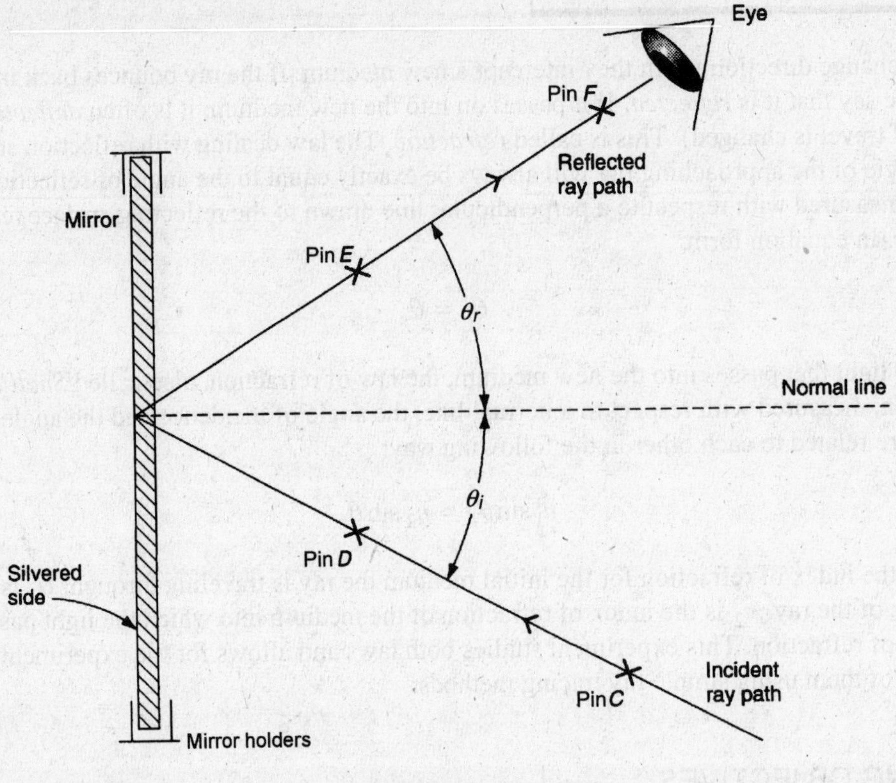

Figure 13.1

Carefully draw a normal (perpendicular) line at right angles to the mirror's surface. All angle measurements should be made with respect to this line. With the protractor, measure angles θ_i and θ_r, and record them in Data Table 13.1. Repeat this procedure for two other angles of incidence (θ_i), and record all additional data in Data Table 13.1.

EXPERIMENT 13 NAME (print) ______________________ DATE ______________
LAST FIRST

LABORATORY SECTION ______ PARTNER(S) ______________________

DATA TABLE 13.1

Trial	θ_i	Estimated Error	θ_r	Estimated Error
1		±		±
2		±		±
3		±		±

Carefully compare the angles θ_i and θ_r for each of your three trials. In the space below write the general law for reflection of a light ray from the surface of a smooth plane mirror.

Considering your drawing ability and the accuracy of your measuring instruments, estimate the error in your measurement of θ_i and θ_r. Since we have taken only a few readings, these will be only estimates, but try to be as realistic as possible. Express your error estimates as plus or minus some number of degrees or tenths of degrees. (Example: 16.3° ± 3.5°.) Now make a statement indicating how well your experimental work has proved or disproved the law of reflection.

PROCEDURE 2

In this part of the experiment you will attempt to use parallax (depth perception) to locate the position of a virtual image behind a mirror. This is a rather difficult procedure, and your lab instructor may need to help you get started. You may either use two long objects such as pencils and view the back one (*B*) *over* the mirror or use two pins and view the back one (*B*) *under* the mirror. The idea is to position object *A* in front of the mirror and then locate the position of its virtual image *behind* the mirror by moving object *B* back and forth along the normal line until it shows no parallax (relative movement with respect to the image of object *A* as seen in the mirror) when you move your eye sideways, parallel to the mirror's surface. When no apparent relative motion is observed, object *B* is located precisely at the location of the virtual image of object *A* produced by the mirror.

Place the mirror in the center of a sheet of paper. Outline the front and back surfaces of the mirror, and then construct a normal line through the surface outlines near the center of the mirror. Replace the mirror on the paper. (Lift it slightly in its holder if you decide you would rather look under it, as shown in Fig. 13.2, instead of over it.) Place a pencil or a pin at point *A* (somewhere on the normal line), and observe the image in the mirror. Then adjust the position of a pencil or a pin at point *B* (also on the normal line) behind the mirror so that there is no parallax between it and the image of object *A* in the mirror. When no parallax exists, you can move your eye from side to side with no apparent motion of pin *B* relative to the image of *A* in the mirror. In this case, the distance of the virtual image behind the mirror coincides with the location of object *B*.

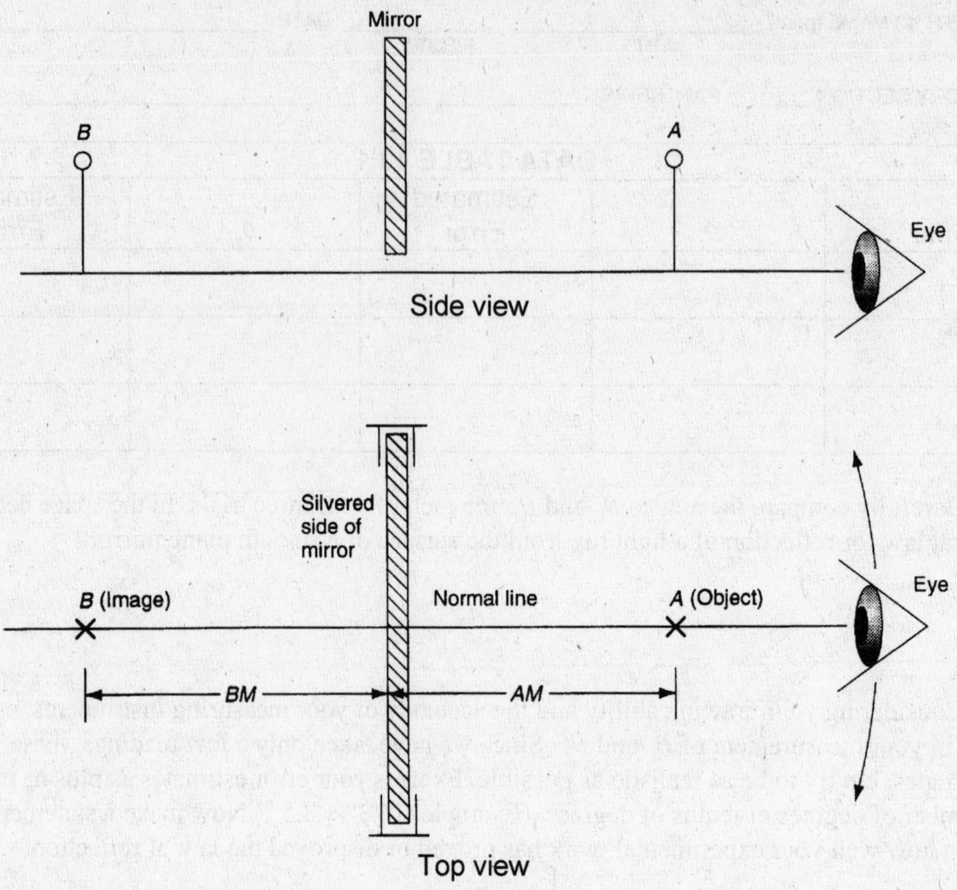

Figure 13.2

Mark the position of objects *A* and *B*, remove the mirror, and measure the distances from *A* (*AM*) and *B* (*BM*) to the *back* of the mirror. This is trial 1.

Repeat these observations for a different position of object *A* (on the normal line). This is trial 2. Record all data in Data Table 13.2.

DATA TABLE 13.2

Trial	Object Distance (*AM*), in meters	Image Distance (*BM*), in meters
1		
2		

EXPERIMENT 13 NAME (print) ______________________________ DATE ______________________
LAST FIRST

LABORATORY SECTION ______ PARTNER(S) ____

What does your data suggest about the relationship between the object distance (*AM*) and the image distance (*BM*) when measured with respect to the reflecting surface of a plane mirror?

Draw a ray diagram showing how the eye sees the virtual image in a plane mirror.

Calculate the percent difference between your two values for the object and image distances in Trials 1 and 2.

Trial 1 % difference ______________________

Trial 2 % difference ______________________

PROCEDURE 3

Place the rectangular glass plate in the center of a sheet of paper as shown in Fig 13.3 on the soft wood base. After placing pins at *A* and *B*, place additional pins at *C* and *D* so that all four pins appear to be in a straight line when *A* and *B* are viewed through the glass plate. Outline the glass plate, remove the glass plate and pins, and draw the light rays *AB* and *CD*. Draw the normal lines perpendicular to the faces of the glass plate at the points where rays *AB* and *CD* intersect the glass surface. Also draw the light path inside the glass plate between the intersection points at the surfaces as shown. Measure the angles θ_{i1}, θ_{r1}, θ_{i2}, and θ_{r2} with a protractor and determine the index of refraction for both pairs of incident and refracted rays. Record the data in Data Table 13.3.

In this case, the index of refraction n_2 of glass is given by Snell's law:

$$\frac{n_2}{n_1} = \frac{\sin\theta_i}{\sin\theta_r}$$

where $n = 1.00$ (to 3 significant digits) because the media outside the glass prism is air.

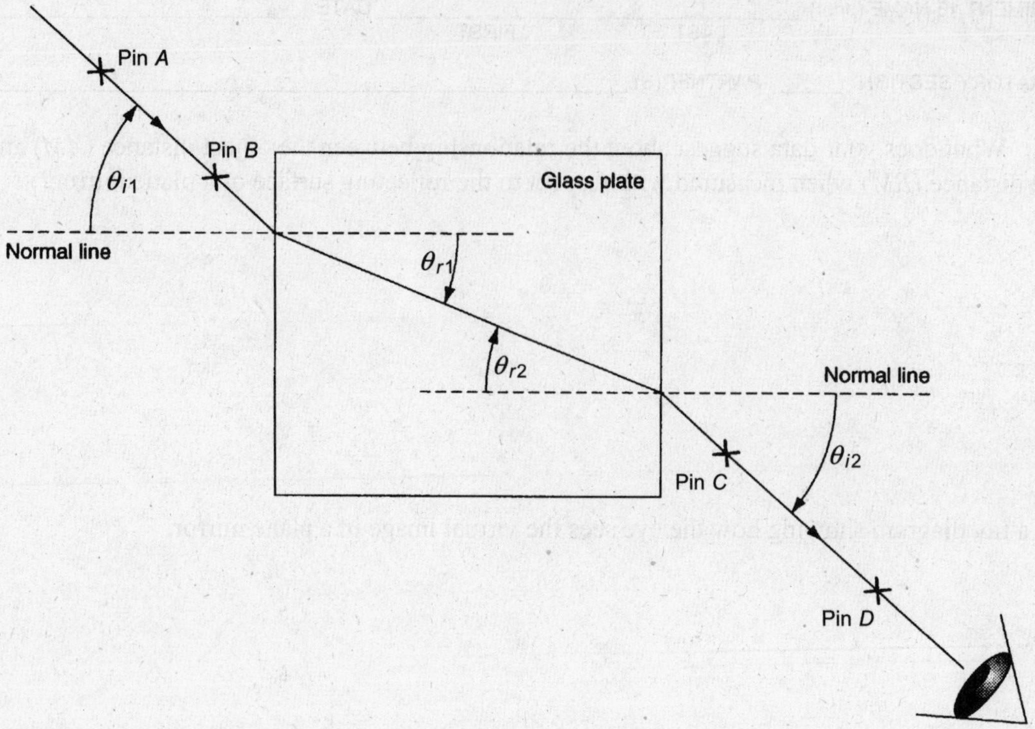

Figure 13.3

DATA TABLE 13.3					
Surface	Angle of Incidence θ_i	$\sin\theta_i$	Angle of Refraction θ_r	$\sin\theta_r$	Index of Refraction $n_2 = \frac{\sin\theta_i}{\sin\theta_r}$
Left (1)					
Right (2)					

Calculate the percent difference between your two values for the index of refraction.

% difference = ________________

EXPERIMENT 13 NAME (print) ______________________________ DATE ______________
LAST FIRST

LABORATORY SECTION ______ PARTNER(S) ______________________________________

QUESTIONS

1. State the law of reflection, and give an example showing its relevance to this experiment.

2. Distinguish between *reflection* and *refraction.*

3. State the law of refraction (Snell's law), and explain how it can be used to find the index of refraction for a glass plate.

4. If the index of refraction for a glass plate is higher than the index of refraction for water, which medium will bend a light beam more when the light beam enters the new material from the air? Explain your choice.

Experiment 14

Mirrors, Lenses, and Prisms

INTRODUCTION

Concave mirrors and convex lenses are optical devices that focus light rays and produce images of objects set in front of them. These optical devices can be studied using actual lenses and mirrors set up on optical benches or by doing ray diagrams that will show the positions and characteristics of the images formed. This experiment will allow you to do both and compare the results of the two methods to see how they work together. In addition, you also will use a prism to split white light into its component colors and observe the various colors produced in this process.

Some instructors prefer to have their students do Procedure 2 before they come to the lab. This helps them to visualize where images that are formed by concave mirrors and convex lenses will be found and also shows them what image characteristics they will observe for different placements of the object (light source) during this experiment. Even if your instructor has not specifically required you to complete these ray diagrams before you come to lab, you might consider doing them early on your own. These diagrams only require a sharp pencil and a straight edge. Once you have done them, you should understand the material in Procedure 1 much better and you will have a better idea where to look for the images formed by the various optical elements studies in this experiment.

LEARNING OBJECTIVES

After completing this experiment, you should be able to do the following:

- Determine experimentally the position, size, and characteristic features of images formed by concave mirrors and convex lens.
- Draw ray diagrams for concave mirrors and convex lens.
- Determine experimentally the colors of the visible part of the electromagnetic spectrum.

APPARATUS

Concave mirror, convex lens, meter stick, candle or other optical light sources, screen (small card), holders for mirror and lens and for the screen.

Caution

Safety Tips

Open flame of candle can be dangerous. Do not leave lighted candle unattended or reach across flame.

EXPERIMENT 14 NAME (print) ______________________ DATE ______________
LAST FIRST

LABORATORY SECTION ______ PARTNER(S) ______________________________

PROCEDURE 1

Step 1 Mount the concave mirror and screen (card) on the meter stick, as shown in Figs. 14.1 and 14.2. Determine the focal length of the mirror as follows: Place the mirror at some fixed position on the meter stick; then move the screen toward or away from the mirror until a clear image of a distant object outside the window, or on the other side of the room, appears on the screen. The distance between the mirror and the screen, when adjusted for a clear image, is the focal length. Take three trials, and calculate the average. Record in Data Table 14.1.

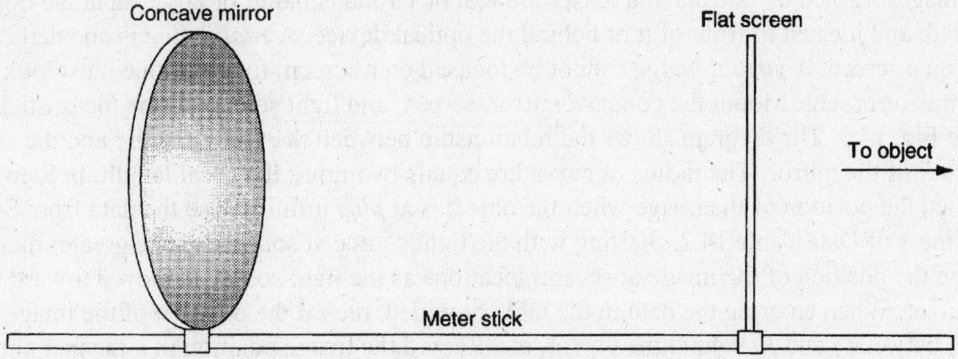

Figure 14.1 Side view of optical bench. The concave mirror is tilted slightly toward the screen. The screen is positioned only on one side of the optical bench. See top view in Fig. 14.2.

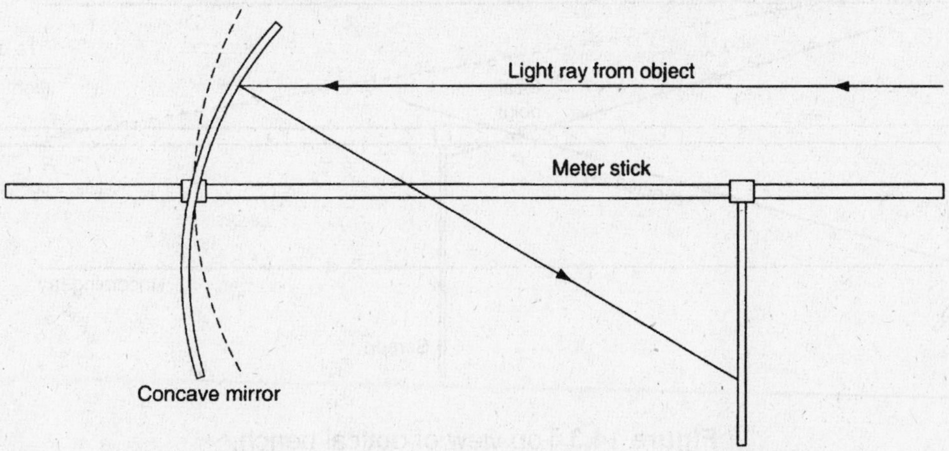

Figure 14.2 Top view of optical bench. The mirror is tilted slightly toward the screen.

DATA TABLE 14.1	
Trial	Focal Length of Mirror
1	
2	
3	
Average	

Step 2 Images formed by mirrors and lenses are real or virtual, smaller or lárger than the object, erect or inverted, and located in front of it or behind the optical device. A *real image* is one that can be focused on a screen. A *virtual image* cannot be focused on a screen; to see it, one must look directly into the mirror or lens. Mount the concave mirror, screen, and light source on the meter stick, as shown in Fig. 14.3. The diagram shows the relationship between the focal length *f* and the radius of curvature *R* of the mirror. The radius of curvature equals two times the focal length. In Step 1, you determined the position of the image when the object is at *plus* infinity. Use the data from Step 1 to answer line 1 of Data Table 14.2. Starting with the light source at some distance greater than *2R*, determine the position of the image at several locations as the light source is moved toward the face of the mirror. When entering the data in the table provided, record the position of the image as being at *f*, at *R*, between *f* and *R*, behind the mirror, etc. Record the image position in a manner similar to the way the light source's position is given.

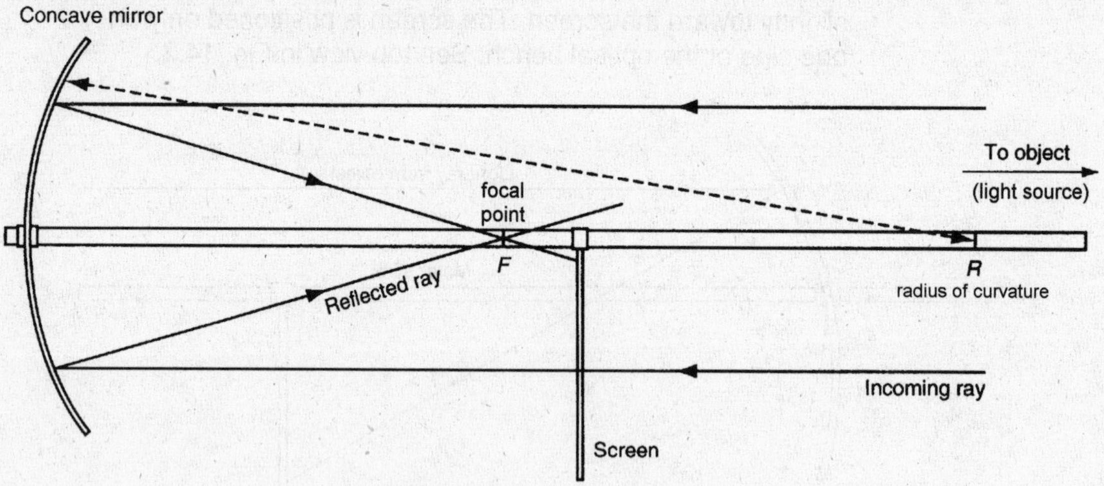

Figure 14.3 Top view of optical bench.

EXPERIMENT 14 NAME (print) ______________________________ DATE ____________________
LAST FIRST

LABORATORY SECTION ______ PARTNER(S) ______________________________

DATA TABLE 14.2 Data Table for Mirror

Position of Light Source	Information on Image			
	Position of Image	Real or Virtual	Erect or Inverted	Larger or Smaller
+ Infinity (at least 8*f*)				
At 4*f*				
At $2f = R$				
Between *R* and *f*				
At *f*				
Between *f* and mirror				

Step 3 The focal length of a lens is determined by the radius of curvature and the index of refraction of the glass. Mount the double convex lens as shown in Fig. 14.4. Determine the focal length of the lens using some object outside the window or across the room and adjusting the screen for a clear image. Consider this object to be at plus infinity, and record the data on line 1 of Data Table 14.3. Place the meter stick on support holders on the laboratory table. Position the light source at the distance in front of the lens called for in the data table. Place the screen in its holder and position it on the other side of the lens from the light source. Adjust the position of the screen until a clear image of the light source is produced on it. Record the data; then move the light source toward the lens and focus again, obtaining information to complete the data table.

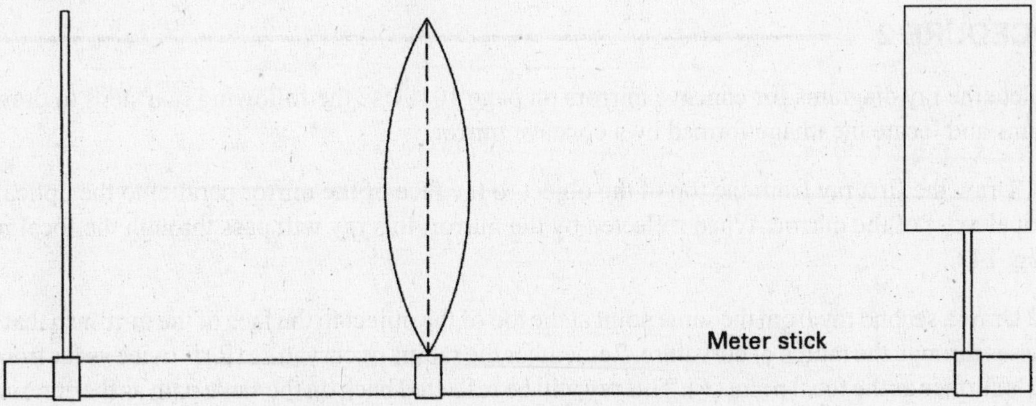

Figure 14.4 Side view of optical bench.

DATA TABLE 14.3 Data Table for Lens				
	Information on Image			
Position of Light Source	Position of Image	Real or Virtual	Erect or Inverted	Larger or Smaller
+ Infinity (at least 8*f*)				
At 4*f*				
At 2*f*				
Between 2*f* and *f*				
At *f*				
Between *f* and lens				

Step 4 Position the prism with respect to the light source as shown in Fig. 14.5, and record the colors observed in the spectrum.

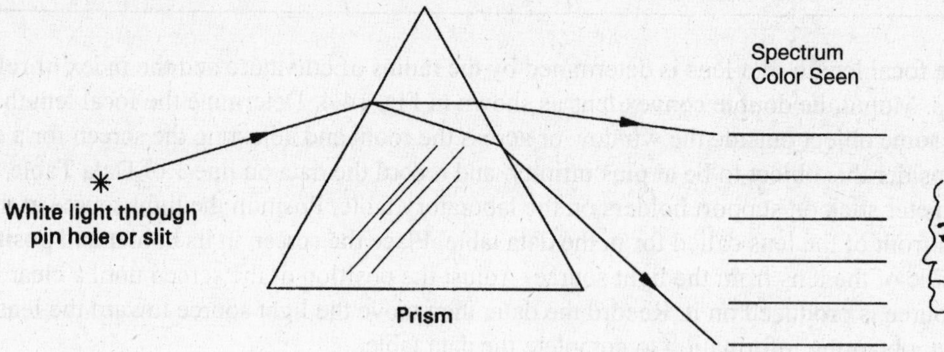

Figure 14.5

PROCEDURE 2

Complete the ray diagrams for concave mirrors on page 101. Use the following two steps to draw ray diagrams and locate the image formed by a concave mirror.

Step 1 Draw the first ray from the top of the object to the face of the mirror parallel to the optical axis (principal axis) of the mirror. When reflected by the mirror, this ray will pass through the focal point. See Fig. 14.6.

Step 2 Draw a second ray from the same point at the top of the object to the face of the mirror so that the ray passes through the radius of curvature. Remember the radius of curvature (R) is twice as far from the mirror's surface as the focal point (*F*). This ray will be reflected back on the same path as the entering ray since the entering ray came in along the normal line. The angle of incidence is zero, therefore the reflected angle is zero and this ray bounces right back along the incoming line. The image is located at the intersection of the two rays that you drew in Step 1 and Step 2.

EXPERIMENT 14 NAME (print) ______________________________ DATE ____________________

LAST FIRST

LABORATORY SECTION ______ PARTNER(S) ________________________________

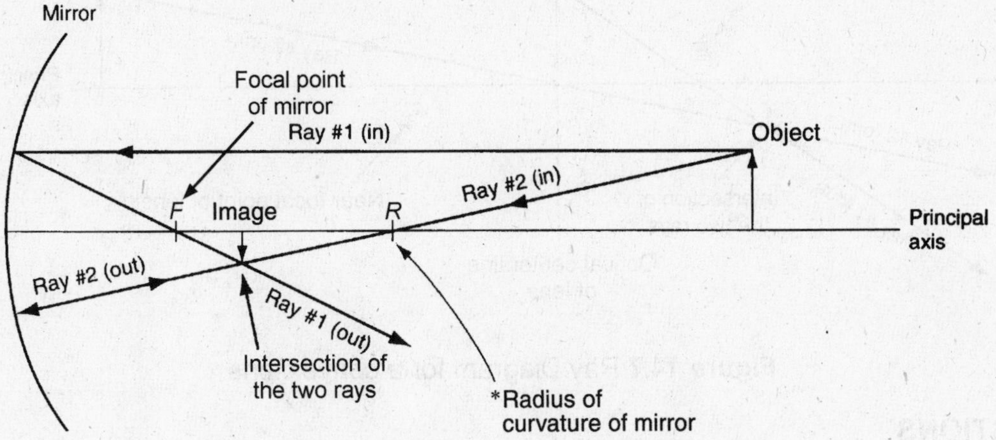

Figure 14.6 Ray diagram for a concave mirror.

*For any spherical concave mirror *R* is twice as far from the mirror's surface as *F*, the focal point.

If you have trouble, refer to your textbook (Section 7.5) or ask your instructor about drawing ray diagrams when the object is between the focal point and the mirror. The procedure is the same but the outgoing rays do not converge on the same side of the mirror as the object, so you must extend each outgoing ray *behind* the mirror to locate their point of intersection. This makes the resulting image virtual instead of real because this image cannot be focused on a screen or card.

PROCEDURE 3

Complete the ray diagram for convex lenses on page 102. Use the following two steps to draw ray diagrams to locate images formed by a convex lens.

Step 1 Draw the first ray from the top of the object to the lens parallel to the optical (principal) axis. For these drawings the rays go half way through the lens and then change direction at the optical center line, except for the ray which goes through the actual geometric center of the lens which travels completely through the lens in a straight line and does not change direction at all. Extend the first ray line from the center line of the lens through the focal point on the opposite side of the lens. See Fig. 14.7.

Step 2 Draw a second ray from the same point at the top of the object to the lens such that the ray passes through the optical center of the lens. This ray will not be deviated from its original path. In other words, draw a straight line from the point on the object through the center of the lens and on out the other side of the lens. The image is located at the intersection of the two rays. Again if the object is between the lens and the near focal point, the rays drawn in steps 1 and 2 will not converge on the opposite side of the lens from the object. You must again extend these rays backward, behind the lens on the same side as the object, to find their crossing point and locate the virtual image formed in this case.

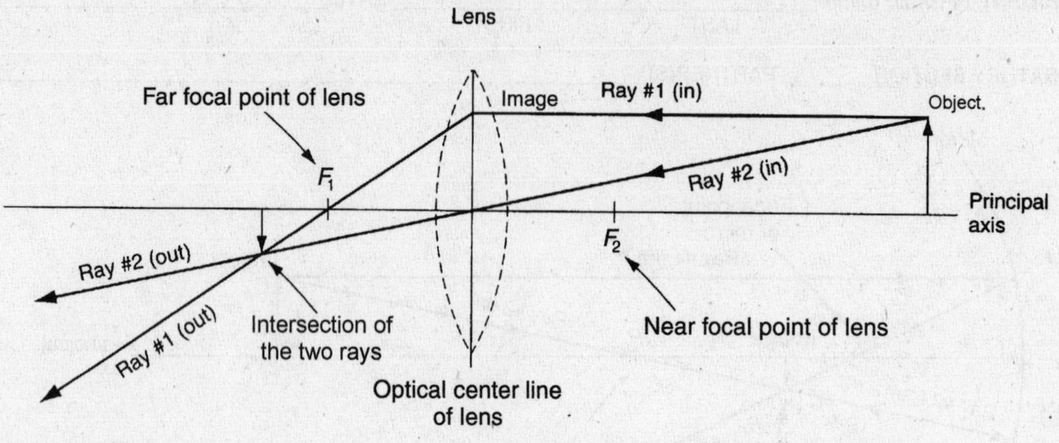

Figure 14.7 Ray Diagram for a convex lens.

QUESTIONS

1. Describe the difference between a real and a virtual image.

2. Where must the object be located with respect to a concave mirror and its focal point to produce a *real* image?

3. Where must the object be located with respect to a lens and its focal point to produce a *virtual* image?

4. At what distance must the object be placed on the principle axes in front of a concave mirror for the image to appear the same size as the object? Draw a ray diagram to illustrate your answer.

EXPERIMENT 14 NAME (print) ______________________________ DATE ______________
LAST FIRST

LABORATORY SECTION ______ PARTNER(S) ______________________________

Ray Diagrams for the Concave Mirror
(See page 101 for instructions)
R = radius of curvature; F = focal point

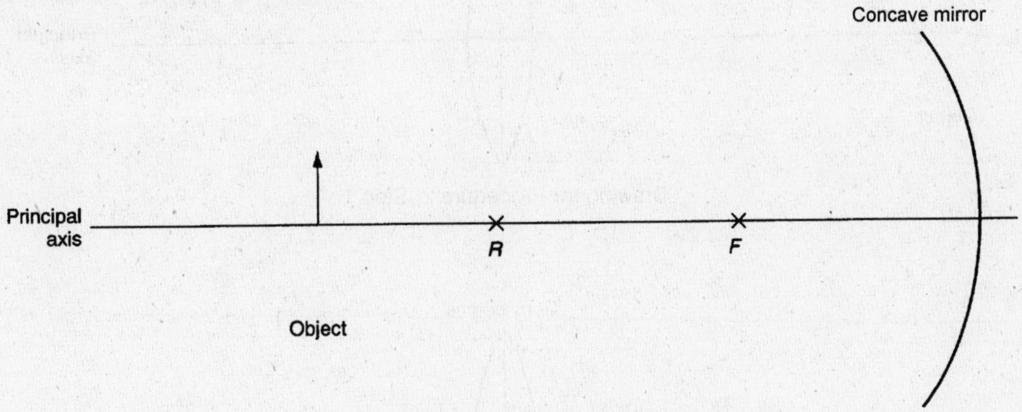

Drawing for Procedure 1, Step 1

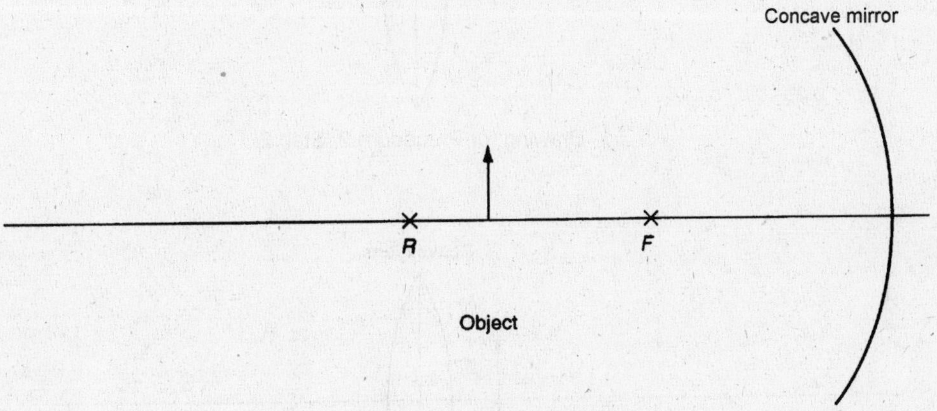

Drawing for Procedure 1, Step 2

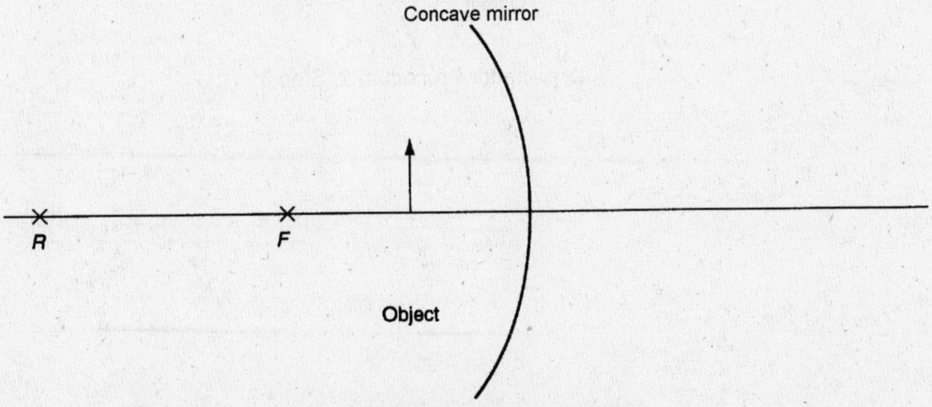

Drawing for Procedure 1, Step 3

Ray Diagrams for the Convex Lens
(see page 102 for instructions)
× = focal points.

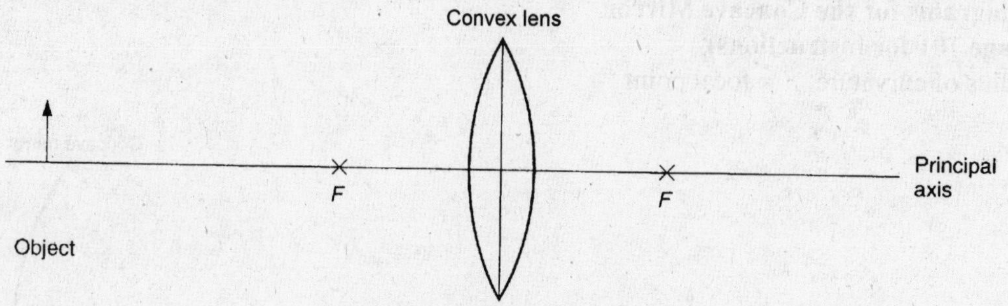

Drawing for Procedure 2, Step 1

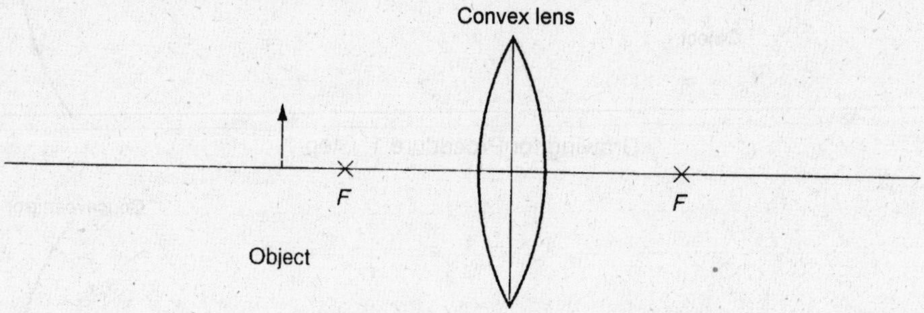

Drawing for Procedure 2, Step 2

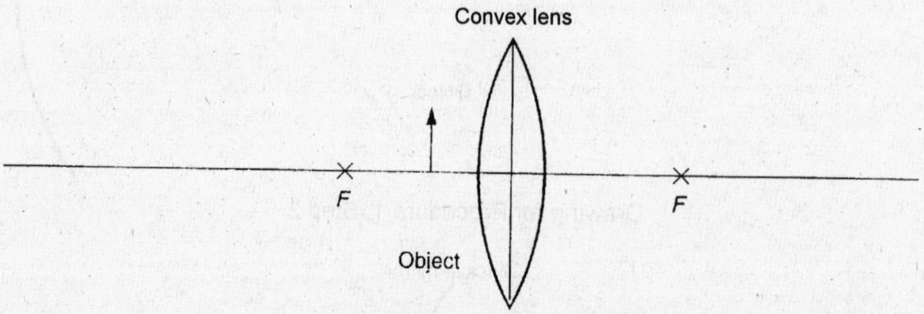

Drawing for Procedure 2, Step 3

Experiment 15

The Refracting Telescope

INTRODUCTION

Lenses made of glass refract light. That is, the lens changes the direction in which light rays travel because the velocity of light changes when it passes through a different medium. Parallel rays of light entering one face of a glass lens will be refracted and converge to a point (called the *focus* or *focal point*) on the back side of the lens. See Fig. 15.1. When the entering rays are parallel to the principal or optical axis of the lens (Fig. 15.la), then the distance from the center of the lens to the focal point is known as the *focal length* of the lens. Thus, if a card or screen were placed on the back side of the lens at the focal point, an image of some distant object located in front of the lens would be seen on the screen. However, if you put your eye at the focal point, you would not see the image of the object. For example, if the object were the Moon, your eye would only see a bright lens. To see an image, the converging rays of light again must be changed to parallel rays so that your eye's lens will be able to focus them on the retina. This can be accomplished by using an additional lens.

Figure 15.2 illustrates a simple refracting telescope. Light from some distant object enters the telescope through lens L_1, called the *objective lens,* forming an inverted image at the focal plane. The diameter of this lens is usually referred to as the *aperture.* The greater the aperture, the greater is the light-gathering power of the telescope. The objective lens, which has a long focal length (f_o) is fixed in position. Lens L_2 is called the *eyepiece.* This lens is set in a movable tube so that the distance between it and the objective lens can be varied. The eyepiece has a small diameter and a short focal length (f_e) in comparison to the objective lens. The image formed by the objective lens becomes the object for the eyepiece. The eyepiece then forms a clear, distinct, magnified, inverted, and virtual image of the distant object. See Fig. 15.3. This virtual image is the one seen by the eye when you look through the eyepiece.

The magnification of the simple two-lens telescope is a function of the focal length of the two lenses. The relationship can be written as

$$\text{Magnification} = \frac{\text{focal length of the objective lens}}{\text{focal length of the eyepiece lens}} = \frac{f_o}{f_e} \qquad \text{Eq. (15.1)}$$

Another way to find the magnification of a telescope is to find the ratio of the actual image height (h_i) to the object height (h_o)

$$\text{Magnification} = \frac{\text{image height}}{\text{object height}} = \frac{h_i}{h_o} \qquad \text{Eq. (15.2)}$$

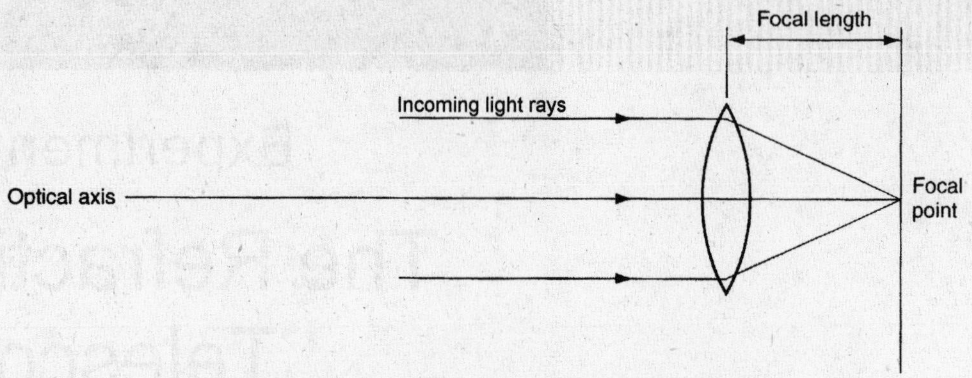

Figure 15.1 Diagram illustrating the converging of parallel rays of light by a convex lens when the incoming parallel beam is parallel to the optical axis of the lens.

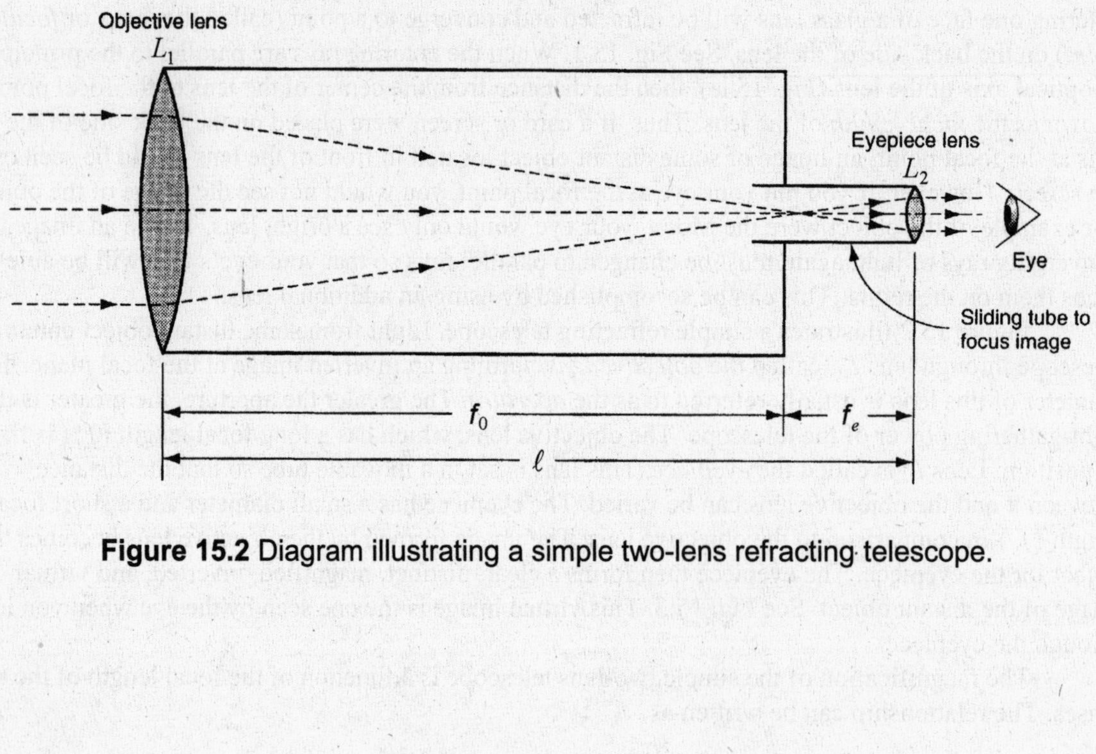

Figure 15.2 Diagram illustrating a simple two-lens refracting telescope.

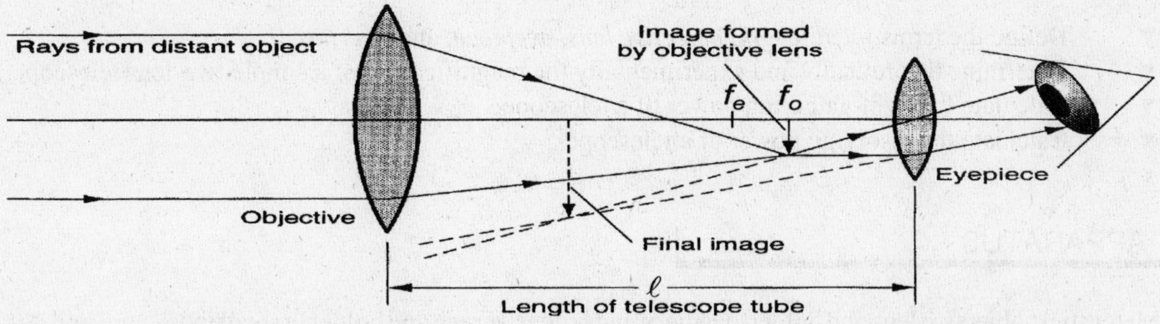

Figure 15.3 Ray diagram showing final image of some distant object.

LEARNING OBJECTIVES

After completing this experiment, you should be able to do the following:

- Define the terms *focal length, objective lens, eyepiece,* and *aperture.*
- Determine theoretically and experimentally the magnification of a simple two-lens telescope.
- Calculate the light-gathering power of a telescope.
- Calculate the resolving power of a telescope.

APPARATUS

Meter stick, objective lens and holder, eyepiece and holder, screen and holder (a small index card will do), a colored filter and holder. The filter is not necessary but is helpful when determining the magnification of the telescope.

EXPERIMENT 15 NAME (print) ______________________________ DATE ______________________
LAST FIRST

LABORATORY SECTION ______ PARTNER(S) ____

PROCEDURE

1. Mount the objective lens (the lens with the longer focal length) in its holder, and place it on the meter stick at the 10-cm mark. Mount the cardboard screen in its holder, and place it on the meter stick near the 30-cm mark. Go to an outside window and point the meter stick toward the outside with the lens toward the window and the screen next to you. Adjust the screen back and forth until you obtain a clear image on the screen of some distant object (tree or building) outside the laboratory. If the experiment is being done at night, choose a street lamp or a neon sign as the object. Once a clear image is obtained, measure the distance from the center of the lens to the screen. This is the focal length (f_o) of the objective lens. Make three independent trials; record them in Data Table 15.1, and calculate their average.
2. In the same way, obtain the focal length (f_e) of the eyepiece lens. Again make three independent trials; record them and calculate their average.
3. Mount the two lenses in their holders on the meter stick. Remove the screen and its holder from the stick entirely. Place the lenses a distance apart nearly equal to the sum of their focal lengths (l), and look through the eyepiece lens and the objective lens at some distant object outside the laboratory window. Move the eyepiece lens slightly until the object is clear and sharp. Now measure and record the distance between lenses. Make two additional trials and record the data. Calculate the average of the three trials.
4. Two students are needed to do this part of the experiment. One student will use the telescope at the end of the laboratory opposite the blackboard. The other student will be at the blackboard. Draw a wide 20-cm vertical line on the blackboard with white chalk. Next draw a horizontal line 5 cm long about 5 cm to the right of the top of the vertical 20-cm line to indicate where the top of the 20-cm line is located. Set the lenses of your telescope the average distance apart found in (3) above and mount the green filter between them. Look at the 20-cm wide vertical line through the telescope with one eye closed. You will see a magnified vertical green line if the filter is used. Otherwise, you will see a magnified white line. Continue looking through the telescope with one eye closed. Now open the other eye. Both eyes are now open. One eye is seeing a magnified green image, and the other eye is seeing the white 20-cm vertical line as drawn on the blackboard. This is not easy for some to do. Keep trying. Next adjust the elevation of the telescope so that the top of the green magnified line is at the same level as the unmagnified vertical white line. With this done, instruct your partner at the blackboard to make a second horizontal line at the bottom of the green magnified image. Your partner may have to draw and erase a few lines before you accomplish this. Once this is done, measure the distance in centimeters between the two horizontal lines that your partner just drew on the blackboard, and record this value as h_i in the data table. The wide white vertical line is the object, and its height has been drawn to 20 cm. This means that h_o is 20 cm, so record this in the data table, and then, using Eq. 15.2, determine the magnification of the telescope you have just built as Calculation 2(a).

   The second value for the magnification can be found by using Eq. 15.2 and the focal length for the two lenses that you determined in Procedures 1 and 2. This will be Calculation 2(b). Now you can find the percent difference between the two magnifications you have just determined as Calculation 2(c).

DATA TABLE 15.1		Trial 1	Trial 2	Trial 3
1.	Position of objective lens on meter stick	______ cm	______ cm	______ cm
	Position of screen on meter stick	______ cm	______ cm	______ cm
	Focal length of objective	______ cm	______ cm	______ cm
	Average focal length of objective (f_o)	______ cm		
2.	Position of eyepiece lens on meter stick	______ cm	______ cm	______ cm
	Position of screen on meter stick	______ cm	______ cm	______ cm
	Focal length of eyepiece	______ cm	______ cm	______ cm
	Average focal length of eyepiece (f_e)	______ cm		
3.	Position of objective lens on meter stick	______ cm	______ cm	______ cm
	Position of eyepiece on meter stick	______ cm	______ cm	______ cm
	Distance between lenses	______ cm	______ cm	______ cm
	Average distance between lenses (l)	______ cm		
4.	(a) Vertical height of image h_i	______ cm		
	(b) Vertical height of object h_o	______ cm		

CALCULATIONS

1. Sum of average focal lengths of lenses (from 1 and 2) $L = f_o + f_e =$ ______ cm

   Actual Distance between lenses (from 3 in Data Table 15.1) $\ell =$ ______ cm

   Percentage difference between L and ℓ % difference = ______ cm

2. (a) Magnification (from 4) $= \dfrac{\text{height of image}}{\text{height of object}} \dfrac{h_i}{h_o} =$ ______

   (b) Magnification (from 1 and 2) $= \dfrac{\text{focal length of objective}}{\text{focal length of objective}} \dfrac{f_o}{f_e} =$ ______

   (c) Percentage difference between magnification values (a) and (b) = ______

EXPERIMENT 15 NAME (print) ______________________________ DATE ______________________
LAST FIRST

LABORATORY SECTION ______ PARTNER(S) ____

3. The brightness of the image of an object increases in direct proportion to the area of the objective lens or the square of its diameter. The objective lens collects light from a large beam of light and then concentrates it into a small beam for the eye to see. The *light-gathering power* of the telescope is defined as the ratio of the light per unit time entering the eye by way of the telescope to the light from the object that enters the unaided eye per unit time. The relationship can be written as

$$\text{Light-gathering power} = \frac{(\text{aperture of the telescope})^2}{(\text{aperture of the eye})^2} = \frac{(\text{diameter of objective lens})^2}{(\text{diameter of lens of eye})^2}$$

The aperture of an average human eye in this case can be determined by using the diameter of the lens opening of the eye which is 0.64 cm.

Measure the diameter of the objective lens of your telescope, and calculate the light-gathering power of the telescope constructed in the laboratory. Show your work.

Diameter of objective lens = ____________ cm

Light-gathering power of your telescope = __________

4. In addition to its light-gathering power, a telescope is rated on its resolving power or its ability to separate two objects that are close together. A relationship for resolving power can be derived from diffraction theory and can be written as

$$\text{Resolving power} = \frac{12.7\ \text{cm}}{\text{diameter of the objective lens in cm}}$$

Calculate the resolving power of the telescope constructed in the laboratory. Show your work.
Resolving power of your telescope = _______________

QUESTIONS

1. What is the simplest way to increase the magnifying power of a two-lens telescope? Hint: The eyepiece lens is often mounted in a short tube that can easily be removed from the main part of the telescope.

2. Why is it better if a distant object outside the window of the laboratory is used when the focal lengths of the two lenses were determined?

3. Calculate the light-gathering power of the 508-cm (200-in) telescope on Mount Wilson. Show your work.

4. Calculate the resolving power of the 508-cm (200-in) telescope on Mount Wilson. Show your work.

Experiment 16

Color

INTRODUCTION

Color is a perception made possible by the complex structure of the human eye. The eye, as you know, is sensitive to light, which allows us to see objects. We detect light either directly from glowing objects or reflected from the surface of an object that does not itself glow. The color that we see is dependent on the wavelengths of the light that are produced by the glowing object or that are reflected from the nonglowing object. White is the presence of all colors (i.e., all wavelengths of light) in the same proportional amount as in the light produced by our nearest star, the Sun. Black is the absence of all color; if an object appears black, it does not produce or reflect any visible light.

In this experiment you will look at several aspects of color by doing six short miniexperiments ranging from simple observation to the mixing of pigments. Because each stands alone, the miniexperiments can be done in any order. In most cases you will move from one experimental setup to another, performing one miniexperiment at each location.

LEARNING OBJECTIVES

After completing this experiment, you should be able to do the following:

- ▼ Describe the spectrum of white light after it has passed through a diffraction grating.
- ▼ List the three primary colors of light and tell what secondary colors are produced by the various combinations of these colors.
- ▼ Explain what occurs when plastic color filters of various colors are combined.
- ▼ Show how to best match color paint chips to cloth swatches of various colors.
- ▼ Mix paint pigments to form the secondary colors in the artist's subtractive color theory, and learn about tints, tones, and shades.
- ▼ Observe and explain the effects of colored lighting on a variety of colored objects.

APPARATUS

White-light source and diffraction grating, Singerman light box or Fisher black-ray box with color filters for the primary colors of light, set of 12 to 20 plastic color filters, several paint chips from a paint store in colors that approximate 4 to 6 pieces of colored cloth, 4 to 6 pieces of colored cloth, colored tempera paints (liquid, in red, blue, green, yellow, black, and white), wooden mixing sticks, and brushes, plastic mixing sheets, newspapers to paint over, 6 to 8 small cups or beakers for water, one high-intensity lamp with red, blue, and green plastic transmission filters.

Caution

Safety Tips

Paint may be hard to remove from clothes. Even though paint is water-based, it can still stain light-colored clothing. It is a good idea *not* to wear good clothes when you come in to do this lab.

EXPERIMENT 16 NAME (print) ______________________________ DATE ______________
LAST FIRST

LABORATORY SECTION ______ PARTNER(S) ______________________________

Note: The following procedures can be done in any order so the equipment can be shared and the painting done early so that it can dry.

PROCEDURE 1

Look at the white-light spectrum produced by a diffraction grating. List below the hues that you can see in this spectrum.

PROCEDURE 2

Turn on each of the three light sources one at a time and record the color produced by each one, as seen on a white surface or on the screen of the Singerman light box.

List the three *primary* colors of light.

1. ____________________

2. ____________________

3. ____________________

Now turn on two of the light sources at a time and record the color observed on the screen. List the *secondary* colors of light that you see.

red + blue = ____________________

green + blue = ____________________

red + green = ____________________

Turn on all three light sources at once and record what you observe on the screen.

red + green + blue = ____________________

Carefully take the green filter out and turn on each of the other light sources, one at a time, with the white light. Record the results below.

red + white = ____________________

blue + white = ____________________

PROCEDURE 3

You will find several plastic colored filters on the table. Combine these in pairs on top of a white sheet of paper. For six pairs, record the color you observe in the region where the two color filters overlap. Record the results below.*

Trial	Color 1	Color 2	Combination Color
1			
2			
3			
4			
5			
6			

* The color filters used probably are not assigned identifying names so make up your own, light green, grass green, and so on. You can also name the combination color. Have some fun! There are no right answers. The idea is to see the color effects produced.

PROCEDURE 4

Keeping the cloth and color chart well separated, take a sample of cloth and look at its color. Now look at the paint chip provided and pick out the color that you think best matches the cloth. Record its designation number below.

Now place the cloth sample and the paint chip side by side and see if you still agree with your original choice. Record your second choice below. (Your lab partner may not always agree with your choice of paint color chip so record your own choice below. Seeing color is always somewhat subjective so there is no absolutely correct answer here.)

EXPERIMENT 16 NAME (print) ______________________________ DATE ____________________
LAST FIRST

LABORATORY SECTION ______ PARTNER(S) ______________________________________

Why do you think there might be a difference between your choices?

Do the same for the other cloth samples provided. Record the closest color-code number for each.

Sample 2. Sample 3. Sample 4.

Did everyone in your group agree on the choice of color-code number for the paint chip that they thought best matched the cloth? Explain why not. (If your group is typical, you probably did not all agree.)

PROCEDURE 5

1. On the Color Work Sheet provided, fill in circles with the appropriate color or combination of colors. Place one or more *small* dabs of paint on a plastic mixing sheet with the wooden sticks provided, and then mix it with a brush to form a uniform color. This color can then be applied to the appropriate circle. (*Note:* The circles do *not* need to be completely filled in.)

   When making blue tints, start with pure blue and add a little more white to each succeeding dot (Row 3).

   When making red shades, start with pure red and add a little more black to each succeeding dot (Row 4).

   Fill in the circles on Row 5 with a series of gray tones. To make gray tones, start with pure white and add a *little more* black to each succeeding dot. Add the black in *very small* amounts as this darkens quickly.

2. Mix up a medium gray tone and paint a dot in the center of a piece of white paper and another dot using the same paint mixture in the center of a piece of black paper.

Question: Do there appear to be any differences in color between the two identical gray paint dots that you applied to the white paper and to the black paper? Explain.

PROCEDURE 6

1. (a) Place a red-colored filter over a high-intensity lamp, and allow the light to fall on the piece of white paper on which you painted the dots in Procedure 5. Note any major changes in the colors of the dots. Remove and replace the filter several times while watching the color dots, and pick out the ones that show the most dramatic color changes and describe these changes below.
(*Note*: Your instructor may provide you with a set of previously painted dots or other color wheel diagram to use because your paint may not be dry from your work in Procedure 5.)

1. (b) Why do you think changes are observed in the different shades of gray when the red filter is placed in front of the white light source?

2. Now place a blue filter over the light and record any changes in the colors of the dots.

3. Repeat the procedure with a green filter and record any color changes of the dots.

EXPERIMENT 16 NAME (print) ______________________ ______________________ DATE ______________________
LAST FIRST

LABORATORY SECTION ______ PARTNER(S) ______________________

1. Define what it meant when a surface color is said to be "black."

2. What colors of paint must be mixed to create a black surface coating?

3. The secondary colors in the subtractive color theory can be formed by mixing two of the pure primary colors together in exactly equal amounts. Name the two primary colors that must be mixed to create the following colors.

Yellow	1. ____________	2. ____________
Cyan	1. ____________	2. ____________
Magenta	1. ____________	2. ____________

4. Why are the secondary colors described in question 3, as mixed on your COLOR WORK SHEET, not exactly the color you would expect them to be in the true subtractive color theory for yellow, cyan, and magenta?

5. What is meant by the terms TINTS, SHADES, and TONES and how can they be created using paint? Use the pure hue RED as the basic color in your explanation.

TINTS:

SHADES:

TONES:

EXPERIMENT 16 NAME (print) ________________________________ DATE ____________________
LAST FIRST

LABORATORY SECTION ______ PARTNER(S) ____

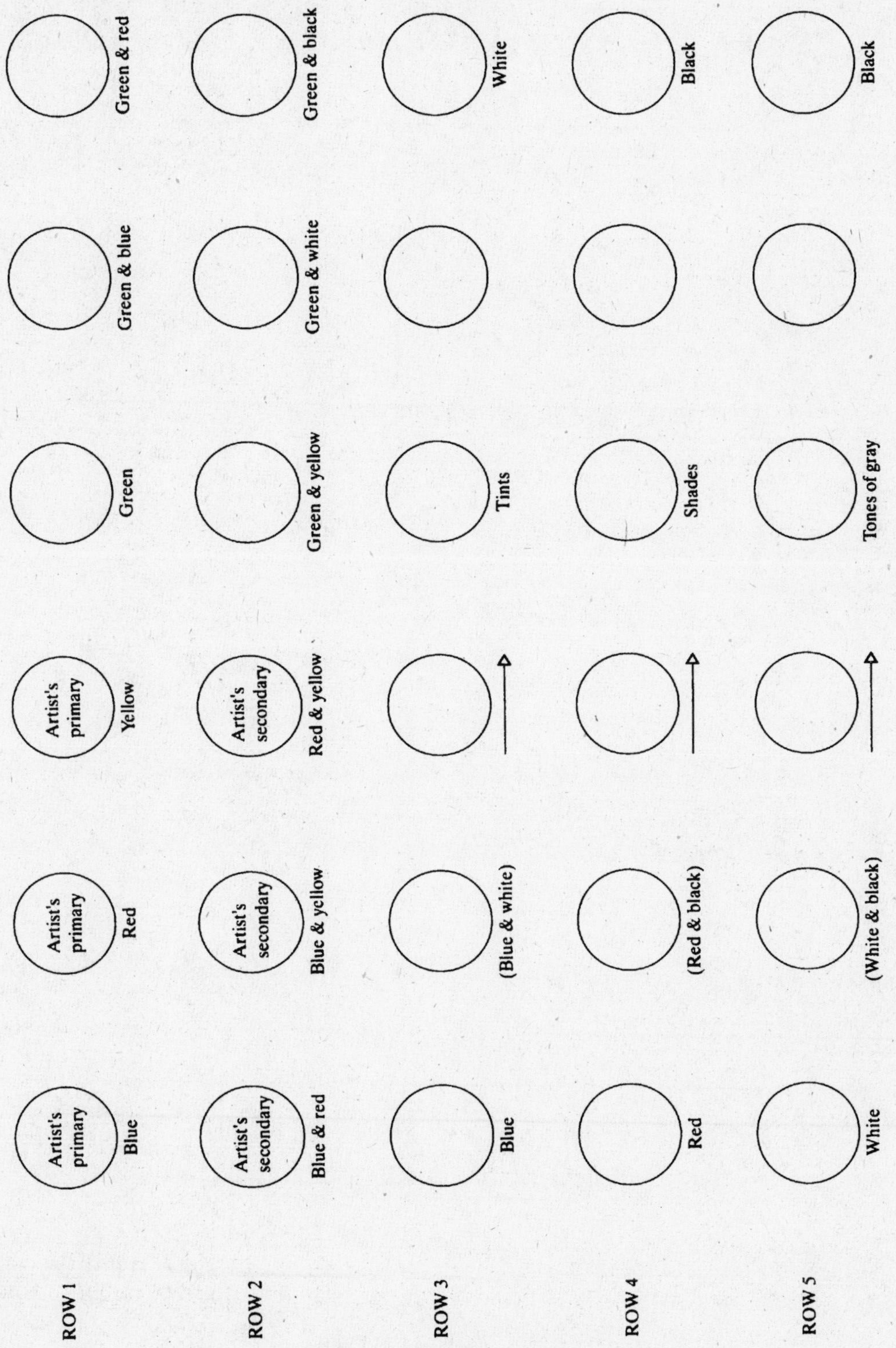

COLOR WORK SHEET

Experiment 17

Static Electricity

INTRODUCTION

According to atomic theory, all matter consists of very small particles called *atoms* that can be described as being made up of negatively charged particles called *electrons,* positively charged particles called *protons,* and particles called *neutrons* that carry no electric charge.

Electrons and protons possess the fundamental properties of matter—mass, length, and electric charge. *Charge* refers to the force field that the electron and the proton possess due to this positive and negative nature. To differentiate this force field from a gravitational or nuclear force field, we call it an *electric field.*

The magnitudes of the electric charge on an electron and a proton are exactly equal, but their polarity or sign is different. When the same number of electrons and protons are present, the total net charge is zero. That is, there is no apparent force field present around the neutral body.

Electric charge is measured in units called *coulombs* (C) and is designated by the symbol (q). Negative charges are designated by a — sign and positive charges by a + sign. All electrons carry the negative charge (-1.6×10^{-19} C), and all protons have the same charge ($+1.6\times10^{-19}$ C), but with the opposite sign, positive. All charged bodies found in nature have multiples of this amount of electric charge. Scientists now believe in the existence of subparticles called *quarks.* Quarks have charges of $-\frac{2}{3}$, $-\frac{1}{3}$ etc. In the currently accepted theory, quarks combine (three at a time) to form protons and neutrons. So far, however, no single free quark has ever been found so our experiment will not be effected by these subparticles.

The relationship between charges is given by *Coulomb's law.* The law states:

> The force of attraction between two unlike charges or the force of repulsion between two like charges is directly proportional to the product of the two charges and inversely proportional to the square of the distance between them.

$$F_e = \frac{kq_1q_2}{r^2}$$

where k (electrostatic constant) $= 9.0\times10^{-9}\ \mathrm{Nm^2/C^2}$

q_1 and q_2 are the magnitude of each of the charges
r is the distance between the two charges

Another relationship, the general law of signs for electric charges, states that unlike charges attract and like charges repel. See Fig. 17.1.

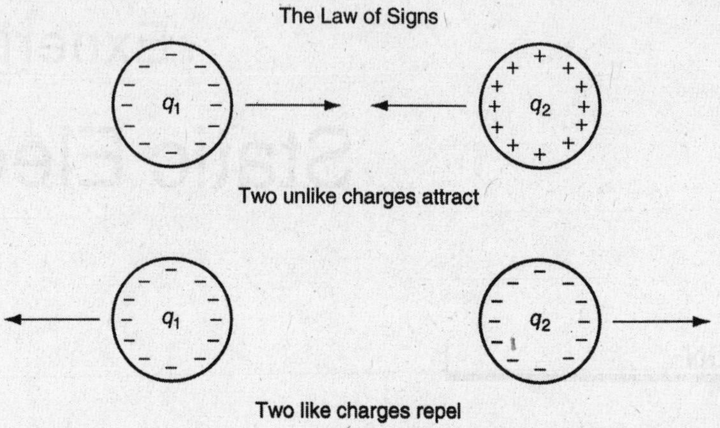

Figure 17.1 Diagram illustrating the law of signs for electric charges.

LEARNING OBJECTIVES

After completing this experiment, you should be able to do the following:

- ▼ State the fundamental properties of the electron and proton.
- ▼ State Coulomb's law for electric charges.
- ▼ State the general law of signs for electric charges.
- ▼ Identify certain properties of static electricity.

APPARATUS

Aluminum or gold leaf electroscope, two hard-rubber or ebonite rods, wool cloth or piece of fur, two glass rods, silk cloth, support stand with clamp, wire cradle to loosely hold rods, short length of light string or thread, two metal spheres mounted on insulated stands, toy balloons (rubber). Note: This experiment may not work well if the air has a high relative humidity as on a day when it is raining.

EXPERIMENT 17 NAME (print) ______________________ DATE ______________
LAST FIRST

LABORATORY SECTION ______ PARTNER(S) ______________________

PROCEDURE 1

Place an electrostatic charge on one of the rubber rods by rubbing it with wool or fur. Suspend the rod in the wire cradle as shown in Fig. 17.2. Place a charge on the other rubber rod by rubbing it in a similar fashion and bring it near one end of the suspended charged rubber rod. Record what you observe as answer A in Data Table 17.1. Bring the second charged rubber rod near the other end of the suspended charged rubber rod. Record what you observe as answer B in Data Table 17.1. Bring the charged fur or wool near the suspended charged rubber rod. Record what you observe as answer C in the data table. Place an electrostatic charge on one of the glass rods by rubbing it with the silk cloth. Bring the charged glass rod near one end of the suspended, rubber rod. Record what you observe as answer D in the data table. Bring the charged glass rod near the other end of the suspended charged rubber rod. Record what you observe as answer E in the data table. Bring the charged silk cloth near the suspended rod. Record what you observe as answer F in Data Table 17.1. What can you conclude from the observations in the above procedure? Record your answer as answer G in Data Table 17.1.

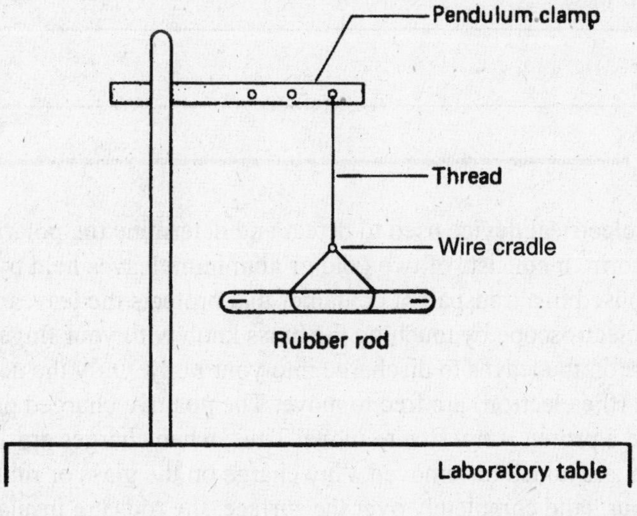

Figure 17.2 Suspended rubber rod free to move in a horizontal plane.

DATA TABLE 17.1	
Answer A	
Answer B	
Answer C	
Answer D	
Answer E	
Answer F	
Answer G	

PROCEDURE 2

The electroscope is an electrical device used to detect and determine the polarity (sign) of charged bodies. In its simplest form, it consists of two gold or aluminum leaves held by a conductor (usually copper or brass) and housed in a transparent container that protects the leaves from air currents.

Discharge the electroscope by touching the brass knob with your finger. A path is thereby provided for any charge on the leaves to discharge into your body. Only the negative charged particles in the nucleus (the electron) are free to move. The positive charged proton is part of the nucleus of an atom, and a proton is not free to move. Thus, when charges are added to or taken from a body, it is electrons that are added or removed. Any charge on the glass or rubber rods can be removed by rubbing your hand completely over the surface; the rods are insulators, and therefore, contact with all the surface is necessary to remove all excess charges. With no charge on the rubber rod, there should be no effect on the leaves of the electroscope when the rubber rod is brought into the vicinity of the electroscope. See Fig. 17.3(a).

A charged electroscope is needed to detect and determine the polarity (sign) of charged objects. One of two methods can be used to charge an electroscope. One method is by contact, or simply touching the aluminum or brass knob of the electroscope with an object carrying a charge. This method is not used in this experiment because it is difficult to control the amount of charge transferred to the leaves of the electroscope.

Caution: Do not touch the brass knob with any charged object.

EXPERIMENT 17 NAME (print) ________________________________ DATE ____________
LAST FIRST

LABORATORY SECTION ______ PARTNER(S) ________________________________

The second method of charging an electroscope is by induction. To charge the electroscope by induction, proceed as follows: Stroke the rubber rod with wool or fur, thereby placing a negative charge (excess electrons) on the rubber rod. Electrons are removed from the wool or fur and transferred to the rubber rod. As you bring the charged rod into the vicinity of the electroscope, you should observe the electroscope leaves diverging. See Fig. 17.3(b). While holding the rod in a position that causes the leaves to spread apart, but not touching the rod to the knob, touch the knob of the electroscope with a finger of your free hand. See Fig. 17.3(c). Electrons on the leaves will flow into your body. Remove your finger from the knob; then remove the rubber rod from the vicinity of the electroscope. Because some electrons on the leaves have been removed, the electroscope will now be positively charged. See Fig. 17.3(d).

As the rubber rod is brought near or taken away from the charged electroscope, you will observe that the leaves of the electroscope spread farther apart or come closer together. As the negatively charged rubber rod is brought close to the knob of the electroscope, electrons on the knob are repelled into the leaves and neutralize part or all of the positive charges. Thus the leaves come closer to one another or collapse completely, depending on how close the rubber rod is brought near the knob of the electroscope. When the rod is taken away, the charge distribution redistributes itself throughout the entire electroscope leaving it charged as it originally was by induction.

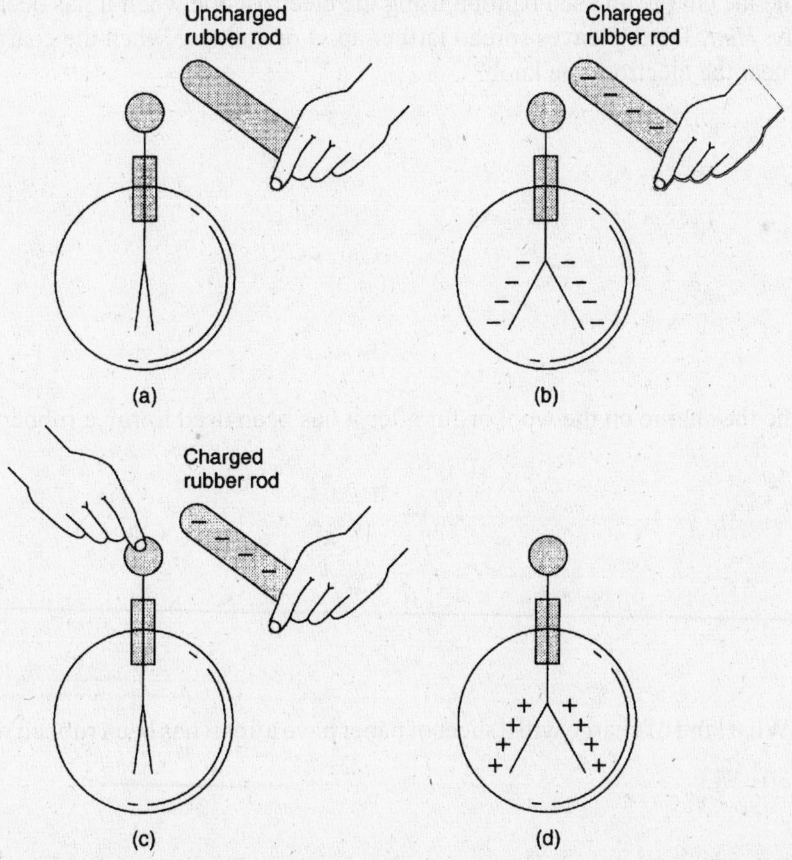

Figure 17.3 Charging an electroscope by induction. See the text for an explanation.

PROCEDURE 3

Using the charged electroscope, complete the following:

1. Determine the charge on a glass rod after rubbing it with silk. The charge on the glass rod is ______________ (positive or negative). The charge on the silk cloth is ______________.

2. Rub each of two inflated toy balloons with wool or fur. Will they attract each other or repel each other? Explain why.

3. Determine the charge on each balloon using the electroscope when it has been charged positively. *Hint:* Do the leaves spread farther apart or collapse when the charged object is brought near the electroscope knob?

4. Determine the charge on the wool or fur after it has been used to rub a rubber rod.

5. (a) What kind of charge will a sheet of paper have after it has been rubbed with wool or fur?

   (b) What kind of charge will a sheet of paper have after it has been rubbed with silk?

EXPERIMENT 17 NAME (print) ______________________________ DATE ______________
LAST FIRST

LABORATORY SECTION ______ PARTNER(S) ______________________________

6. Obtain a few (four or five) pieces of paper 1 or 2 cm in size. Remove any electric charge by touching them with your hands. Test for any charge on the pieces of paper by bringing them near the knob of the charged electroscope. The pieces of paper are neutral. That is, they have zero charge. Place them on the table free of any electric charge.

(a) Place a negative charge on a rubber rod by rubbing it with wool or fur. Bring the charged rubber rod near the neutral pieces of paper. What do you observe? Explain.

(b) Repeat using a glass rod and a silk cloth. What do you observe? Explain.

7. Place the two metal spheres mounted on insulated stands in contact with one another as shown in Fig. 17.4, Part 1. A rubber rod carrying a negative charge is placed near the left side of sphere A. The spheres are then separated while the rubber rod is held near sphere A. See Fig. 17.4, Part 2 below. The spheres will now possess an electric charge if they are separated before the charged rod is removed. Determine what kind of charge will be on each by moving each in turn near the knob of a positively charged electroscope. Explain why each sphere was charged the way it was.

Record the charge you detected on each sphere.

Sphere A ______________________

Sphere B______________________

Note: Separate the spheres *before* you take away the charged rubber rod.

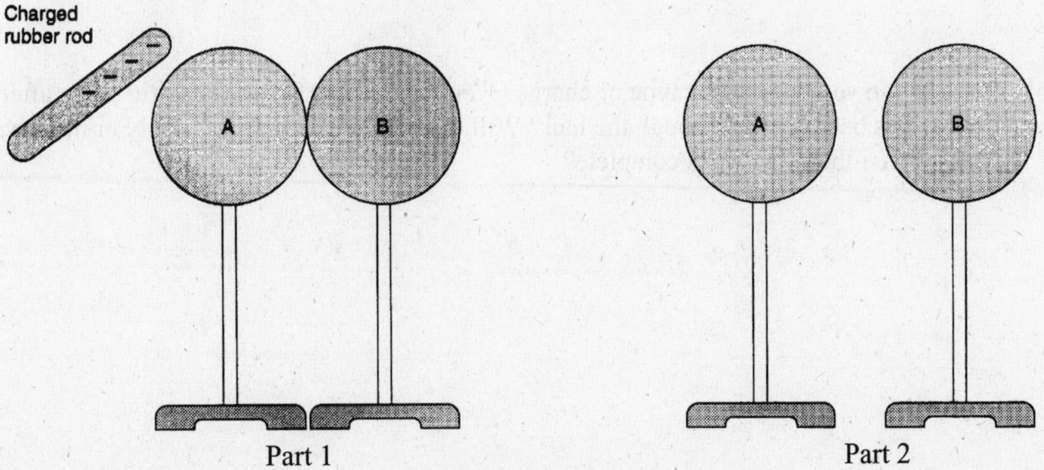

Figure 17.4

QUESTIONS

1. What kind of a charge can be placed on an electroscope by induction using a glass rod rubbed with silk? Explain.

2. How many different kinds of charges did you observe the effects of in this experiment?

3. State the law of signs for electric charges.

4. (a) Sometimes when a person combs their hair on a dry day (low relative humidity), a crackling sound can be heard as the plastic comb passes through the hair. What causes this to happen?

   (b) Can you guess what type of charge (+ or –) will be left on the plastic comb after it has been passed through the hair? Will the hair be charged positively or negatively after the combing is complete?

Experiment 18

Magnetism and Electromagnetism

INTRODUCTION

The history of magnetism began with ancient civilization in Asia Minor in a region known as Magnesia, where rocks were found that would attract each other. These rocks were called *lodestones* or *magnets*.

The origin of the compass is unknown, but it is fairly certain that the first magnetic compass was the lodestone or leading stone used in the latter part of the thirteenth century.

Magnetism is the phenomenon of matter (matter that does not display an electric force field) to attract or repel other matter. This phenomenon is produced by a magnetic force field.

A magnetic force field is generated when electric-charged matter such as an electron, proton, or ion is put in motion. Thus spinning electrons or electrons moving through a conductor generate a magnetic field of force. No magnetic field exists around a stationary charged particle. That is, if the particle is not spinning or moving in some direction, no magnetic field is present. The magnetic force field possessed by the bar magnets used in this experiment is due to the spinning of electrons in the atoms of iron or other elements used to make the magnet.

LEARNING OBJECTIVES

After completing this experiment, you should be able to do the following:

- State the general law for magnetic poles.
- State Coulomb's law for magnetic poles.
- Determine the direction of a magnetic field at a given location.
- Determine the direction of a magnetic field generated by a moving charge.
- Draw on a sheet of paper the magnetic lines of force produced by a bar magnet.
- Measure the angle of dip for Earth's magnetic field.

APPARATUS

Two bar magnets, meter stick, ring stand or a vertical support rod for the laboratory table, wire cradle to support magnet, pendulum clamp (Cenco number 72296 or similar support clamp), thread, dip needle, large sheet of brown wrapping paper, magnetic compass (small size), dry cell (1.5 V), a short piece (25 cm) of number 14 copper wire.

EXPERIMENT 18 NAME (print) ________________________________ DATE ____________________
LAST FIRST

LABORATORY SECTION ______ PARTNER(S) ____

PROCEDURE 1

Suspend one of the bar magnets using the wire cradle as shown in Fig. 18.1 so that it is free to move in a horizontal plane about a vertical axis. After removing any magnets or other items that might influence the movement of the suspended magnet, allow it to swing freely and come to rest. At rest, the suspended magnet has aligned its long axis with Earth's magnetic field. That is, the magnetic field of the suspended magnet and the magnetic field of Earth act on each other, and being free to move, the magnet orients its long axis parallel to Earth's magnetic field. The end of the magnet pointing in the direction of geographic north is called the *north-seeking pole* of the magnet, or simply the *north pole*. The opposite end of the magnet is called the *south pole.*

If the suspended magnet fails to orient in a north—south direction with the end marked (N) pointing north, ask the instructor for assistance. The word *north* should be posted on one wall of the laboratory to inform you of the direction of geographic north.

The suspended magnet, which acts as a compass, may not point directly toward geographic north. In fact, in most places on Earth's surface it will not point directly toward geographic north. The geographic north pole and the magnetic north pole are not located at the same place but are several hundred miles apart and over time the North magnetic pole moves around very slowly within about 50 miles of its average central location. The angle that a compass needle deviates from geographic north is called the *angle of declination.*

Holding the second bar magnet in your hand, bring its north pole near the north pole of the suspended magnet. Record what you observe in Data Table 18.1 as answer A. Next, bring the north pole of the magnet in your hand near the south pole of the suspended magnet. Record what you observe in Data Table 18.1 as answer B. Bring the south pole of the magnet held in your hand first to the north pole of the suspended magnet, then to the south pole. Record what you observe in Data Table 18.1 as answers C and D, respectively. What can you conclude from the four observations concerning the attraction and repulsion of magnetic poles? Record your answer in Data Table 18.1 as answer E.

The general law for magnetic poles states that two unlike magnetic poles attract each other, and two like magnetic poles repel each other. How does this compare with your answer E?

Coulomb's law for magnetic poles states that two unlike magnetic poles attract each other and two like magnetic poles repel each other with a force directly proportional to the product of their pole strength and inversely proportional to the square of the distance between the poles.

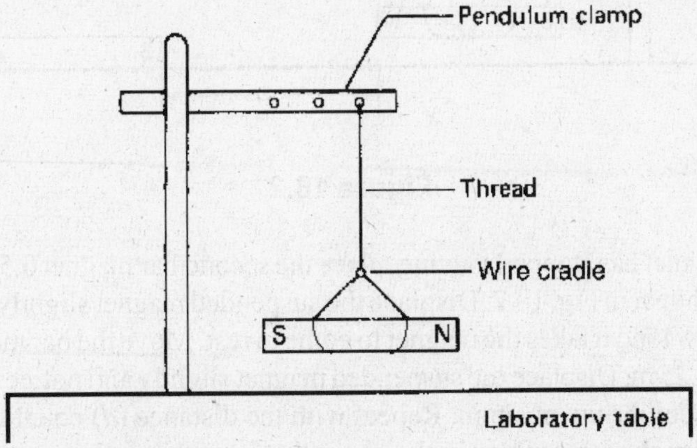

Figure 18.1 Suspended bar magnet free to move in a horizontal plane.

DATA TABLE 18.1 Procedures 1 and 2
Answer A:
Answer B:
Answer C:
Answer D:
Answer E:
Answer F:

Loosen the pendulum clamp and lower the suspended bar magnet until it is just above the surface of the laboratory table, as shown in Fig. 18.2. Allow the suspended magnet to come to rest in Earth's magnetic field. Make sure the second bar magnet is far enough away so that it does not influence the suspended magnet.

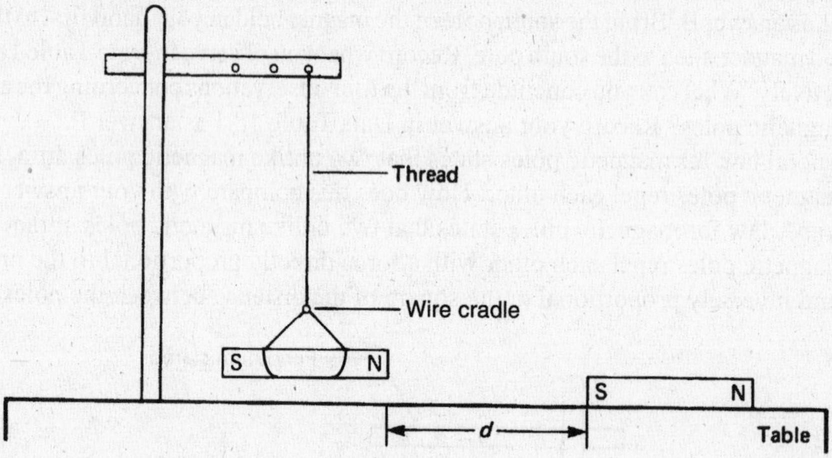

Figure 18.2

After the suspended magnet has stopped moving, place the second bar magnet 0.5 m north of the suspended magnet, as shown in Fig. 18.2. Displace the suspended magnet slightly from its equilibrium position; then notice how long it takes the magnet to come to rest. Move the bar magnet on the table so that the distance (*d*) is 0.25 m. Displace the suspended magnet slightly and notice how long it takes the magnet to come to its equilibrium position. Repeat with the distance (*d*) equal to 0.1 m.

What general conclusion can you make about the force that influences the suspended magnet in relationship to the distance between its magnetic pole and the magnetic pole of the magnet on the table? Record your answer in Data Table 18.1 as answer F.

EXPERIMENT 18 NAME (print) ______________________ DATE ______________
LAST FIRST

LABORATORY SECTION ______ PARTNER(S) ______________________

PROCEDURE 2

A portion of the magnetic field surrounding a bar magnet can be plotted using a large sheet of wrapping paper to record the magnetic lines of force. A very small compass can be used to plot the location and direction of the lines of force. See Fig. 18.3.

A magnetic force field is a vector quantity, and the direction of the force field at any given location, by definition, is the direction a unit north pole would tend to move when placed in the force field at that location. Since like poles repel and unlike poles attract, the direction of the force field is away from the north pole of the magnet producing the field and toward the south pole. See Fig. 18.3.

Position a bar magnet in the center of a large sheet of paper with its long axis in an east—west direction. (One sheet of paper is to be used by each group of students at a laboratory table.) Draw an outline of the magnet with a pen or pencil. Label the poles of the magnet on the outline N and S, respectively. See Fig. 18.3. With the magnet in position, make a dot near one end of the magnet; then place the small compass so that one end of the compass needle coincides with the dot. Make sure the compass needle rotates freely. Tap the top of the compass slightly to free a needle that tends to stick. Place a dot at the opposite end of the compass needle after the needle comes to rest. Move the compass until the end of the needle that originally coincided with the first dot coincides with the second. Continue this process until the compass is back to some part of the magnet or reaches the edge of the paper. See Fig. 18.3. Connect the points with a smooth line and draw arrows on each line to indicate the direction in which the north pole of the compass pointed.

Continue this process for a number of other lines until a symmetrical field is obtained. A diagram similar to that shown in Fig. 18.4 should be obtained. Notice on the diagram that no two magnetic field lines ever cross. The points indicated by (O) are points where, due to the bar magnet, the magnetic field is equal and opposite to the horizontal component of Earth's magnetic field. When a small compass is placed at either of these points, the needle of the compass will rotate freely and not point in any particular direction.

The plot shown on the large sheet of paper illustrates the reaction between the magnetic field of the bar magnet and the horizontal component of Earth's magnetic field.

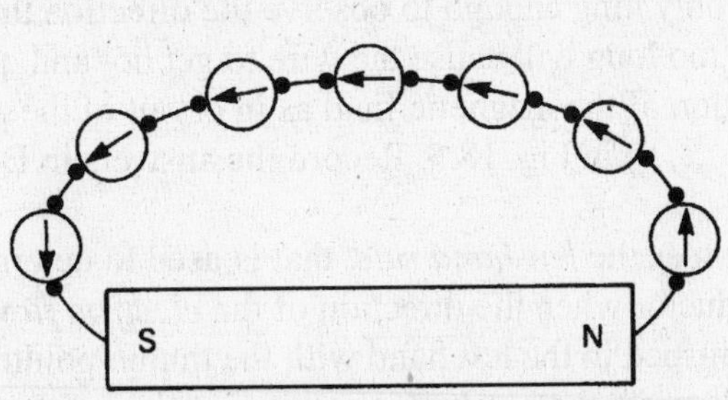

Figure 18.3 Compass used to plot a magnetic field line of a bar magnet.

Note: The compasses are shown at every other plotting position along the field lines.

The true direction of Earth's magnetic field can be obtained with a dip needle. Ask the instructor for a dip needle and proceed as follows:

1. Remove all magnets and magnetic material that will influence the dip needle.
2. Adjust the dip needle so that it will rotate in the horizontal plane.
3. Allow the needle to reach equilibrium in a north—south line.

4. Adjust the dip needle so that it will rotate in the vertical plane.
5. Allow the needle to reach equilibrium.
6. Read the angle of dip. This is the angle Earth's magnetic field makes with Earth's surface.
7. Record the dip angle below:
   Magnetic dip angle at this location on earth is ____________ degrees.

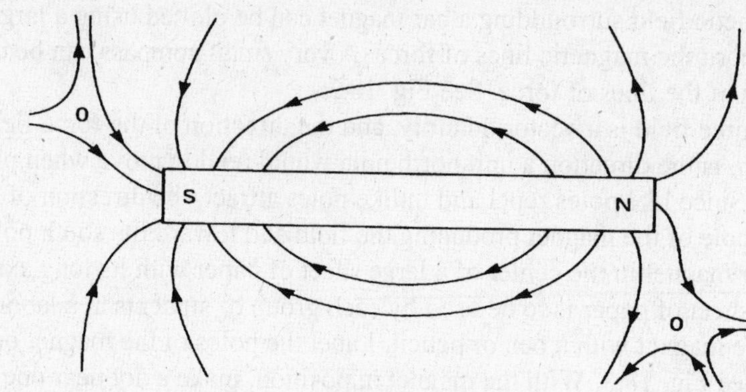

Figure 18.4

PROCEDURE 3

It was stated in the Introduction that a magnetic force field is generated by a moving electric charge. This can be shown experimentally as follows:

1. Bend the number 14 wire to form a square loop as shown in Fig. 18.5.
2. Connect one end of the wire to the positive (+) terminal of the dry cell, but *do not connect* the other end of the wire to the negative (–) terminal.
3. Determine the direction of the magnetic force field around the wire by observing the compass needle when the compass is placed at the positions *A* through *F* as shown in Fig. 18.5. This is done by holding the compass in the labeled position, and then *momentarily* touching the loose end of the wire to the negative (–) terminal of the dry cell. The time of contact with the negative terminal is only long enough to observe the direction the compass needle points. Holding the wire on too long will cause the wire to get hot and quickly discharge the dry cell.
4. Determine the direction of the magnetic field as in or out of the paper at each point around the wire loop (A, B, …, F) in Fig. 18.5. Record the answers in Data Table 18.2.

There is a rule, known as the *left-hand rule,* that is used to determine the direction of a magnetic field around a conductor when the direction of the *electron flow* is known. The rule states that when the conductor is grasped in the left hand with the thumb pointing in the direction of the *electron flow*, the fingers will circle the conductor in the direction of the magnetic field lines. See the section on magnetic fields in your textbook for further information on using the left hand rule. Remember that *electron flow* is always away from the negative (–) terminal of a battery and towards the positive (+) terminal. When the direction of *electron flow* is considered, the left hand rule is used as described above. Another convention that is often used deals with the electric current which by definition flows from the positive (+) terminal to the negative (–) terminal. When electric current is considered, it is convenient to use the right hand rule which is essentially the same except that the thumb points in the direction of the *current flow* instead of the direction of electron flow. The fingers still circle around the conducting wire in the direction of the magnetic field lines. Do both of these rulers agree with your results? If your answer is no, ask your instructor to explain this further.

EXPERIMENT 18 NAME (print) ______________________ LAST ______________________ FIRST DATE ____________

LABORATORY SECTION ________ PARTNER(S) ______________________________

DATA TABLE 18.2 Procedure 3	
Position A	
Position B	
Position C	
Position D	
Position E	
Position F	

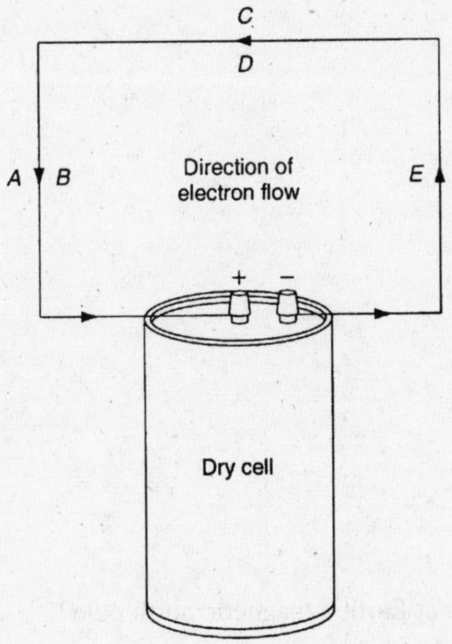

Note: The arrows show the direction of **electron flow** through wire so the left-hand rule is used to determine the orientation of the magnetic field lines around the wire.

Figure 18.5

QUESTIONS

1. State the general rule for the interaction of two magnetic poles.

2. State Coulomb's law for magnetic poles.

3. How is the direction of a magnetic field defined?

4. What is the polarity (north or south) of Earth's magnetic pole which is located near Earth's north geographic pole? Explain.

5. What is *magnetic declination?*

6. What is the *angle of dip?*

7. What is the magnitude of the angle of dip at Earth's magnetic north pole?

8. Explain why the lines of force never cross one another. (*Hint:* Consider the direction of a magnetic line of force.)

Experiment 19

Ohm's Law for Electricity

INTRODUCTION

The study of electricity is the study of static charges and charges in motion. The flow of electrical charges produces the effect called *electricity*. The electron is the smallest stable particle known, and at rest it possesses the physical property of an electric force field. If the electron is moving, then it possesses, in addition to the electric force field, a magnetic force field. The combined effect of these two force fields is called *electricity*. The flow of electrons can be brought about by having an excess of electrons at one position and a deficiency at another. If a conducting path is provided between the two positions, the electrons will flow from the position of the excess to that of the deficiency. In an electric circuit, the electrons flow from the negative terminal (position of excess electrons) through the circuit to the positive terminal (position of deficiency).

Work is done in building up an excess of electrons; therefore, they will possess potential energy due to their position. The potential energy per unit charge is known as the *electrical potential energy* and is measured in volts. The rate of flow of charge, called *current,* is measured in amperes. Opposition to the flow of charge is called *resistance* and is measured in ohms. Ohm's law states the relationship between these concepts. The law can be written symbolically as

$$V = IR$$

where V = potential difference, in volts
I = current, in amperes
R = resistance, in ohms

It is important to learn the distinction between rate of flow of electric charge and the amount of charge present. The number of electrical charges constitutes the amount of charge present, which is measured in coulombs. The rate of flow is the amount of charge passing a particular point per unit time and is measured in amperes. The relationship is expressed as

$$I = \frac{q}{t}$$

where I = current (rate of flow), in amperes
q = amount of electrical charge (charge present) in coulombs
t = time, in seconds

Ohm's law (expressed in an alternative way) states that the ratio of the voltage (V) to the current (I) in an electric circuit is a constant and is equal to the resistance (R). This can be written symbolically as

$$\frac{V}{I} = R$$

The magnitude of an unknown resistance can be determined using this relationship by measuring the voltage across the resistor and the current through it.

LEARNING OBJECTIVES

After completing this experiment, you should be able to do the following:

- Define the terms *voltage, current,* and *resistance.*
- Determine experimentally the voltage, current, and resistance of an electric circuit.
- State Ohm's law.
- Identify a few basic electrical components and connect them to form an electric circuit.
- Use electrical meters to measure voltage and current.

APPARATUS

One 44-ohm rheostat, dc voltmeter (0–10 V), dc milliammeter (0–200 mA), one 40-ohm resistor (5 watt), one unknown resistor, fuse board, connecting wire, low-voltage power source.

Caution Safety Tips

Even low-voltage electricity may cause a shock. Never try any of these experiments with 110 V ac.

EXPERIMENT 19 NAME (print) ______________________________ DATE ______________
LAST FIRST

LABORATORY SECTION ______ PARTNER(S) ______________________________

PROCEDURE 1

Connect the 40-ohm resistor and other electrical components in a series circuit, as shown in Fig. 19.1. The voltmeter is connected across (in parallel with) the 40-ohm resistor. **Caution:** Note the polarity when connecting meters. See Fig. 19.1. Have the instructor check the circuit when you have everything connected. He or she will connect the circuit to the low-voltage direct-current power source; either a battery or a power supply may be used. When this has been done, adjust the variable resistor (the rheostat) to different settings and obtain at least six different values for the current and voltage. Proceed with the taking of data by starting with the rheostat at maximum resistance. This will give minimum current in the circuit. Do not allow excess current to flow in the circuit such that any electrical meter goes off scale. Record the measurements in Data Table 19.1 and calculate the resistance (R) for each of the readings.

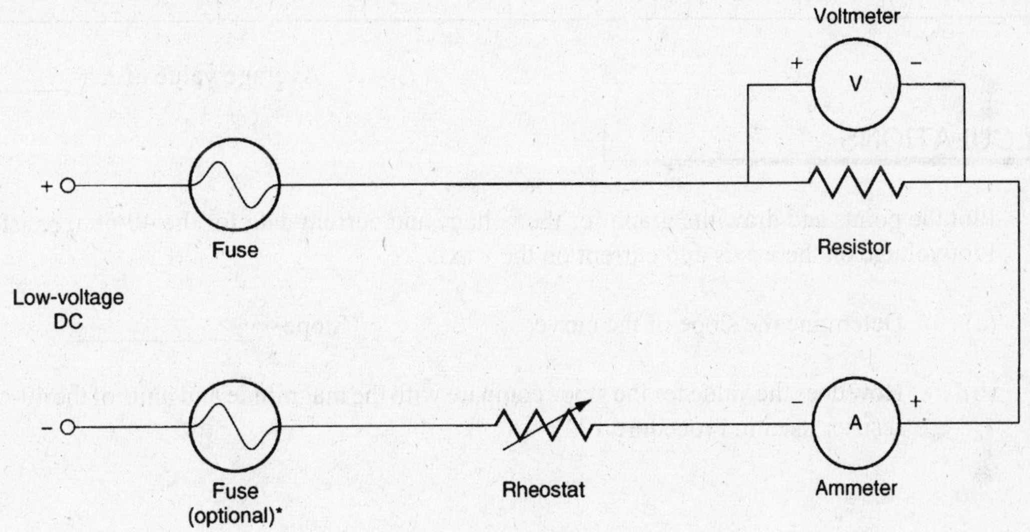

Figure 19.1

DATA TABLE 19.1 40-ohm Resistor						
Rheostat setting	No. 1	No. 2	No. 3	No. 4	No. 5	No. 6
Voltage, in volts (V)						
Current, in amperes (I)						
Resistance $R = \dfrac{V}{I}$						

Average value of R = __________

PROCEDURE 2

Disconnect the circuit from the supply voltage by removing the plug from the outlet socket or disconnecting the battery. Remove the 40-ohm resistor, and replace it with the unknown resistor. Have the instructor check the circuit again. He or she will connect the circuit to the low-voltage power supply. Repeat Procedure 1 and record the six voltage and current measurements in Data Table 19.2 and calculate the resistance (*R*) for each set of readings.

DATA TABLE 19.2 Unknown Resistor

Rheostat setting	No. 1	No. 2	No. 3	No. 4	No. 5	No. 6
Voltage, in volts (V)						
Current, in amperes (I)						
Resistance $R = \dfrac{V}{I}$						

Average value of R = ________

CALCULATIONS

1. Plot the points and draw the graph for the voltage and current data for the 40-ohm resistor. Plot voltage on the y axis and current on the x axis.

   (a) Determine the slope of the curve. Slope = ________

   (b) How does the value for the slope compare with the magnitude and units of the 40-ohm resistor used in Procedure 1?

   (c) How does the value of the slope compare with the average value of R found for Data Table 19.1?

2. Plot the points and draw the graph for the voltage and current data for the unknown resistor. Plot voltage on the y axis and current on the x axis.

   (a) Determine the slope of the curve. Slope = ________

   (b) What is the value of the unknown resistor? R_x = ________

   (c) How does the value found in (b) from the slope compare to the average value of R found for Data Table 19.2?

EXPERIMENT 19 NAME (print) ______________________________ DATE ____________________
LAST FIRST

LABORATORY SECTION ______ PARTNER(S) ____

QUESTIONS

1. If you connect the circuit as shown in Fig. 19.1 with either resistor in place, but one fuse is removed from the circuit, this represents a circuit with a burned-out fuse. What is the voltage across the resistor? Explain your answer.

   $V =$ ____________

2. What is the voltage across the fuse that was left in the circuit? Explain your answer.

   $V =$ ____________

3. What is the voltage across the terminals of the fuse holder without the fuse? Explain your answer.

   $V =$ ____________

4. A voltmeter reads 80 volts when connected in parallel with an unknown resistor that has 125 milliamperes flowing through it. What is the resistance of the unknown resistance? Show your work.

   $R_x =$ ____________

5. Distinguish between an open and a closed circuit. Give an example of each.

6. A 60-watt light bulb operates on regular house voltage of 120 volts. Normal current through the bulb is 0.5 amp. Determine the resistance of the bulb's filament. Show your work.

R_x = ____________

Experiment 20

Electric Circuits

INTRODUCTION

The concepts of electrical energy (measured in volts), the rate of charge flow as current (measured in amperes), and the opposition to the flow of charge or resistance (measured in ohms) were introduced and discussed in the last experiment. We will continue our study of electric circuits in this experiment.

An *electric circuit* is a closed path (loop) of the flow of electric charge (current). A source of electrical potential energy (a low-voltage power source) is required to produce such flow of current in a conductor such as a metal wire. An electric circuit may be composed of one or more components connected in one of two basic patterns designated as *series* or *parallel*, or a more complex circuit may be a combination of the two.

These two basic circuits are defined with respect to the way that charge flows through the circuit elements. A *series circuit* is one in which the current flows through each component in turn and is the same in all components throughout the circuit. See Fig 20.1. A *parallel circuit* is one in which an entering current can divide proportionally between the circuit components and thus follow different paths through the circuit. See Fig. 20.2. In this latter case the current will in general not be the same in the various elements making up the circuit.

In this experiment we will be concerned with circuits involving steady-state or dc currents and voltages. The flow of charge is direct, or in one direction, so dc stands for *direct current*. We find dc current in battery-operated devices such as flashlights, portable radios, portable computers, and automobile headlights. In batteries, the flow of charge (moving electrons) is away from the negative terminal and toward the positive terminal, but for historical reasons, the *current* is considered to flow from positive to negative.

LEARNING OBJECTIVES

After completing this experiment, you should be able to do the following:

- ▼ Describe series and parallel circuits.
- ▼ Make the correct electrical connections for circuit elements in both series and parallel circuits.
- ▼ Connect voltmeters and ammeters properly in electric circuits to measure the voltage and current in these circuits.
- ▼ Determine experimentally and mathematically the values of current, voltage, and resistance in both series and parallel circuits.

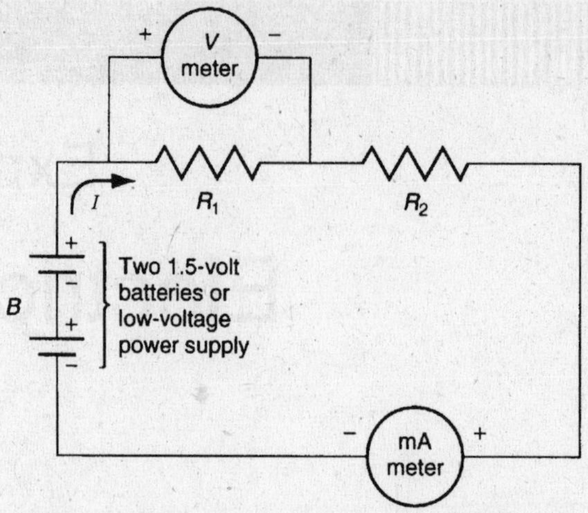

Figure 20.1 A series circuit with two resistors showing the placement of the milliammeter to read I_T, total current, and the voltmeter to read V_1.

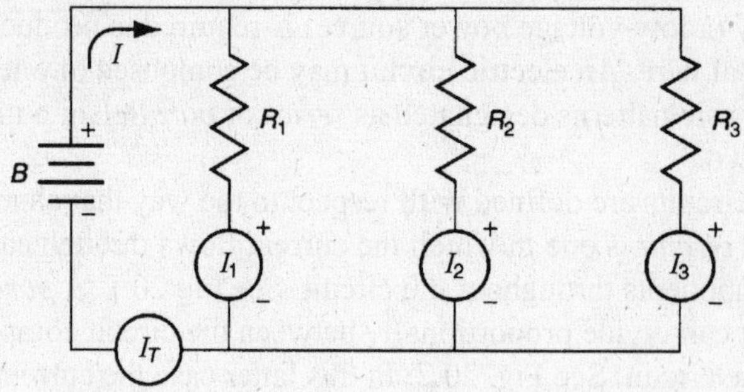

Figure 20.2 A parallel circuit of three resistors showing placement of milliammeters to read I_T and the current through each resistor, I_1, I_2, and I_3.

APPARATUS

One high-resistance dc voltmeter (0–10 V), one low-resistance dc milliammeter (0–200 mA or 0–500 mA) (four milliammeters can be used if available in the parallel circuit), two 1.5-volt dry cells (with battery holder if possible) or a low-voltage power supply, several resistors of low wattage (typical values are 10, 22, 47, and 82 ohms because they are readily available from Radio Shack), about 10 wire connectors with alligator clips on each end.

Even low-voltage electricity may cause a shock. Never try any of these experiments with 110 V ac.

EXPERIMENT 20 NAME (print) ______________________________ DATE ____________________
LAST FIRST

LABORATORY SECTION ______ PARTNER(S) ______________________________________

GENERAL PROCEDURES FOR ELECTRIC CIRCUITS

When setting up electric circuits, it is always necessary to exercise care in handling the equipment and connecting the wires. Even a low-voltage battery can become extremely hot or even explode if a direct short is placed across it. All electric meters are delicate and must be handled with care to prevent unnecessary damage.

The circuits used in this *Laboratory Guide* were tested using connection leads with alligator clips on both ends. This makes it possible to clip directly to the leads on resistors and to meter terminals without having to construct any special resistor boards or holders. It also eliminates the need for separate switches because the connection to the positive terminal of the battery holder or low-voltage power supply can be attached or removed easily to control the flow of current into the circuit.

A two-cell battery holder was chosen to simplify the way that the two D cells were connected to the circuits under study. The ends of the power leads were simply stripped back about an inch and clips used to connect them to the rest of the circuit components. The values of the resistors were chosen to coincide with those available from Radio Shack in packets of 6 for about 50 cents.

Once the resistors have been removed from their packages, it is necessary to read their color codes to keep track of their resistive values. To do this, the first two color bands are read as a number, and the third color band tells how many zeros must be placed behind that two-digit number. The fourth color band indicates the precision to which the resistor was manufactured. The colors correspond to the following numbers:

0 = black	1 = brown	2 = red
3 = orange	4 = yellow	5 = green
6 = blue	7 = violet	8 = gray
9 = white	5% = gold	10% = silver

As an example, a resistor having color bands of red, red, black, and silver would be a 22 + no zeros with 10 percent tolerance, or a 22-ohm resistor. If the third band had been orange, the value would have been 22 + 3 zeros, or 22,000 ohms.

When taking voltage readings, the voltmeter can be clipped directly across the circuit element for which the voltage must be measured. The milliammeter must be connected in series, so the wires must be unclipped and reconnected if the meter is moved around to read the current in different loops of a parallel circuit. Remember that if a milliammeter is used, the readings must be converted to amperes before calculations using Ohm's law or the power equation can be done.

Remember, be careful when doing any electrical experiments, and do not assume that since you have done simple experiments with low-voltage power sources you can play with high-voltage circuits and not get hurt. High-voltage electricity, especially ac, can be extremely dangerous and must not be treated casually.

PROCEDURE 1

(a) Connect the milliammeter in series with a 47-ohm resistor R_1 to the power source as shown in Fig. 20.3. If you do not have a 47-ohm resistor, use another with a similar value.

R_1 = ________________ ohms

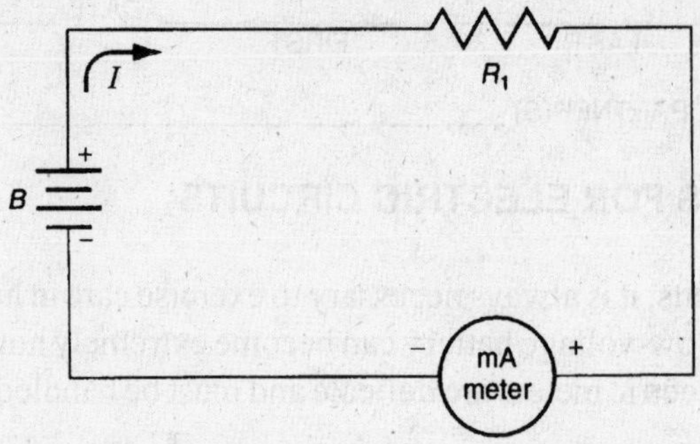

Figure 20.3 Series circuit with one resistor, a milliammeter, and connected to a power source.

(b) Measure the total current flowing in the series circuit.

$I_T =$ __________ mA = __________ amps

(c) Measure the voltage across the power source.

$V_B =$ __________ volts

(d) Measure the voltage across the resistor R_1.

$V_1 =$ __________ volts

(e) Calculate the resistance of R_1 using Ohm's law.

$$R_1 = \frac{V_1}{I_T} = \frac{\_\_\_\_\_\_\_\_\_\_ \text{ volts}}{\_\_\_\_\_\_\_\_\_\_ \text{ amps}} = \_\_\_\_\_\_\_\_\_\_ \text{ ohms}$$

QUESTIONS

1. How does the value of the voltage across the power source V_B compare with the voltage across the resistor R_1?

2. Find the percentage error between the value printed on resistor R_1 and the value calculated in Procedure 1(e).

EXPERIMENT 20 NAME (print) ______________________________ DATE ______________________________
LAST FIRST

LABORATORY SECTION ______ PARTNER(S) ______________________________

PROCEDURE 2

(a) Add a second 47-ohm resistor R_2 in series with R_1 and the milliammeter. If a 47-ohm resistor is not used, fill in a different value for R_2. $R_2 =$ __________ ohms. See Figure 20.1.

(b) Measure the total current.

$I_T =$ ___________ mA = ___________ amps

(c) Measure the voltage across each resistor.

$V_1 =$ __________ volts

$V_2 =$ __________ volts

(d) Measure the voltage across *both* resistors at once. *Note:* This is the same as the voltage supplied by the voltage source.

$V_{1+2} = V_B =$ __________ volts

(e) Calculate the resistances R_1, R_2, and R_{1+2}.

$R_1 = \frac{V_1}{I_T} =$ _________ ohms

$R_2 = \frac{V_2}{I_T} =$ _________ ohms

$R_{1+2} = \frac{V_B}{I_T} =$ _________ ohms

(f) Calculate the value of the series combination of resistors R_1 and R_2 with the series resistor equation using the values marked on the resistors. $R_{1+2} = R_1 + R_2$.

$R_{1+2} =$ ________ ohms

QUESTIONS

1. How well do the values for the resistors R_1 and R_2 calculated in Procedure 2(e) compare with the value marked on them by the manufacturer?

2. Calculate the percentage difference between the value calculated for R_{1+2} in Procedure 2(e) and the value calculated for the series combination of these two resistors R_{1+2} found in Procedure 2(f).

PROCEDURE 3

(a) Add a third 47-ohm resistor R_3 in series with R_1, R_2, and the milliammeter. If a 47-ohm resistor is not used, fill in a different value for R_3,

$R_3 =$ _________ ohms

(b) Measure the total current.

$I_T =$ _______ mA = _________ amps

(c) Measure the voltage across each resistor.

$V_1 =$ _______ volts

$V_2 =$ _______ volts

$V_3 =$ _______ volts

(d) Measure the voltage across *all three* resistors in series at once, and record this value below. *Note:* This is also the value for the voltage supplied by the voltage source, V_B.

$V_{1+2+3} = V_B =$ _______ volts

(e) Calculate the resistances R_1, R_2, R_3, and R_{1+2+3}.

$R_1 = \dfrac{V_1}{I_T} =$ _________ ohms

$R_2 = \dfrac{V_2}{I_T} =$ _________ ohms

EXPERIMENT 20 NAME (print) ______________________________ DATE ______________
LAST FIRST

LABORATORY SECTION ______ PARTNER(S) ______________________________

$R_3 = \frac{V_3}{I_T} =$ __________ ohms

$R_{1+2+3} = \frac{V_B}{I_T} =$ __________ ohms

(f) Calculate the value of the series combination of resistors R_1, R_2, and R_3 with the series resistor equation using the values marked on the resistors. $R_{1+2+3} = R_1 + R_2 + R_3$

$R_{1+2+3} =$ ________ ohms

QUESTIONS

1. Calculate the percentage difference between the values calculated for R_{1+2+3} in Procedure 3(e) and the value calculated for the series combination of these three resistors R_{1+2+3} found in Procedure 3(f).

2. Add the voltages measured across the three resistors V_1, V_2, and V_3 together and compare this sum with the value measured for V_{1+2+3} in Procedure 3(d).

3. Calculate the *IR* drops across the three resistors in the series circuit and compare them with the voltages measured across each resistor.

$I_T R_1 =$ ___________ volts $V_1 =$ ___________ volts

$I_T R_2 =$ ___________ volts $V_2 =$ ___________ volts

$I_T R_3 =$ ___________ volts $V_3 =$ ___________ volts

4. Add the three values for the *IR* drops together and compare them with the value for the voltage across the entire series circuit.

Sum of 3 *IR* drops = __________ volts $V_B =$ ___________ volts

PROCEDURE 4

(a) Replace two of the resistors in the series circuit used in Procedure 3 with a 10-ohm resistor and a 22-ohm resistor. This makes $R_1 = 10$ ohms, $R_2 = 22$ ohms, and $R_3 = 47$ ohms. Again, if you do not have resistors of these exact values, use similar ones and fill in their values below.

$R_1 =$ ________ ohms

$R_2 =$ ________ ohms

$R_3 =$ ________ ohms

(b) Measure the total current.

$I_T =$ ________ mA= ________ amps

(c) Measure the voltage across each resistor.

$V_1 =$ ________ volts

$V_2 =$ ________ volts

$V_3 =$ ________ volts

(d) Measure the voltage across *all three* resistors at once.

$V_{1+2+3} = V_B =$ ________ volts

(e) Calculate the resistances R_1, R_2, R_3, and R_{1+2+3}.

$R_1 = \frac{V_1}{I_T} =$ ________ ohms

$R_2 = \frac{V_2}{I_T} =$ ________ ohms

$R_3 = \frac{V_3}{I_T} =$ ________ ohms

$R_{1+2+3} = \frac{V_B}{I_T} =$ ________ ohms

(f) Calculate the value of the series combination of resistors R_1, R_2, and R_3 with the series resistor equation using the values marked on the resistors. $R_{1+2+3} = R_1 + R_2 + R_3$

$R_{1+2+3} =$ ________ ohms

EXPERIMENT 20 NAME (print) ________________________________ DATE ______________
LAST FIRST

LABORATORY SECTION ______ PARTNER(S) ________________________________

QUESTIONS

1. Calculate the percentage difference between the values calculated for R_{1+2+3} in Procedure 4(e) and the value calculated for the series combination of these two resistors R_{1+2+3} found in Procedure 4(f).

2. Add the voltages measured across the three resistors V_1, V_2, and V_3 together and compare this sum with the value measured for V_{1+2+3} in Procedure 3(d).

3. Calculate the *IR* drops across the three resistors in the series circuit and compare them with the voltages measured across each resistor.

$I_T R_1 =$ ________________ volts $V_1 =$ ________________ volts

$I_T R_2 =$ ________________ volts $V_2 =$ ________________ volts

$I_T R_3 =$ ________________ volts $V_3 =$ ________________ volts

4. Add the three values for the *IR* drops together and compare them with the value for the voltage across the entire series circuit.

Sum of 3 *IR* drops = ____________ volts $V_B =$ ____________ volts

5. In general, how do you think the sum of all of the *IR* drops in a series circuit compares with the total voltage across the entire circuit based on your results in Procedures 2, 3, and 4?

6. Why didn't you have to take separate measurements for the individual currents through resistors R_1, R_2, and R_3 in Procedures 2, 3, and 4?

7. Do you think that this experiment confirms the validity of the series resistor equation used in Procedures 2(f), 3(f), and 4(f)?

PROCEDURE 5

(a) Connect three 47-ohm resistors in parallel with the milliammeter as shown in Fig. 20.4. If 47-ohm resistors are not used, fill in a different value in the space below.

$R_1 = R_2 = R_3 =$ _________ ohms

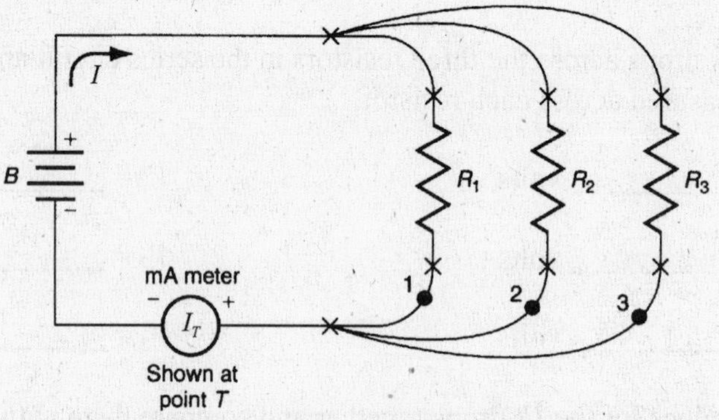

Figure 20.4 Parallel circuit shown as it might look with three resistors connected using clip leads (ends marked with x). Ammeter can be moved to points 1, 2, and 3 to measure current through the individual resistors.

(b) Measure the total current.

$I_T =$ ________ mA= ___________ amps

(c) Measure the voltages across each resistor.

$V_1 =$ ________ volts

$V_2 =$ ________ volts

$V_3 =$ ________ volts

EXPERIMENT 20 NAME (print) ______________________________ DATE ______________
LAST FIRST

LABORATORY SECTION ______ PARTNER(S) ____

(d) Measure the voltage across the power source.

$V_B =$ __________ volts

(e) Now carefully move the milliammeter from point *T* in the parallel circuit to point 1, and measure the current through resistor R_1 and record this value below. Now repeat by moving the milliammeter to point 2 and to point 3. Keep all three parallel resistors connected while you make each measurement. If you happen to have four milliammeters available, you can set up the circuit with meters at *T*, 1, 2, and 3 all at once and take the current readings in that way, but you will get acceptable results if you use one meter and carefully move it to each location in turn. Now calculate the resistances R_1, R_2, R_3, and R_{1+2+3}.

$I_1 =$ _______ mA = ________ amps $R_1 = \frac{V_1}{I_1} =$ ________ ohms

$I_2 =$ _______ mA = ________ amps $R_2 = \frac{V_2}{I_2} =$ ________ ohms

$I_3 =$ _______ mA = ________ amps $R_3 = \frac{V_3}{I_3} =$ ________ ohms

$R_{1+2+3} = \frac{V_B}{I_T} =$ ________ ohms

(f) Calculate the value of the parallel combination of resistors R_1, R_2, and R_3 with the parallel resistor equation using the values marked on the resistors.

$$\frac{1}{R_{1+2+3}} = \frac{1}{R_1} + \frac{1}{R_2} + \frac{1}{R_3}$$

$R_{1+2+3} =$ _______ ohms

QUESTIONS

1. How well do the values calculated for the resistors R_1, R_2, and R_3 compare with the value marked on them by the manufacturer?

2. Calculate the percentage difference between the values calculated for R_{1+2+3} in Procedure 5(e) and the value R_{1+2+3} calculated for the parallel combination of these three resistors found in Procedure 5(f).

3. Add the currents I_1, I_2, and I_3 that you measured flowing through the three resistors in Procedure 5(e), and compare this sum with the value measured for I_T in Procedure 5(b).

4. Calculate the *IR* drops across the three resistors in the series circuit and compare them with the voltages measured across the individual resistors V_1, V_2, and V_3.

I_1R_1 = __________ volts V_1 = __________ volts

I_2R_2 = __________ volts V_2 = __________ volts

I_3R_3 = __________ volts V_3 = __________ volts

PROCEDURE 6

(a) Replace one of the resistors in the parallel circuit used in Procedure 5 with an 82-ohm resistor and take the third one out of the circuit completely. This makes $R_1 = 47$ ohms and $R_2 = 82$ ohms. Again, if you do not have resistors of these exact values, use similar ones.

R_1 = __________ ohms R_2 = __________ ohms

(b) Measure the current flowing through *both* resistors at once.

I_T = __________ mA = __________ amps

(c) Measure the voltage across each resistor.

V_1 = _______ volts

V_2 = _______ volts

(d) Measure the voltage across the voltage source.

V_B = ______ volts

EXPERIMENT 20 NAME (print) ______________________________ DATE ______________
LAST FIRST

LABORATORY SECTION ______ PARTNER(S) ____

(e) Measure the current flowing through each resistor I_1 and I_2. You may have to move the milliammeter around in the circuit if you only have one meter. Then calculate the resistances R_1, R_2, and R_{1+2}.

I_1 = __________ mA = __________ amps $R_1 = \frac{V_1}{I_1}$ = __________ ohms

I_2 = __________ mA = __________ amps $R_2 = \frac{V_2}{I_2}$ = __________ ohms

$R_{1+2} = \frac{V_B}{I_T}$ = ____________ ohms

(f) Calculate the value of the parallel combination of resistors R_1 and R_2 with the parallel resistor equation using the values marked on the resistors.

$$\frac{1}{R_{1+2}} = \frac{1}{R_1} + \frac{1}{R_2}$$

R_{1+2} = ________ ohms

QUESTIONS

1. Calculate the percentage difference between the values calculated for R_{1+2} in Procedure 6(e) and the value calculated for the parallel combination of these two resistors R_{1+2} found in Procedure 6(f).

2. Add the currents you measured flowing through the two resistors I_1, and I_2 together and compare this sum with the value measured for I_T in Procedure 6(b).

3. Calculate the *IR* drops across the two resistors in the parallel circuit and compare them with the voltages measured across each resistor V_1 and V_2.

I_1R_1 = __________ volts V_1 = __________ volts

I_2R_2 = __________ volts V_2 = __________ volts

4. After reviewing the data from Procedures 5 and 6, in general how do you think the sum of all the currents flowing through the resistors in a parallel circuit compares with the total current flowing in the circuit?

5. After reviewing the data from Procedures 5 and 6, in general how do you think the voltages across individual elements in a parallel circuit compare with the total voltage supplied to the circuit?

6. Why do you have to take separate measurements for the individual currents through resistors in Procedures 5 and 6?

7. Does this experiment confirm the validity of the parallel resistor equation used in Procedures 5(f) and 6(f)?

Experiment 21

Electromagnetic Waves

INTRODUCTION

When a charged particle such as an electron, proton, or ion is accelerated, energy in the form of electromagnetic radiation emanates in all directions from the accelerated particle. An example of electromagnetic radiation is visible light, which consists of oscillating electric and magnetic fields that travel through vacuum at a speed of 3×10^8 m/s (186,000 mi/s). Your awareness of these printed words is due to the sensitivity of your eyes to electromagnetic radiation, which is reflected from the page of the *Laboratory Guide.*

Electromagnetic radiation consists of oscillating electric and magnetic fields (vector quantities) perpendicular to each other and also perpendicular to the direction of the wave motion. The fields are in phase. That is, their minimum and maximum values occur at the same points in the oscillating cycle. Figure 21.1 illustrates the relationship between the force fields and the direction of the wave motion.

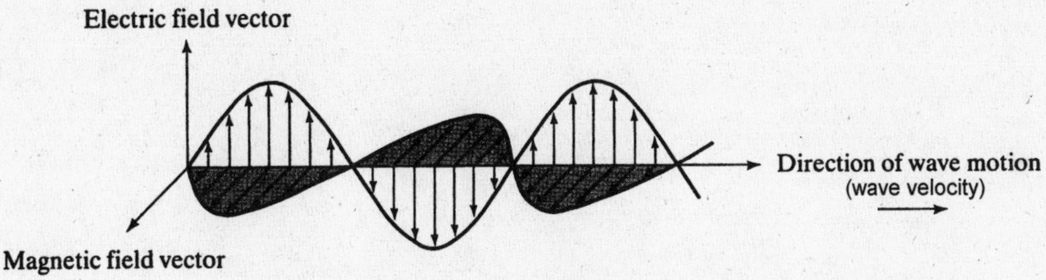

Figure 21.1 Electromagnetic waves. Electromagnetic waves consist of oscillating electric and magnetic fields. Note that the fields are at right angles to each other and also at right angles to the direction of the wave motion.

All electromagnetic radiation travels through empty space with a speed *c*. The speed *c*, frequency *f*, and the wavelength λ (lambda) are related by the equation

$$c = f\lambda \qquad \text{Eq. (21.1)}$$

where c = speed of the wave = 3.00×10^8 m/s

f = frequency in hertz (Hz)

λ = wavelength in meters (m)

Very short wavelengths are often expressed in angstrom or nanometer units. One angstrom equals 10^{-10} m; one nanometer equals 10^{-9} m. To convert angstroms to nanometers, move the decimal point one place to the left.

LEARNING OBJECTIVES

After completing this experiment, you should be able to do the following:

- Explain the physical characteristics of electromagnetic waves.
- State how electromagnetic waves of different frequencies are generated and detected.
- Calculate the frequency or wavelength of electromagnetic waves when either the frequency or wavelength is given.
- Give the relationship between frequency, wavelength, and the speed of electromagnetic waves.

APPARATUS

A hand calculator is useful but not necessary.

EXPERIMENT 21 NAME (print) ______________________ DATE ______________
LAST FIRST

LABORATORY SECTION ______ PARTNER(S) ______________________________

PROCEDURE 1

Electromagnetic radiation sensitive to the human eye is called *visible light* and has frequencies ranging from 4.3×10^{14} Hz to 7.5×10^{14} Hz. Use Eq. 21.1 and calculate the wavelengths (in meters) for the frequencies given in Data Table 21.1. Record your answers in the space provided. Use the conversion information given above and record the wavelengths in angstrom and nanometer units. Use Fig. 21.2 to obtain the color for the calculated wavelengths. Record the color for each wavelength in Data Table 21.1.

DATA TABLE 21.1 Calculated Wavelengths for Visible Light Frequencies

Frequency f (Hz)	Wavelength λ (m)	Wavelength λ (Å)	Wavelength λ (nm)	Color
4.3×10^{14}				
5.0×10^{14}				
6.0×10^{14}				
7.5×10^{14}				

The electromagnetic spectrum (an ordered arrangement of frequencies and/or wavelengths) is shown in Fig. 21.2. The spectrum is continuous; it ranges from low frequency radio waves to extremely high frequency gamma rays. There is no sharp dividing line between one type of electromagnetic wave and its neighbor. In fact, there is an overlapping of frequencies, as shown in Fig. 21.2. Electromagnetic waves differ from one another only in their frequency or wavelength.

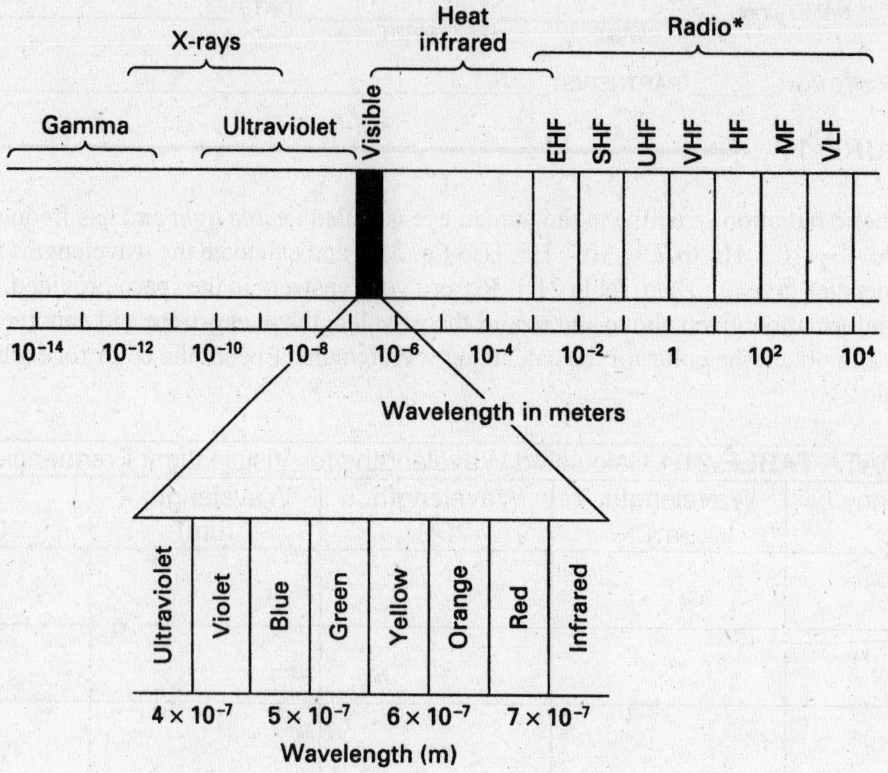

Figure 21.2 The electromagnetic spectrum

*E, extremely; F, frequency; H, high; L, low; M, medium; S, super; U, ultra; V, very.
Example: UHF means ultra high frequency radiation

PROCEDURE 2

Plot a graph that shows the relationship between frequency and wavelength. Use the data in Data Table 21.1. Plot wavelength on the x axis. (If necessary, refer to Experiment 1 in this *Laboratory Guide* for information on how to plot and label a graph.)

QUESTIONS

1. State in words the relationship between frequency and wavelength that your graph illustrates.

2. Refer to Experiment 1 in this *Laboratory Guide* and give the name that describes the curve in your graph. (*Hint:* What is the equation for the curve?)

EXPERIMENT 21 NAME (print) ______________________ DATE ______________
LAST FIRST

LABORATORY SECTION ______ PARTNER(S) ______________________________

PROCEDURE 3

Table 21.2 lists the entire frequency range for each type of electromagnetic radiation. The values for the corresponding wavelengths are not given. Use the equation $c = f\lambda$ and calculate the wavelengths for each of these frequencies. Record the calculated values in the appropriate place in Table 21.2.

Electromagnetic waves are generated and detected in different ways and by different devices, as listed in Table 21.2. Some of these may be unfamiliar to you. It is not feasible to give an explanation for all terms listed in columns one and two of Table 21.2. If you have a question concerning one or more terms, ask your instructor for assistance, or refer to an outside source.

The next paragraph gives a brief explanation of how radio waves are generated by an LC circuit. They are also detected by the same type of circuit.

An LC circuit consists of a capacitor C (two conductors separated by an insulator) and an inductor L (a conductor in the form of a coil). If the capacitor has an initial charge (stored electrons) and the switch S is closed (see Table 21.2, column two), the capacitor will begin to discharge. The total energy in the circuit, which is stored in the electric field between the two conductors, will be transferred by the flow of electrons to the magnetic field of the inductor. The magnetic field cannot sustain itself and collapses. This induces an electron flow in the circuit that places an opposite charge back on the capacitor. The process then repeats itself in the opposite direction. Thus electrons are accelerated back and forth between the capacitor and the inductor, and electromagnetic waves are generated. Because there is resistance (opposition to the flow of the electrons) in the circuit, energy must be continually supplied to the circuit, or else the oscillations will gradually die out.

PROCEDURE 4

Electromagnetic radiation has a dual nature. Sometimes it acts as a wave; sometimes it acts as a bundle of energy. Max Planck, a German physicist, introduced the idea that a frequency generator could have only discrete or certain amounts of energy. The energy of the generator depends on its frequency according to the relationship

$$E = hf$$

where h (called *Planck's constant*) $= 6.67 \times 10^{-34}$ J·s
f = frequency in Hz
E = a quantum of energy (called a photon) in joules

Calculate the energy of a photon in Joules with an associated frequency listed below:

(a) 2×10^{14} Hz

(b) 4.3×10^{14} Hz

(c) 7.5×10^{14} Hz

(d) 3×10^{18} Hz

DATA TABLE 21.2 Electromagnetic Radiations

Name of Radiation	Radiation Generated by	Radiation Detected by
Radio waves AM broadcast band FM broadcast band TV broadcast band Microwaves	Accelerated electrons in an LC circuit 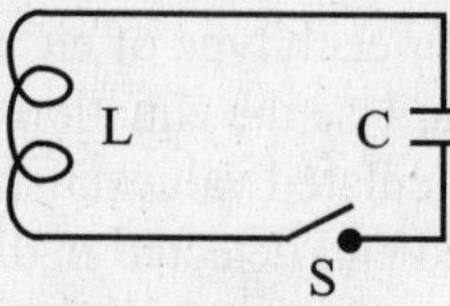	Electronic LC circuit Charged-coupled device (CCD)
Infrared (heat waves)	Hot bodies (disturbed molecules)	Skin Thermometers Thermocouples Radiometers
Visible light	Hot bodies (disturbed electrons in atoms and molecules)	Eye Photographic film Photocell
Ultraviolet	Electric arcs Special lamps (disturbed electrons in atoms)	Photographic film Photocell
X-rays	Special vacuum tubes (disturbed electrons in atoms)	Photographic film Geiger counter Ionization chamber
Gamma rays	Radioactive nuclei (disturbed nucleus in atoms)	Geiger counter Ionization chamber Scintillation chamber

Name of Radiation	Frequency Range Hertz (cycles/s) Approximate Values	Wavelength Range Meters to be Calculated by Student
Radio waves AM broadcast band FM broadcast band TV broadcast band Microwaves	AM 0.55×10^6 to 1.60×10^6 FM 88×10^6 to 108×10^6 TV 54×10^6 to 890×10^6 MW 1×10^9 to 1×10^{11}	
Infrared (heat waves)	1×10^{11} to 4.3×10^{14}	
Visible light	4.3×10^{14} to 7.5×10^{14}	
Ultraviolet	7.5×10^{14} to 3×10^{17}	
X-rays	3×10^{17} to 3×10^{19}	
Gamma rays	Greater than 3×10^{19}	

EXPERIMENT 21 NAME (print) ________________________ DATE ____________
LAST FIRST

LABORATORY SECTION ______ PARTNER(S) ______________________________

QUESTIONS

1. What is the basic explanation for the origin of electromagnetic waves?

2. All electromagnetic waves have the same physical characteristics. They differ only in frequency and wavelength. How are they classified into different types?

3. How do you account for the fact that a sharp dividing line does not exist between one type of electromagnetic wave and its neighbor?

4. How are radio waves generated? How are they detected? What do you think limits the highest radio frequency that can be generated?

5. Why is it impossible to generate radio waves with a frequency of 4×10^{14} Hz?

Use the data given in Table 21.2, Fig. 21.1, and Fig. 21.2 to answer Questions 6, 7, and 8.

6. How are X-rays generated? Give three methods for detecting them.

7. What is the minimum frequency generated by radioactive nuclei? The maximum wavelength? What is the name of this radiation?

8. Name the type of radiation that has the greatest range of frequencies. Give the range width.

9. Electromagnetic waves are transverse waves. State the relationship between the electric, magnetic, and wave velocity vectors.

10. Calculate the difference in energy (joules) between red and violet light.

11. Does a photon associated with blue light have more or less energy than a photon associated with red light? Give the reason for your answer.

Experiment 22
Temperature

INTRODUCTION

Temperature can be defined as the degree of hotness or coldness of a body; temperature also can refer to that property of a body that determines the direction of heat flow by conduction. Temperature is also proportional to the average kinetic energy of the random motion of the particles in matter.

Regardless of how temperature is defined, one common way of measuring it is with a mercury-in-glass thermometer. The method of the thermometer's functioning is based on the expansion and contraction effect of heat. Mercury exists in the liquid phase over a large temperature range (melting point, –38.87°C; boiling point, 356.9°C). When a mercury-in-glass thermometer is placed in a gas such as the air, heat is transferred from this gas to the thermometer or from the thermometer to the gas, depending on which is at the higher energy level.

If heat flows from the gas into the glass and the mercury, the glass and the mercury expand to a larger volume, and a higher temperature is recorded by the thermometer. If the heat flow is in the reverse direction, the glass and mercury contract, and the thermometer records a lower temperature. The coefficient of cubical expansion of mercury is large compared with that of glass; that is, mercury expands much more than glass for each degree change in temperature, so the mercury rises and falls in a narrow glass tube which thus indicated the current temperature.

Other materials can also be used as the liquid in glass thermometer tubes. Because mercury vapor is poisonous, thermometers that were filled with mercury are becoming less common in situations where a narrower temperature range will work such as in school laboratories. For that reason we recommend alcohol-in-glass thermometers for student use whenever they are available.

A thermometer must be calibrated before a temperature can be recorded accurately. This is done by determining two fixed points (usually the melting point and boiling point of water) on the thermometer, choosing an arbitrary unit of measurement, and marking a scale on the glass tube. Two primary temperature scales are in use today. The Celsius scale sets 100°C as the boiling point of pure water and 0°C as its freezing point. The distance between these points is then divided into 100 equal increments. The other scale was developed by Fahrenheit and although it was not originally defined in this way, it too can be set up using the boiling point of water as 212°F and its freezing point is 32°F with 180 equal divisions between the two points.

LEARNING OBJECTIVES

After completing this experiment, you should be able to do the following:

- ▼ Define and explain *temperature* and state its units of measurement.
- ▼ Calibrate an alcohol-in-glass or mercury-in-glass thermometer.
- ▼ Measure temperature with your newly calibrated thermometer.

APPARATUS

Nongraduated thermometer (alcohol-in-glass or mercury-in-glass), one liter Pyrex glass beaker, tripod base to hold glass beaker, ice cubes, Bunsen burner, marker for making temporary mark on glass, ruler.

Safety glasses or goggles are recommended. Open flame can be dangerous. Do not leave a lighted burner unattended or reach across flame. Hot water can scald; handle with care. Whenever possible use a safer alcohol-in-glass thermometer in case of accidental breakage.

EXPERIMENT 22 NAME (print) ______________________ DATE ______________
LAST FIRST

LABORATORY SECTION ______ PARTNER(S) ______________________

PROCEDURE

1. Fill the glass beaker about half full with tap water. Place the thermometer in water, and heat the water to boiling point.
2. Allow the system to reach equilibrium; that is, wait until the liquid in the thermometer reaches its highest point in the glass tube and remains there for a few minutes. This is called the *steam point.* Lay the thermometer on a paper towel on the table to cool somewhat before the next step.
3. Mark the steam point with the glass marking pen provided by the instructor. This should be done carefully and quickly. Lay the thermometer on a paper towel on the table to cool somewhat before the next step.
4. Place a mixture of ice and water in the plastic beaker, and position the thermometer in the mixture. Make sure that there is sufficient water in the beaker that the entire volume of liquid in the thermometer will be completely submerged.
5. Allow the system to come to equilibrium; that is, wait until the liquid reaches its lowest point in the glass tube and remains there for a few minutes. This is called the *ice point.*
6. Mark the ice point with a glass marker. Make the mark carefully and quickly, and be careful not to break the thermometer in doing so. Be certain that the mark is placed on the glass tube where you observe the ice point.

   Note: Normally, in calibrating an alcohol (or mercury)-in-glass thermometer, the ice and steam points are determined with pure water at standard atmospheric pressure. We have not called for either of these conditions for this experiment so the calibration of your thermometer will not be perfect.
7. Since the expansion of alcohol or mercury is fairly linear from 0°C to 100°C, a linear scale can be marked on the glass tube over this temperature range. Using the ruler, mark off a scale on the thermometer between the two fixed points. Assign 0° Celsius to the ice point and 100° Celsius to the steam point. Divide the distance between the two marks into 10 equal sections and then mark carefully 10 divisions between 20°C and 30°C. This range on the thermometer will be used in determining the existing air temperature.
8. Determine the existing air temperature in the laboratory with your newly calibrated thermometer.

   ______________°C

9. Ask the instructor for the location in the laboratory of a standard thermometer. Read and record the existing air temperature of the laboratory.

   ______________°C

QUESTIONS

1. What is the least count of your hand calibrated thermometer?

2. What is the percent error of your newly calibrated thermometer in your determination of the existing air temperature in the laboratory? Use the reading on the standard commercially calibrated thermometer as the accepted value. Show your work.

Percent error = ____________

3. Would you expect the steam point on your calibrated thermometer to be higher or lower than a regular standard thermometer? Why?

4. What are the disadvantages of using a water-in-glass thermometer to measure outside air temperature?

5. When a mercury-in-glass thermometer is placed in boiling water, the mercury level falls slightly at first and then begins to rise. Explain this effect. If you have lab time, perform the experiment. Obtain the lab instructor's permission.

Experiment 23

Specific Heat

INTRODUCTION

The law of conservation of energy states that the energy in any isolated system remains constant. This law provides a means for determining the specific heat of a substance.

Experiments have shown that the amount of heat H absorbed or lost by a substance undergoing a temperature change ΔT is directly proportional to the change in the temperature, the mass m of the substance, and the type of substance. We write these relations as follows:

Since: $$H \propto \Delta T \text{ and } H \propto m$$

then: $$H \propto m\Delta T$$

and: $$H = cm\Delta T$$

where H = the heat absorbed or lost in calories
ΔT = change in temperature in degrees Celsius, $\Delta T = T_f - T_i$
m = the mass of the substance in grams
c = the proportionality constant called the specific heat

Solving the equation $H = cm\Delta T$ for c yields

$$c = \frac{H}{m\Delta T}$$

The specific heat of a substance is thus seen to be the amount of heat required to raise one unit mass of a substance one degree in temperature. In terms of cgs units of measurement, the specific heat of a substance is the number of calories required to change the temperature of one gram of the substance by one degree Celsius. By definition, the specific heat of water is 1.00 calorie per gram degree Celsius. Because 4.18 joules equals one calorie, the specific heat of water in standard SI energy units is 4.18 joules/gram degree Celsius.

In obtaining the experimental value of the specific heat c, it would appear at first to be a simple task to measure the number of calories required to raise one gram of the substance one degree C. True, the mass can be determined by using a balance and the temperature by a thermometer, but we have no direct method of measuring the amount of heat gained or lost. Because the heat cannot be measured directly, we resort to a procedure known as the *method of mixtures,* which is based on the law of conservation of energy.

The method of mixtures consists of bringing together a known mass of a substance at a known high temperature and a known mass of water at a known lower temperature and then determining the resulting temperature of the mixture. In the process, the heat lost by the substance is absorbed by the water and the container holding the water. If we assume that no heat is lost to the surroundings or gained from them, we can express this relation as follows:

$$\text{Heat lost by substance} = \text{heat gained by water and container}$$

The substance in this experiment will be copper or aluminum metal. Therefore, we can formulate the equation

$$H_{\text{metal}} = H_{\text{water}} + H_{\text{container}}$$

Because $H = cm\Delta T$ for any substance, we can substitute for H as follows:

$$(cm\Delta T)_{\text{metal}} = (cm\Delta T)_{\text{water}} + (cm\Delta T)_{\text{container}}$$

or

$$c_{\text{mt}}m_{\text{mt}}(T_{\text{mt}} - T_{\text{f}}) = c_{\text{w}}m_{\text{w}}(T_{\text{f}} - T_{\text{w}}) + c_{\text{cn}}m_{\text{cn}}(T_{\text{f}} - T_{\text{cn}})$$

where: T = temperature in °C
m = mass in grams
c = specific heat in $\dfrac{\text{cal}}{\text{g}\cdot°\text{C}}$

and the subscripts used refer to: mt = the symbol for the metal
w = the symbol for the water
cn = the symbol for the container
f = the symbol for final

Solving for the specific heat of the metal c_{mt},

$$c_{\text{mt}} = \frac{c_{\text{w}}m_{\text{w}}(T_{\text{f}} - T_{\text{w}}) + c_{\text{cn}}m_{\text{cn}}(T_{\text{f}} - T_{\text{cn}})}{m_{\text{mt}}(T_{\text{mt}} - T_{\text{f}})} \qquad \text{Eq. (23.1)}$$

All the quantities on the right side of the equals sign in the equation above are known or can be measured so this allows us to determine the specific heat of the metal.

LEARNING OBJECTIVES

After completing this experiment, you should be able to do the following:

- ▼ Define *specific heat* and state its units of measurement.
- ▼ Determine the specific heat of a metal using the method of mixtures.

APPARATUS

Calorimeter (Cenco or coffee-cup type), steam generator, Bunsen burner, thermometer, balance, aluminum and copper metal pellets.

Safety glasses or goggles are recommended. Open flame can be dangerous. Do not leave gas turned on with an unlighted burner, and light with care. Do not leave a lighted burner unattended or reach across flame. Hot water can scald; handle with care.

EXPERIMENT 23 NAME (print) ______________________ DATE ______________
LAST FIRST

LABORATORY SECTION ______ PARTNER(S) ______________________

PROCEDURE 1

1. Fill the steam generator with tap water to about one-third of its capacity. Light the Bunsen burner, and place the flame under the steam generator. Do this first so the water will be heating while you are doing other tasks.
2. Fill the dipper about two-thirds full of the metal. Carefully insert the thermometer in the metal pellets by tipping the dipper to loosen the pellets and shaking slightly as you gently insert the thermometer tip. Place the dipper in the steam generator and let the water boil until the thermometer reaches its highest temperature and remains constant for at least one minute.
3. While the metal is heating, use the balance to obtain the combined mass of the calorimeter and stirrer. Record the mass in Data Table 23.1.
4. In the introduction to this experiment it was stated that we take a metal at a known *high* temperature and water at a known *low* temperature and bring the two together. The cold water obtained from the cold-water tap will warm to room temperature if allowed to remain standing on the laboratory table. Therefore, do not draw the cold water until the metal has reached and remains constant at its highest temperature. Once the metal has reached this condition, measure and record the temperature of the hot metal in the data table. Remove the thermometer from the metal pellets and allow it to cool for use in determining the cold-water temperature. Keep the flame under the steam generator. This will keep the metal at the same high temperature until needed.
5. Fill the inner cup of the calorimeter about two-thirds full of cold water. Do not fill the cup full. Remember, the metal is to be added to the cold water. Determine the mass of the cup (with stirrer) plus the water. Record the mass in the data table. Also measure and record the temperature of the cold water in the data table.
6. The metal and water are now ready to be brought together. With the calorimeter cup positioned in its housing, the thermometer and stirrer placed through the lid of the housing, and the lid slightly removed from the top, pour the metal into the water. Quickly restore the lid and stir the water carefully. Do not lose any water by spilling or splashing. Observe the thermometer closely, and obtain the temperature of the mixture when it has reached its highest equilibrium value. This is the final temperature of the mixture.
7. Remove the thermometer, then determine the combined mass of the calorimeter cup, water, and metal.
8. Complete the data table. Use Eq. 23.1 to calculate the value of the specific heat c_{mt} of the metal.

Table of Specific Heats

Substance	Specific Heat
Water	1.00 cal/g·°C
Steam	0.50 cal/g·°C
Ice	0.48 cal/g·°C
Aluminum	0.22 cal/g·°C
Glass	0.16 cal/g·°C
Iron	0.105 cal/g·°C
Copper	0.092 cal/g·°C

DATA TABLE 23.1			
Metal used:			
Mass of the calorimeter cup and stirrer (1)	m_{cn}		g
Mass of the calorimeter cup, stirrer, and water (2)			g
Mass of the calorimeter cup, stirrer, water, and metal (3)			g
Mass of the water [(2) − (1)]	m_w		g
Mass of the metal [(3) − (2)]	m_{mt}		g
Temperature of the hot metal	T_{mt}		°C
Temperature of the cold water	T_w		°C
Temperature of final mixture	T_f		°C

CALCULATIONS

1. Determine the specific heat c_{mt} of the metal. Use Eq. 23.1, and show your work. The specific heat of the calorimeter cup (made of aluminum) is 0.22 cal/g·°C

2. From the accepted value of the specific heat for the metal used in the experiment, determine the percent error. Show your work.

EXPERIMENT 23 NAME (print) ______________________________ DATE ______________
LAST FIRST

LABORATORY SECTION ______ PARTNER(S) ______________________________

QUESTIONS

1. Distinguish between *heat* and *temperature.* Compare definitions and units of measurements.

2. State the specific heat of water in terms of standard SI energy units (joules).

3. The heat capacity of a material is defined as: heat capacity = mass of the material times its specific heat. What is the heat capacity in calories for:

   (a) 100 g of iron?

   (b) 2,500 g of water?

4. If the final temperature was determined incorrectly (a value greater than the true value was obtained), how would this affect the calculated value of the specific heat? Explain your answer.

5. Where, do you feel, is the greatest possible source of error in this experiment? Why?

Experiment 24

Heat of Fusion

INTRODUCTION

The amount of heat necessary to change one gram of a solid into a liquid (a change of phase) at the same temperature is called the *latent* (hidden) *heat of fusion*. This change of phase is called *melting*. In this experiment the heat of fusion of ice will be determined by using the method of mixtures. Ice, a solid at 0°C will be mixed with water, a liquid, at a temperature of 30° to 35°C.

The law of conservation of energy may be stated in many different ways. For example: The total energy of a system remains constant; energy may be neither created nor destroyed; in changing from one form to another, energy is conserved. This experiment, using the method of mixtures to determine the latent heat of fusion of ice, is based on the law of conservation of energy.

The calorie is the unit of heat (energy) in the metric system. One calorie is defined as the amount of heat necessary to raise the temperature of one gram of pure water one degree Celsius. The specific heat of a substance is the amount of heat, measured in calories, necessary to raise the temperature of one gram of the substance one degree Celsius. Thus the specific heat of pure water is one calorie per gram degree Celsius.

The definition of specific heat provides an equation for the amount of heat necessary to change the temperature of a given substance. The equation is

$$H = mc(\Delta T) \qquad \text{Eq. (24.1)}$$

where H = heat gained or lost by the substance
m = mass of the substance
c = specific heat of the substance
ΔT = change in temperature of the substance in °C or K units

As mentioned above, the phase change from a solid to a liquid is called *melting*. The heat necessary to change a solid to a liquid is provided by the following equation:

$$H = mH_\text{f} \qquad \text{Eq. (24.2)}$$

where H = heat added to the substance
m = mass of the substance
H_f = latent heat of fusion of the substance

Equations 24.1 and 24.2 will be used to calculate the heat of fusion of ice. Examine the terms in the equation $H = mH_f$. Notice that in order to calculate H_f heat of fusion, we must know the value of *m* and the value of *H*. The mass (*m*) we can measure with a balance, but we do not have any direct method for measuring the heat (*H*). Since we cannot measure the heat (*H*), we devise a method to eliminate it from the calculation. This is where the method of mixtures is used.

From the law of conservation of energy we know that when two substances at different temperatures are mixed, the heat lost by one will equal the heat gained by the other.

In this experiment the heat lost by the water and the calorimeter cup holding the water will be equal to the heat gained by the ice plus heat gained by the water produced when the ice melts. This can be stated in words and symbols as follows:

Heat lost by water + heat lost by calorimeter cup = heat gained by ice to melt ice + heat gained by the water from melted ice

$$\text{Heat lost} = \text{Heat gained}$$

$$(mc\Delta T)_{\text{water}} + (mc\Delta T)_{\text{cup}} = H_f\, m_{\text{ice}} + (mc\Delta T)_{\text{water from melted ice}}$$

Rearranging to solve for H_f

$$H_f = \frac{(mc\Delta T)_{\text{water}} + (mc\Delta T)_{\text{cup}} - (mc\Delta T)_{\text{water from melted ice}}}{m_{\text{ice}}} \qquad \text{Eq. (24.3)}$$

LEARNING OBJECTIVES

After completing this experiment, you should be able to do the following:

- Define *latent heat of fusion* and give an example.
- State the law of conservation of energy.
- Give an experimental value for the latent heat of fusion of ice.

APPARATUS

Balance, thermometer (Celsius –10°C to 110°C), calorimeter, stirring rod, ice cubes, paper towels. *Note:* If a calorimeter is not available, a substitute can be made using a small aluminum can (a soft drink can will do) placed in a Styrofoam cup plus a cardboard lid with holes for the thermometer and the stirring rod.

EXPERIMENT 24 NAME (print) ______________________________ DATE ______________
LAST FIRST

LABORATORY SECTION ______ PARTNER(S) ______________________________

PROCEDURE

1. Using the balance, determine the mass of the calorimeter cup (inner container) and the stirrer. Record in Data Table 24.1.
2. Fill calorimeter cup about half full of water at a temperature between 30° to 35°C.
3. Determine the mass of the cup, stirrer, and the water. Record in the data table.
4. Place the calorimeter cup with water and stirrer in the calorimeter housing. Stir the water, and determine the temperature to the nearest tenth of a degree. Record in the data table.
5. Dry two ice cubes with paper towels to remove any water, and place them immediately in the calorimeter cup, being careful not to splash any water from the cup.
6. Begin stirring, fairly fast, but be careful not to splash. Stir until all the ice is melted; then determine the temperature to the nearest tenth of a degree. If the temperature has not dropped at least 20 degrees C below your starting temperature, repeat steps 5 and 6. Once the final temperature has reached a value of 20 degrees C or more below your starting temperature stop adding ice and, after the system has reached equilibrium, measure and record this final temperature in Data Table 24.1.
7. Remove the calorimeter cup, stirrer, and water from the calorimeter housing, and determine their total mass. Record in the data table.
8. Make the calculation necessary to complete the data table.
9. Calculate the value for the heat of fusion of ice using Eq. 24.3. Show your work. Record the value of H_f in the data table.

DATA TABLE 24.1		
Mass of calorimeter cup and stirrer (1)	m_{cup}	g
Mass of calorimeter cup, stirrer, and water (2)		g
Mass of water [(2) − (1)]	m_w	g
Mass of cup, stirrer, water, and melted ice (3)		g
Mass of ice [(3) − (2)]	m_i	g
Initial temperature of water, cup, and stirrer	T_w	°C
Final temperature of water, cup, and stirrer	T_f	°C
$\Delta T(T_{initial} - T_{final})$		°C
Specific heat of water	c_w	1.00 cal/g·°C
Specific heat of calorimeter cup and stirrer (obtain from instructor)	c_{cup}	cal/g·°C
Heat of fusion of ice (experimental value)	H_f	cal/g

CALCULATIONS AND QUESTIONS

1. The accepted value for the heat of fusion of ice is 80 cal/g. This is the amount of energy that must be transferred into one gram of ice at its melting temperature (0°C) in order to change it into liquid water. Also, 80 cal/g is the amount of energy that must be removed from one gram of water at 0°C in order to change liquid water to ice. Calculate the percent error for your experimental value of the heat of fusion of ice using this value as the standard. Show your work.

2. How will using wet ice (some water at 0°C is added along with the ice) affect the experimental value for the latent heat of fusion of ice? Explain.

3. If the calorimeter is not insulated, how will heat from the air in the laboratory affect the experimental value for the latent heat of fusion of ice? Explain.

4. How do you think citrus growers in Florida might use heat of fusion to protect against frost damage?

5. Is it possible to add heat energy to a substance and not increase its temperature? Give an example.

Experiment 25

Heat of Vaporization of Water

INTRODUCTION

The amount of heat necessary to change one gram of vapor into a liquid (a change in phase) at the same temperature is called the *latent* (hidden) *heat* of vaporization. This change of phase is called *condensation.* In this experiment, the heat of vaporization of water will be determined by using the method of mixtures. Steam, a vapor, at 100°C will be added to water, a liquid, at a temperature of 15° to 25°C.

The law of conservation of energy may be stated in many different ways. For example: The total energy of a system remains constant; energy may be neither created nor destroyed; in changing from one form to another, energy is conserved. This experiment, using the method of mixtures to determine the latent heat of vaporization of water, is based on the law of conservation of energy.

The calorie is the unit of heat (energy) in the metric system. One calorie is defined as the amount of heat necessary to raise the temperature of one gram of pure water one degree Celsius. The specific heat of a substance is the amount of heat, measured in calories, necessary to raise the temperature of one gram of the substance one degree Celsius. Thus the specific heat of pure water is one calorie per gram degree Celsius.

The definition of specific heat provides an equation for the amount of heat necessary to change the temperature of a given substance. The equation is

$$H = mc(\Delta T) \qquad \text{Eq. (25.1)}$$

where $H =$ heat gained or lost by the substance
$m =$ mass of the substance
$c =$ specific heat of the substance
$\Delta T =$ change in temperature of the substance in °C or K units

As mentioned above, in the method of mixtures, the law of conservation of energy requires that the heat given up by the steam be equal to the heat gained by the water. The heat given up by the steam in condensing into water is provided by the following equation:

$$H = m_s H_v \qquad \text{Eq. (25.2)}$$

where $H =$ the heat given up by the steam
$m_s =$ mass of the steam
$H_v =$ the latent heat of vaporization

After the steam condenses into water, the mass (m_s) will continue to decrease in temperature due to mixing with the cool (15° to 25°C) water in the calorimeter cup. The complete process can be stated as follows:

Heat lost = Heat gained

[Condensation of steam] + [cooling of condensed steam = [warming up of cool water in calorimeter cup plus the cup]

$$(mH_v)_{\text{steam}} + (mc\Delta T)_{\text{water}} = (mc\Delta T)_{\text{water}} + (mc\Delta T)_{\text{cup}}$$

$$m_s H_v + m_s c_w (T_s - T_f) = m_w c_w (T_f - T_w) + m_{\text{cup}} c_{\text{cup}} (T_f - T_w)$$

Rearranging to solve for H_v *yields*

$$H_v = \frac{m_w c_w (T_f - T_w) + m_{\text{cup}} c_{\text{cup}} (T_f - T_w) - m_s c_w (T_s - T_f)}{m_s} \qquad \text{Eq. (25.3)}$$

where the symbols are defined in Data Table 25.1.
(*Note:* Here T_s is the average temperature in degrees Celsius of steam at the elevation where the experiment is performed. The steam point temperature varies with altitude above sea level. The value for T_s is usually about 99°C in most cases.)

LEARNING OBJECTIVES

After completing this experiment, you should be able to do the following:

- Define *latent heat of vaporization* and give an example.
- State the law of conservation of energy.
- Give an experimental value for the latent heat of vaporization of water.
- Explain the method of mixtures, and explain how the method can be used to determine the latent heat of vaporization.

APPARATUS

Balance, boiler with hose and steam trap, boiler stand, ring stand, clamp, Bunsen burner or other heat source (an electric boiler unit may be used), small beaker, calorimeter, stirring rod, a thermometer (–10° to 110°C).

Safety glasses or goggles are recommended. Open flame can be dangerous. Do not leave gas turned on with an unlighted burner, and light with care. Do not leave a lighted burner unattended or reach across flame. Hot water can scald; handle with care.

EXPERIMENT 25 NAME (print) ______________________ DATE ______________
LAST FIRST

LABORATORY SECTION ______ PARTNER(S) ______________________

PROCEDURE

1. Assemble the apparatus as shown in Fig. 25.1. Keep the steam outlet hose in the beaker instead of in the calorimeter.
2. Fill the boiler about one-third full of water, and place it over the heat source so that the water will heat and boil. While you are waiting for the water to boil, complete Steps 3–6.
3. Use the balance and determine the mass of the calorimeter cup (inner container) and the stirrer. Record in Data Table 25.1.
4. Fill the calorimeter cup about half full of water at a temperature of approximately 15° to 25°C.
5. Determine the mass of the calorimeter cup, stirrer, and the water. Record in the data table.
6. Place the calorimeter cup with water and stirrer in the calorimeter housing. Stir the water, and determine the temperature to the nearest tenth of a degree. Record in the data table.
7. When a heavy flow of steam is coming from the outlet base, carefully place the end of the delivery hose in the inner container of the calorimeter through the hole in the top. Make sure the end of the delivery hose is submerged so that the steam will bubble through the water already in the cup. The thermometer also should be in the cup while steam is being added. (*Note:* Boil the water *very gently* over a low flame so that excessive pressure does not build up in the boiler or hoses.)

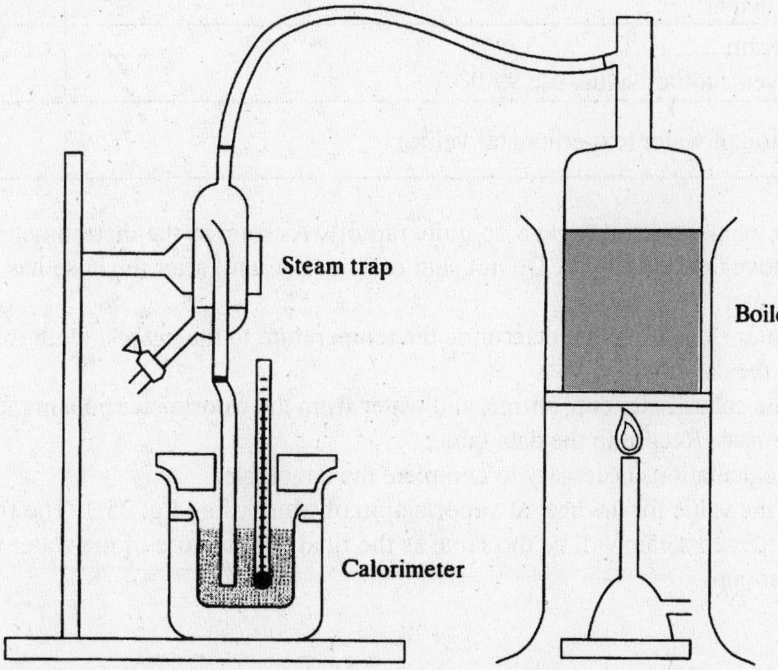

Figure 25.1 Apparatus for determining the heat of vaporization.

DATA TABLE 25.1		
Mass of calorimeter cup and stirrer (1)	m_{cup}	g
Mass of calorimeter cup, stirrer, and water (2)		g
Mass of water[(2 – 1)]	m_w	g
Mass of cup, stirrer, water, and condensed steam (3)		g
Mass of steam [(3) – (2)]	m_s	g
Initial temperature of water, cup, and stirrer	T_w	°C
Final temperature of water, cup, and stirrer	T_f	°C
$\Delta T(T_f - T_w)$		°C
Specific heat of water	c_w	1.00 cal/g·°C
Specific heat of calorimeter cup and stirrer (obtain from instructor)	c_{cup}	cal/g·°C
Temperature of steam *If you are not given another value, use 99.0°C	T_s	°C
Heat of vaporization of water (experimental value)	H_v	cal/g

8. Watch the temperature; it will go up quite rapidly. As soon as the thermometer reads 40° to 45°C, remove the steam hose. Do not shut off the heat until after the hose has been removed from the water.
9. Stir the water carefully; then determine the temperature to the nearest tenth of a degree. Record in the data table.
10. Remove the calorimeter cup, stirrer, and water from the calorimeter housing, and determine their total mass. Record in the data table.
11. Make the calculations necessary to complete the data table.
12. Calculate the value for the heat of vaporization of water using Eq. 25.3. The final temperature of the condensed steam will be the same as the final temperature of the water in the calorimeter cup.

EXPERIMENT 25 NAME (print) ______________________________ DATE ____________
LAST FIRST

LABORATORY SECTION ______ PARTNER(S) ______________________________________

CALCULATIONS AND QUESTIONS

1. The accepted value for the latent heat of vaporization is 540 cal/g. This is the amount of energy that must be removed from one gram of steam at its boiling temperature (100°C) in order to change it into liquid water. Also, 540 cal/g is the amount of energy that must be added to one gram of water at 100°C in order to change liquid water to steam. Calculate the percentage error in your experimental value of H_v, using this value as the standard. Show your work.

2. How will the experimental value be affected if some water (liquid) comes through the hose into the calorimeter cup along with the steam?

3. If the calorimeter is not insulated, how will heat from the air in the laboratory affect the experimental value for the latent heat of vaporization?

4. How can a steam turbine use the heat of vaporization to increase the output of electrical energy it is generating?

5. Is it possible to remove heat energy from a substance and not change its temperature? Explain, and give an example.

6. Calculate the percentage increase in the mass of the water in the calorimeter cup with the addition of the condensed steam. Use mass data from Table 25.1 for your calculation.

Experiment 26
Radiation

INTRODUCTION

There are three main kinds of radiation that may be given off by a radioactive substance. The emitted radiations are alpha particles (helium-4 nuclei), beta particles (electrons), and gamma rays (high-energy, high-frequency electromagnetic radiation). These three kinds of radiation have different properties. For instance, the gamma rays are more penetrating, and the alpha particles and electrons are deflected by magnetic fields.

In our day-to-day life we are constantly exposed to natural radiation. This is radiation due to cosmic rays bombarding our atmosphere or due to the natural decay of radioactive substances present in the Earth. There are also unnatural radioactive substances due to nuclear explosions conducted in the atmosphere and other man-made activities including the use of radioactive isotopes.

The intensity of the radiation received from a radioactive source varies as the inverse square of the distance from the source. Thus, if the distance from the source is doubled, the radiation received by a given area is reduced by one-fourth.

LEARNING OBJECTIVES

After completing this experiment, you should be able to do the following:

- ▼ State the properties of alpha, beta, and gamma radiation.
- ▼ Determine experimentally some of the properties of radiation.

APPARATUS

Radiation sources, detector with power supply, timer, meter stick, magnets, lead shielding.

Even low-level radiation can be hazardous. Follow your instructor's directions exactly, and do not play with a radiation source or try to open its protective covering.

EXPERIMENT 26 NAME (print) ______________________ DATE ______________
LAST FIRST

LABORATORY SECTION ______ PARTNER(S) ______________________

PROCEDURE

1. First investigate the background radiation. With all radioactive sources far removed from the detector, determine the number of counts for 5 minutes. Do this three times.

Compute the counts/min.

Background radiation: #1 ________
#2 ________
#3 ________
Average background ________

2. Next examine radioactive source number 1. Determine the intensity of the radiation in counts per minute produced by the source for four different distances from the detector. Subtract the background count from each reading and then record your data. Plot the intensity versus the distance on your *graph paper*.

Intensity (Counts/Minute) After you subtract background	Distance from Source (d)	$\frac{1}{d^2}$

3. Examine the penetrating effects of rays from the different sources. Measure the radiation produced first with no shielding and then with a sheet of paper, a piece of wood, and a lead shield placed between the radiation source and the detector.

		Intensity (Counts/Minute)			
Source	Distance	No Shielding	Paper	Wood	Lead Shield
1					
2					

4. Examine the effects of the different sources when the radiation travels through a magnetic field.

		Intensity (Counts/Minute)	
Source	Distance	With No Magnetic Field	With a Magnetic Field
1			
2			

Note: Alpha particles have a charge of +2 and are about 7500 times more massive than beta particles. Beta particles have a charge of –1. Gamma rays have no mass or charge. To detect alpha particles a special detector is needed because they are easily absorbed by matter. The case of your detector would absorb the alpha particles before they could arrive at the sensitive part of the detector.

5. From the properties of beta particles and gamma rays, determine what kind of radiation is given off by sources 1 and 2.

   Source 1 ___________

   Source 2 ___________

QUESTIONS

1. Why aren't the three readings for background radiation in Procedure 1 identical?

2. Does your data for Procedure 2 suggest that there is a relation between $1/d^2$ and the intensity? If so, what? Should there be a relationship? Discuss sources of error in this part of the experiment.

3. Why are alpha particles so easily stopped by even very thin sheets of insulation material?

Experiment 27

Spectroscopy

INTRODUCTION

A sodium vapor lamp contains a gas at low pressure. Under this condition of low pressure, the gas atoms are excited by fast-moving electrons from the lamp's negative electrode. When the energy transferred to an atom is sufficient, an outer (or valence) electron is excited, and the atom changes from its normal state of energy E_1 to the first excited state of energy E_2, which is greater than E_1. The electron in the excited state then radiates energy as it subsequently returns to its normal or ground state. The frequency of the radiation is related to the energy difference by the following relationship:

$$hf = E_2 - E_1 \qquad \text{Eq. (27.1)}$$

where E_1 = energy of the electron in its normal or ground state, in joules
E_2 = energy of the electron in its excited state, in joules
f = frequency of the radiation in cycles per second
h = Planck's constant, which has a value of 6.67×10^{-34} J · s

The relationship between the velocity, c, of electromagnetic radiation, the wavelength, λ, and the frequency, f, is

$$f = \frac{c}{\lambda}$$

Substituting in Eq. 27.1, we obtain

$$\frac{hc}{\lambda} = E_2 - E_1$$

The difference between the two energies can be determined if λ is measured, since h and c are known constants.

Many different frequencies or wavelengths are generated from an electrical discharge because there are many energy levels in an atom and the electrons may be transferred between any of these discrete levels. Also, the electron may return to its normal or ground state through a series of transitions rather than through one transition.

When electromagnetic radiation, in this case visible light, is allowed to fall on a diffraction grating, the frequencies or wavelengths are resolved into their individual components. A diffraction grating is made by cutting on the surface of a piece of glass very fine, narrow, parallel grooves spaced very close together. Where the grooves are cut, the glass will be opaque, but the spaces between grooves will allow the radiation to pass through. Some gratings are made with as many as 10,000 spaces or lines per centimeter. Each of the narrow slit openings acts as a new discrete source of radiation, and the radiation from each opening interferes with that from all the other openings. Constructive interference occurs along a line perpendicular to the plane of the grating and also along lines that are at angles with the perpendicular, according to the following equation:

$$n\lambda = d \sin\theta \qquad \text{Eq. (27.2)}$$

where λ = wavelength
d = distance between any two grooves of the grating, known as the *grating spacing*
θ = angle of diffraction measured with respect to the normal line.
n = an integer (any whole number, not zero)

If $n = 1$, an image will appear at some angle. For $n = 2, 3, 4, \ldots$, a similar image will appear at greater angles. These images are called *orders of the spectrum.* See Fig. 27.1.

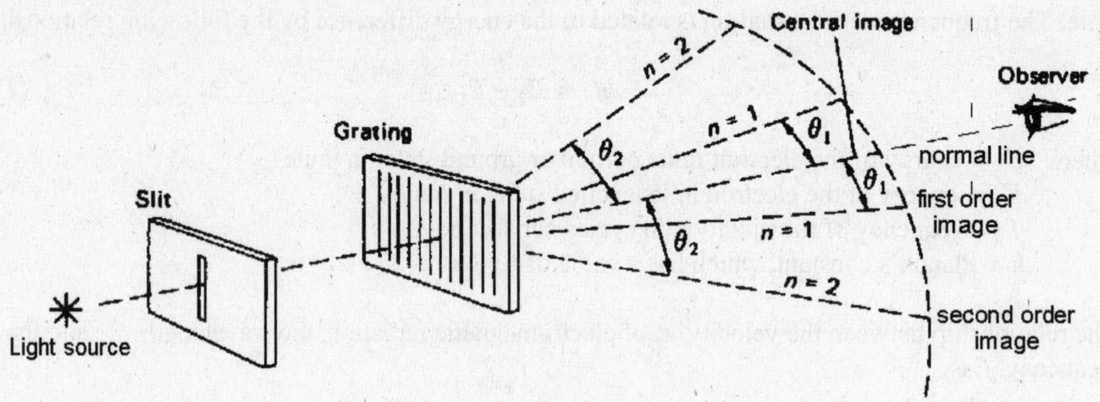

Figure 27.1

LEARNING OBJECTIVES

After completing this experiment, you should be able to do the following:

- ▼ State the relationship between velocity, frequency, and wavelength for electromagnetic radiation.
- ▼ Determine theoretically and experimentally the wavelength of light produced by a sodium vapor lamp as other sources of electromagnetic radiation.
- ▼ Calculate the energy necessary to produce the electromagnetic radiation emitted by the sodium atom.

APPARATUS

Diffraction grating, grating holder, sodium vapor lamp or mercury arc lamp, two meter sticks, two supports for meter sticks, slit with attached scale (Cenco 86260-2).

Ultraviolet light from a mercury lamp can be harmful to your vision. Do not stare at the mercury lamp or read by the light of the mercury source because reflection from the paper surface also contains ultraviolet light.

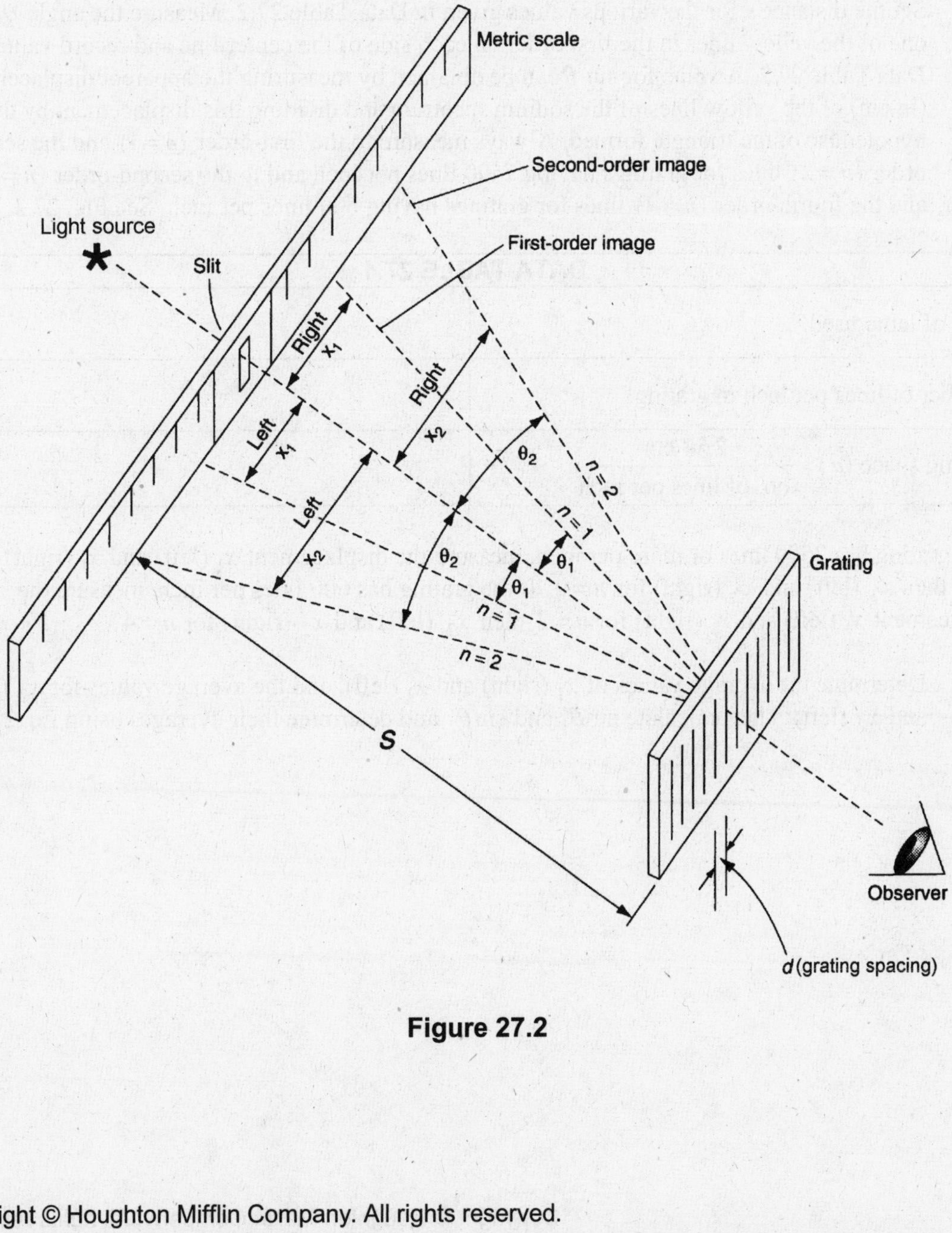

Figure 27.2

PROCEDURE

1. Plug in the sodium vapor lamp and adjust its height so that the lamp is centered in front of the slit.
2. Position the sodium lamp, the slit, and the grating in a straight line, with the plane of the grating perpendicular to the rays of light from the slit. Observe the pattern with a distance (s) between the slit and the screen of 40, 60, 80, and 100 cm. Note the difference between the pattern positions x_1 and x_2 for the first two orders in each case. See Fig. 27.2.
3. Look through the grating and observe the diffracted images on both sides of the centerline. If the grating space d is 2.54/7500 cm, you will see two yellow lines and perhaps other colored lines of the spectrum. If the grating space d is 2.54/600 cm, only one yellow line will appear. The 600-lines-per-inch grating cannot resolve the two yellow lines.

   The images of the slit should appear at equal distances on either side of the centerline. If they do not, rotate the grating slightly until they do.
4. Set the distance s for the various values given in Data Table 27.2. Measure the angle θ for one of the yellow lines in the first order on each side of the center line and record values in Data Table 27.2. A value for sin θ can be obtained by measuring the apparent displacement (in cm) of the yellow lines of the sodium spectrum and dividing this displacement by the hypotenuse of the triangle formed. Always measure to the first-order ($n=1$) and the second-order ($n=2$) lines for gratings having 7500 lines per inch and to the second-order ($n=2$) and the fourth-order ($n=4$) lines for gratings having 600 lines per inch. See Fig. 27.2.

DATA TABLE 27.1	
Type of lamp used	
Number of lines per inch in grating	
Grating space $(d) = \dfrac{2.54 \text{ cm}}{\text{no. of lines per inch}}$	cm

If the grating has 7500 lines or more per inch, measure the displacement x_1 (left) and x_1 (right) for $n=1$, then x_2 (left) and x_2 (right) for $n=2$. If the grating has 600 lines per inch, measure the displacement x_1 (left) and x_1 (right) for $n=2$ then x_2 (left) and x_2 (right) for $n=4$.

5. Determine the average values of x_1 (right) and x_1 (left), and the average values for x_2 (right) and x_2 (left). Then calculate $\sin\theta_1$ and $\sin\theta_2$ and determine their averages using Eq. 27.3.

EXPERIMENT 27 NAME (print) ______________________________ DATE ______________
LAST FIRST

LABORATORY SECTION ________ PARTNER(S) ______________________________

DATA TABLE 27.2

Distance, s Between Grating and Slit (cm)	x_1 Left (cm)	x_1 Right (cm)	Average of x_1 Left and x_1 Right	x_2 Left (cm)	x_2 Right (cm)	Average of x_2 Left and x_2 Right	* $\sin\theta_1$	* $\sin\theta_2$
40								
60								
80								
100								
				Average values of $\sin\theta$				

* For any value of $\sin\theta$: $\sin\theta = \dfrac{\text{side opposite angle } \theta \text{ (use average value for } x)}{\text{hypotenuse}} = \dfrac{x}{\sqrt{x^2 + s^2}}$ Eq. (27.3)

CALCULATIONS

1. Using Eq. 27.2 solve for the average wavelength of the yellow lines of sodium. The accepted value is 5893 angstroms. If a mercury vapor light is used instead of sodium the prominent mercury spectral lines in angstroms are: violet — 4358; green — 5461; red — 6907. (*Note:* 1 angstrom $= 1 \times 10^{-8}$ cm). For $n = 1$ use the average value of $\sin\theta_1$. For $n = 2$ use the average value of $\sin\theta_2$.

2. Use $hc/\lambda = \Delta E$ to calculate ΔE, the energy of the yellow photons produced in the electron's transition from the first excited state to the ground state.

QUESTIONS

1. What is a diffraction grating?

2. What is the grating spacing?

3. In the first-order spectrum of mercury, if available for you to observe, is the red line or the violet line closest to the central image?

Experiment 28

Density of Liquids and Solids

INTRODUCTION

Density is a physical property of matter. It is a derived quantity that is defined as the mass per unit volume. In symbolic notation:

$$\text{Density} = \frac{\text{mass in grams (g)}}{\text{volume in milliliters (mL)}} \qquad \text{Eq. (28.1)}$$

In SI units, density is measured in kg/m^3. Other units of density are g/cm^3 or g/mL (remember that for liquids, one cm^3 is equal to one mL) and lb/ft^3. In this experiment units of g/cm^3 or g/mL will be used. This is so because most laboratory equipment is still calibrated in cgs units (that is, centimeters, grams, and seconds), even though most scientific texts are now using the SI system of units.

The maximum density of pure, air-free water at atmospheric pressure, which occurs at 3.98°C, is defined to be exactly 1.0000 g/cm^3. As the temperature of water increases above or decreases below 3.98°C, its density decreases. Data Table 28.1 gives the density of pure water as a function of temperature in the region of 3.98°C and near room temperature (15–25°C).

DATA TABLE 28.1 Density of Pure Air-Free Water at Atmospheric Pressure

Temperature °C	Density g/mL	Temperature °C	Density g/mL
0	0.99987	24	0.99733
3.98	1.00000	25	0.99708
10	0.99973	26	0.99681
15	0.99913	27	0.99654
20	0.99823	28	0.99626
21	0.99802	29	0.99598
22	0.99780	30	0.99568
23	0.99757	35	0.99423

Other substances can have densities less than or greater than that of pure water. For example, the most dense solid known is osmium, which has a density of 22.5 g/cm^3, while gasoline (a liquid) has a density of 0.68 g/cm^3, and balsa wood (a solid) has a density of only 0.12 g/cm^3. As you can see, the phase of a sample of material is not always a good indication of its density.

The density of any substance can be determined if we can accurately determine its mass and volume. The mass of a solid or liquid can be measured on a pan balance, which allows us to compare the mass of the unknown material with a standard set of masses that are experiencing the same gravitational force. You also might use a triple-beam balance, on which the positioning of known masses along a preset scale will indicate the mass of the sample under study.

The volume of a liquid can be found with a precision of about ±1 ml by using a graduated cylinder. A more precise determination can be made using a pycnometer (pik-'näm-ətər), which is a bottle-like container having a fixed definite volume. Both procedures will be explained in this experiment, and either or both can be performed, depending on the time and equipment available.

The volume of a solid that has a regular shape can be determined by directly measuring its dimensions. The volume can then be calculated using one of the equations given in Data Table 28.2. In many instances it is not possible to measure the dimensions of an irregularly shaped solid, and an indirect method must be used. One such method is to measure the volume of water (or other liquid) that is displaced when a solid sample is completely immersed in liquid. If the displaced liquid is captured in a graduated cylinder, its volume can then be determined. The volume of the displaced liquid will be equal to the volume of the submerged solid object. It is also possible to completely submerge the sample in a partially filled graduated cylinder. The difference in volumes before and after the material is submerged will give the volume of the sample.

LEARNING OBJECTIVES

After completing this experiment, you should be able to do the following:

- Define and give an example of *density*.
- Explain how the density of a solid or a liquid can be calculated.
- Experimentally determine the density of any small sample of liquid or solid using simple laboratory equipment.

APPARATUS

Laboratory balance with a sensitivity of at least 0.01 grams (0.001 grams is preferred), pycnometer (volumetric flask with stopper), graduated cylinder (100 mL or larger), thermometer, ruler or meter stick, calipers or micrometers, water (distilled, if possible), an "unknown" liquid, an "unknown" regularly shaped solid material, such as a block, cylinder, or sphere, an irregular shaped sample, such as a rock or a piece of metal, and a bulk sample (aluminum shot, copper shot, lead shot, or small glass marbles). For best accuracy, these samples need to be small enough to be placed in the volumetric flask or the graduated cylinder.

DATA TABLE 28.2 Equations for Calculating Volume

Volume of a block	Length × width × height
Volume of a cylinder	$\pi \times$ radius^2 × height (where $\pi = 3.1416$)
Volume of a sphere	$4/3 \times \pi \times$ radius^3

EXPERIMENT 28 NAME (print) ______________________________ DATE ______________________
LAST FIRST

LABORATORY SECTION ______ PARTNER(S) ______________________________________

PROCEDURE 1

The purpose of this procedure is to determine the exact volume of the volumetric flask.

1. Weigh the clean dry flask and stopper on the balance. Record this mass, to the nearest 0.001 g if possible, in Data Table 28.3.
2. Determine the temperature of the water. Ideally, the water temperature should be the same as the air temperature in the laboratory. Record this temperature in Data Table 28.3.
3. Use Data Table 28.1 to obtain the exact density of the water at the temperature that you previously measured. Record this density in Data Table 28.3.
4. Fill the flask with water, and *carefully* insert the stopper to remove excess water and any air bubbles. Make sure the stopper is seated properly in the flask and that the flask and stopper neck are completely filled with water.
5. Using a dry paper towel, carefully remove any water that may be on the stopper or on the outside of the flask.
6. Weigh the flask + stopper + water to the nearest 0.001 g, if possible. Record this mass in Data Table 28.3.
7. Determine the mass of the water in the flask by subtracting:

   (Mass of flask + stopper + water) – (mass of flask + stopper)

8. Calculate the exact volume of the flask using Eq. 28.1, the exact density of water and the mass of water measured in Step 7. The volume of the flask is the same as the volume of the water you calculated using Equation 28.1. Be careful to round off your answers to the appropriate number of significant digits. Show your calculations on a separate sheet of paper. Record this volume in Data Table 28.3.

PROCEDURE 2

The purpose of this procedure is to determine the density of an unknown liquid.

1. Empty the water from the flask. Dry the flask and stopper thoroughly. Ask the instructor where to empty the water.
2. Fill the flask with an unknown liquid.
3. Carefully place the stopper in the flask. Dry the outside of the stopper and flask thoroughly.
4. Weigh the flask + stopper + unknown liquid and record the mass, to the nearest 0.001 g, if possible, in Data Table 28.3.
5. Calculate the mass of the unknown liquid. Record this mass in Data Table 28.3. Show your calculations on a separate sheet of paper. This mass may be found by subtracting:

   (Mass of flask + stopper + liquid) — (mass of flask + stopper)

6. Calculate the density of the unknown liquid using Eq. 28.1. The volume of the unknown fluid will be the same as the volume of the flask that you determined in Procedure 1. Record the density in Data Table 28.3. Show your calculations on a separate sheet of paper. Be sure all calculations contain the correct number of significant digits.

7. Empty the unknown liquid from the flask and wash and dry the flask and stopper. Ask the instructor where to empty the unknown liquid.

DATA TABLE 28.3 Procedure 1	
Mass of flask + stopper (also used in Procedure 2 and 3)	g
Temperature of water	°C
Density of water from Data Table 28.1	g/cm^3
Mass of flask + stopper + water	g
Mass of water in flask	g
Calculated volume of flask	cm^3

DATA TABLE 28.3 Procedure 2	
Mass of flask + stopper + unknown liquid	g
Calculated mass of unknown liquid	g
Volume of unknown liquid*	cm^3
Calculated density of unknown liquid	g/cm^3

* This is equal to the calculated volume of flask found in Data Table 28.3 Procedure 1.

PROCEDURE 3

The purpose of this procedure is to determine the density of an unknown solid using the displaced-water method.

This is a less-accurate method for obtaining the density of a solid, but it may be helpful in showing the student a more direct method of determining volume. This procedure may also be used when volumetric flasks are not available, in which case the experiment should be repeated several times.

1. Determine the mass of an unknown solid by using a laboratory balance. Record this mass in Data Table 28.3.
2. Fill the graduated cylinder approximately half full of water. Carefully read the level of the water in the graduated cylinder and record this value in the data table. Attach a piece of fine thread (25 to 30 cm long) to the solid sample. Lower the sample into the graduated cylinder until the sample is completely submerged. Read the new level of the water in the graduated cylinder, and record this value in the data table. The difference between the two readings will be the volume of displaced water and is, therefore, exactly equal to the volume of the solid. Record this volume in Data Table 28.3.

EXPERIMENT 28 NAME (print) ______________________ LAST FIRST DATE ______________

LABORATORY SECTION ______ PARTNER(S) ______________________

If the solid sample happens to float in the water, you will have to use a long thin rod, such as a pencil, to force the sample completely under water, or your determination of density will not be correct. Any portion of the rod that is submerged will also effect your volume determination, so ensure that only the very tip of the rod is in the water. *Note:* If the solid sample is too large to go into the graduated cylinder, you can fill a suitable container, such as a large beaker, completely to the top. Be careful to catch all displaced water in the empty graduated cylinder as you carefully submerge the sample in the beaker, because any drops of water lost will cause a large error. Read the volume of displaced water now in the graduated cylinder.

3. Calculate the density of the unknown solid using Eq. 28.1. Record the density in Data Table 28.3. Use the appropriate number of significant digits in all calculations. Show your calculations on a separate sheet of paper.

DATA TABLE 28.3 Procedure 3

Mass of unknown solid (measured on the balance)	g
Volume of water in graduated cylinder	cm^3
Volume of water + solid in graduated cylinder	cm^3
Volume of solid (found by displaced-water method)	cm^3
Density of unknown solid	g/cm^3

PROCEDURE 4

The purpose of this procedure is to determine the density of various solid objects by measuring their mass and volume directly.

Measure the diameter and height of the two cylinders, the diameter of the sphere, and the three dimensions of the wooden block. Use the micrometer caliper on all dimensions small enough to fit into the instrument. Use the vernier caliper for all other dimensions. Take several measurements at various points on the objects, and obtain an average. Record average dimensions in centimeters in the tables below.

Now obtain the mass of each of the four objects using the beam balance. Measure the mass to the nearest 1/100 g if possible. Make sure the balance stands at zero and is level before using it. Record all masses in grams in the data tables.

DATA TABLE 28.4 Aluminum cylinder

Radius (cm)	Height (cm)	Volume (cm^3)	Mass (g)	Density (g/cm^3)	Accepted Density (g/cm^3)
					2.7

DATA TABLE 28.4 Copper or brass cylinder					
Radius (cm)	Height (cm)	Volume (cm^3)	Mass (g)	Density (g/cm^3)	Accepted Density (g/cm^3)
					7.0 or 8.9*

* Density of copper = 7.0 g/cm^3, density of brass = 8.9 g/cm^3

DATA TABLE 28.4 Iron sphere				
Radius (cm)	Volume (cm^3)	Mass (g)	Density (g/cm^3)	Accepted Density (g/cm^3)
				7.8

DATA TABLE 28.4 Wooden block					
Length (cm)	Width (cm)	Height (cm)	Volume (cm^3)	Mass (g)	Density† (g/cm^3)

† Most wood has a density less than one.

CALCULATIONS

The density of an object may be defined as the mass per unit volume and in the cgs system of units:

$$\text{Density} = \frac{\text{Mass in grams}}{\text{Volume in cm}^3} = \frac{m}{v}$$

Calculate the volume of each object, being careful to retain the proper number of significant figures. Record your values for each volume in Table 28.4 and then calculate the density of each object and record that as well.

EXPERIMENT 28 NAME (print) ______________________________ DATE ______________
LAST FIRST

LABORATORY SECTION ______ PARTNER(S) ______________________________

QUESTIONS

1. Define *density,* and give an example, using SI units.

2. Calculate the volume of 4.250 g of pure, air-free water at atmospheric pressure and at the following temperatures. Make these calculations to *four significant figures*.

   (a) 10°C

   (b) 22°C

3. State the maximum density of pure water and give the temperature at which this maximum density occurs.

4. Determine the mass of a solid that has a volume of 6.39 cm^3 and a density of 5.4 g/cm^3. Be sure that your answer contains the correct number of significant digits.

5. Determine the volume of a liquid that has a mass of 24.6 g and a density of 0.82 g/cm^3. Be sure the answer has the correct number of significant digits.

6. Explain how the volume of the human body could be determined experimentally.

Experiment 29

Oxygen

INTRODUCTION

The element we now call oxygen was discovered in 1774 by Joseph Priestley (1733–1804), an English minister, who obtained the gas from mercuric oxide. The mercuric oxide was obtained by heating mercury with the "burning glass" (magnifying lens). The gas was distinguished as an element and named oxygen by A. L. Lavoisier.

Oxygen is the most abundant element in Earth's crust. The atmosphere contains 23 percent oxygen by weight, and water contains almost 89 percent oxygen by weight. Oxygen has an atomic number of 8, and because oxygen combines readily with many elements, it was chosen as the standard to which the weights of other elements were compared. In 1961 oxygen was dropped as a standard, and the most abundant isotope of carbon was taken as the new standard. The standard value assigned to this isotope of carbon is defined to be exactly 12 atomic mass units.

LEARNING OBJECTIVES

After completing this experiment, you should be able to do the following:

- ▼ Prepare a volume of oxygen from the compound potassium chlorate.
- ▼ State several physical and chemical properties of oxygen.

APPARATUS

A mixture of about 80 percent potassium chlorate ($KC1O_3$) and 20 percent manganese dioxide (MnO_2), one Pyrex test tube, ring stand and clamp, Bunsen burner, gas lighter, one-hole rubber stopper, delivery tube (rubber hose), pneumatic trough, four collecting bottles (500 mL), large shallow tray to catch water overflow from pneumatic trough, wood splints, sulfur (pinch), deflagrating spoon, iron picture wire (15 cm in length), four glass plates.

Caution Safety Tips

Safety glasses or goggles are recommended. Open flame can be dangerous. Do not leave a lighted burner unattended or reach across flame. Do not leave the gas turned on with an unlighted burner, and light with care.

PROCEDURE

Fill the pneumatic trough with water. Completely fill four bottles with water, cover them with glass plates, invert them, submerge their necks in the water in the pneumatic trough, and remove the glass covers. Allow no air bubbles. *Be sure that the pneumatic trough is set in the shallow tray provided for your table.*

You will need a mixture of about 80 percent potassium chlorate ($KClO_3$) and 20 percent manganese dioxide (MnO_2). Your instructor may provide these compounds already premixed in the proper proportions. If you prepare your own mixture, observe the following warning.

Caution: Explosions can occur when combustibles such as carbon, sulfur, or rubber come into contact with fused potassium chlorate. Heat the manganese dioxide well before using, and inspect the potassium chlorate for any foreign materials.

Place 2 to 4 g of the $KClO_3$-MnO_2 mixture in a Pyrex test tube. Insert the stopper and delivery tube, arranging the assembly as shown in Fig. 29.1.

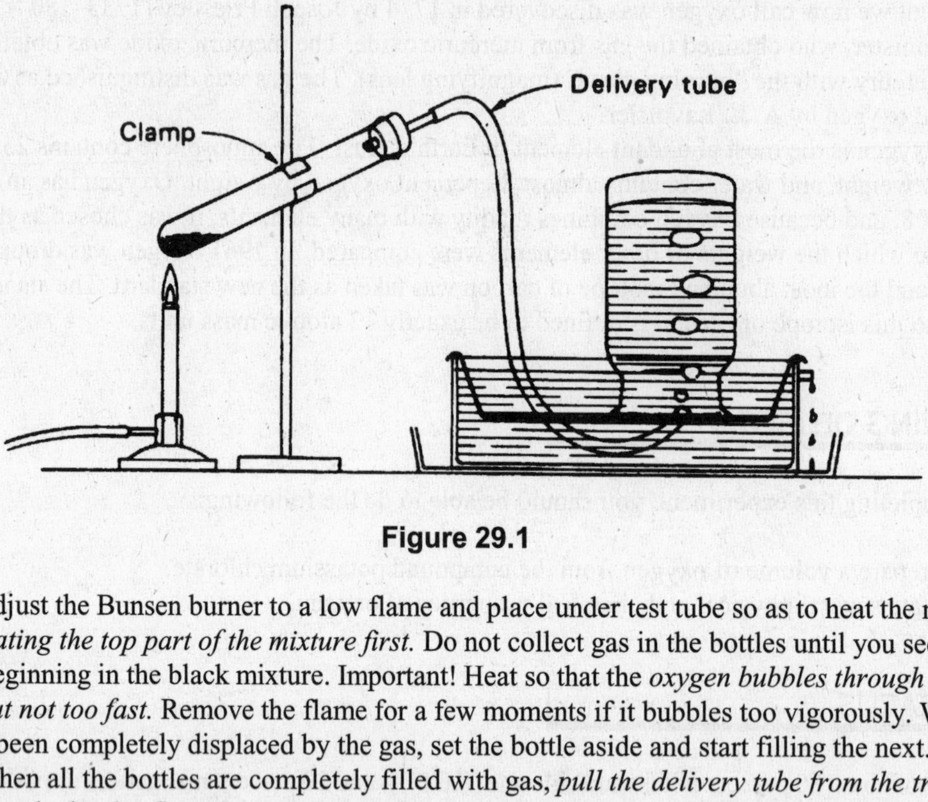

Figure 29.1

Adjust the Bunsen burner to a low flame and place under test tube so as to heat the mixture *slowly, heating the top part of the mixture first.* Do not collect gas in the bottles until you see the reaction beginning in the black mixture. Important! Heat so that the *oxygen bubbles through the water steadily but not too fast.* Remove the flame for a few moments if it bubbles too vigorously. When the water has been completely displaced by the gas, set the bottle aside and start filling the next.

When all the bottles are completely filled with gas, *pull the delivery tube from the trough* and then remove the bottles from the trough as follows. Raise the bottle enough to slip a glass plate over the mouth under water; hold the glass plate tightly against the mouth of the bottle; lift the bottle out and set it upright on the table. *Keep the bottle covered* until the oxygen is needed.

What has happened is this: The heat of the flame decomposes the potassium chlorate ($KClO_3$) producing potassium chloride (KCl) and oxygen gas (O_2); thus

$$2\,KClO_3 \rightarrow 2\,KCl + 3O_2$$

The manganese dioxide has acted as a catalytic agent; that is, it has undergone no chemical change itself but it has promoted a chemical reaction, in this case, the decomposition of the potassium chlorate.

EXPERIMENT 29 NAME (print) ______________________ DATE ______________
LAST FIRST

LABORATORY SECTION ______ PARTNER(S) ______________________

TESTS WITH OXYGEN

1. Ignite a wood splint by means of your Bunsen burner. When the splint has burned for a moment, blow out the flame, leaving the splint still glowing. Plunge the glowing splint into a bottle of oxygen. Perform this procedure at least one more time. Describe what happens.

2. Heat a *small* pinch of sulfur in a deflagrating spoon until the sulfur begins to burn. Observe the color and size of the flame. Now, lower the spoon of burning sulfur into a bottle of oxygen. Compare the size and color of the flame in air and in oxygen. (Put out the flame by submerging the spoon and sulfur in water as soon as the experiment is finished.)

3. Heat the tip of the iron picture wire directly above the tip of the blue inner cone of the Bunsen burner flame until it becomes red hot. *Quickly* plunge it into a bottle of oxygen. Record your observations.

4. Take a pinch of sulfur on a dry glass plate. Again, heat the tip of the iron picture wire red hot, and at once dip it into the pinch of sulfur. While the sulfur is still burning, quickly thrust this end of the wire into a jar of oxygen. What do you observe? A reaction should take place. If you fail to see any reaction, notify the instructor.

QUESTIONS

1. What do you think holds the water in the bottles that, after being filled, are inverted in the pneumatic trough?

2. From the results of Tests 1 and 2 in this experiment, what statement could be made about the effect of oxygen on the process of combustion?

3. When sulfur is oxidized (combined with oxygen), of what chemical elements is the product composed?

4. In Test 4, the sulfur does not become part of the final product. Comparing Tests 3 and 4 of this experiment, what was the purpose of the sulfur in Test 4?

5. In Test 4, what happened to the iron wire? Of what chemical elements was the product of the reaction composed?

Experiment 30

Percentage of Oxygen in Potassium Chlorate

INTRODUCTION

The percentage composition of a compound can be calculated if the formula for the compound is known. For example, the percentage composition of sulfuric acid (H_2SO_4) is calculated as follows:

1. Write the chemical formula:

$$H_2SO_4$$

2. Look up relative weights and place these values above each element's symbols:

H	S	O
1	32	16

3. Determine total relative weight of each element present by multiplying relative weight by subscript:

2 H	S	4 O
2(1) = 2	32	4(16) = 64

4. Add total relative weights and indicate each element as a fraction of the total:

$$\underset{2\,H}{2/98} + \underset{S}{32/98} + \underset{4\,O}{64/98} = 98/98$$

5. Express each fractional part as a percentage by multiplying the fraction by 100:

$$2/98 \times 100 = 2.04\%$$
$$32/98 \times 100 = 32.65\%$$
$$64/98 \times 100 = 65.31\%$$

LEARNING OBJECTIVES

After completing this experiment, you should be able to do the following:

- ▼ Determine the percentage composition of a compound.
- ▼ Determine the theoretical and experimental values of oxygen in the compound potassium chlorate.

APPARATUS

Potassium chlorate, manganese dioxide, Pyrex test tube, ring stand, test-tube clamp, Bunsen burner, gas lighter, beam balance.

Safety glasses or goggles are recommended. Open flame can be dangerous. Do not leave a lighted burner unattended or reach across flame. Do not leave the gas turned on with an unlighted burner, and light with care.

EXPERIMENT 30 NAME (print) ______________________________ DATE ____________________
LAST FIRST

LABORATORY SECTION ______ PARTNER(S) ______________________________

PROCEDURE

Weigh a clean, dry test tube with a few grains of manganese dioxide (MnO_2) in it.* Record weight in Data Table 30.1. Place in the tube about 1 gram of potassium chlorate ($KClO_3$), and weigh again. Put the test tube in the clamp provided on the stand, and cover its mouth with a crucible cover. Using a clean, nonluminous flame, heat the test tube, including the sides, gently for a few minutes, then more strongly. Avoid the occurrence of white fumes in the tube. Continue the heating until the reaction stops. Cool the tube, and weigh it again. Resume heating for a few minutes; again cool, and reweigh. If the two weights are close in value, the process is complete. If not, heat until a constant weight is obtained on two successive weighings. The difference between the weight before heating and the constant weight after heating represents the weight of oxygen driven off.

Calculate from your data the percentage of oxygen in potassium chlorate.

DATA TABLE 30.1	
Weight of test tube and MnO_2	g
Weight of test tube, MnO_2, and $KClO_3$ before heating	g
Weight of test tube and residue after heating	g

COMPUTATIONS

Weight of $KClO_3$	g
Weight of oxygen driven off	g
Percentage of oxygen in $KClO_3$	g

From the formula $KClO_3$ and the known atomic weights of its elements, calculate the theoretical percentage of oxygen in potassium chlorate. Compare this with the value obtained in your experiment by computing percent error; the percent error should be small.

Theoretical percentage of oxygen in $KClO_3$ __________ %

Percent error __________ %

* The instructor will dispense the MnO_2.

QUESTIONS

1. What is the purpose of the manganese dioxide in this experiment?

2. What is the theoretical percentage of Cl in $KClO_3$? Show your calculations.

3. What is the theoretical percentage of K in $KClO_3$? Show your calculations.

4. What do you think is the greatest source of error in this experiment? Why?

5. Balance the chemical equation:

$$\_\_\_\_\_\_\ KClO \rightarrow \_\_\_\_\_\ KCl + \_\_\_\_\_\_\ O_2$$

Experiment 31

Percentage of Oxygen and Nitrogen in the Air

INTRODUCTION

The atmosphere is composed almost entirely of chemically uncombined nitrogen and oxygen. There are also small amounts of argon, krypton, neon, water vapor, carbon dioxide, and dust. During chemical oxidation processes in which metal oxides are formed in the air, oxygen is used up, leaving nitrogen as the chief constituent of the remaining air. Such a process, carried out carefully, can be used to determine the percentage of oxygen and nitrogen in the air.

The general procedure is to pass a measured quantity of air *slowly* over a *red-hot metal,* and after the oxygen is removed in the formation of an oxide, measure the quantity of nitrogen gas that is left. Copper metal serves the purpose very well. Hot copper combines with the oxygen of the air to form a black copper oxide:

$$2\ Cu + O_2 \rightarrow 2\ CuO$$

LEARNING OBJECTIVES

After completing this experiment, you should be able to do the following:

- Determine the percentage of oxygen and nitrogen in the air.
- Show how to remove all oxygen from the air using a simple oxidation reaction.
- Measure the volume of displaced gas in a collection bottle.
- Determine the actual percentage of oxygen and nitrogen in the air.

APPARATUS

Combustion tube, Bunsen burner, wing tip, thistle tube, one 500-mL bottle, one 300-mL bottle, pneumatic trough, large shallow tray, two ring stands, two burette clamps, two one-hole rubber stoppers, one two-hole rubber stopper, rubber hose, pinch clamp, copper turnings, graduated cylinder.

Safety glasses or goggles are recommended. Open flame can be dangerous. Do not leave a lighted burner unattended or reach across flame. Do not leave the gas turned on with an unlighted burner, and light with care.

EXPERIMENT 31 NAME (print) ________________________ DATE ____________
LAST FIRST

LABORATORY SECTION ______ PARTNER(S) ________________________

PROCEDURE

1. *Preparation:* Fill the pneumatic trough nearly full of water. Fill a bottle to the brim with water, and place it in the trough. Be certain that there is no air in the bottle before the experiment is started.

Assemble the apparatus as shown in Fig. 31.1. *Be sure that the delivery tube is not under the collecting bottle* B_2 *when the experiment begins.*

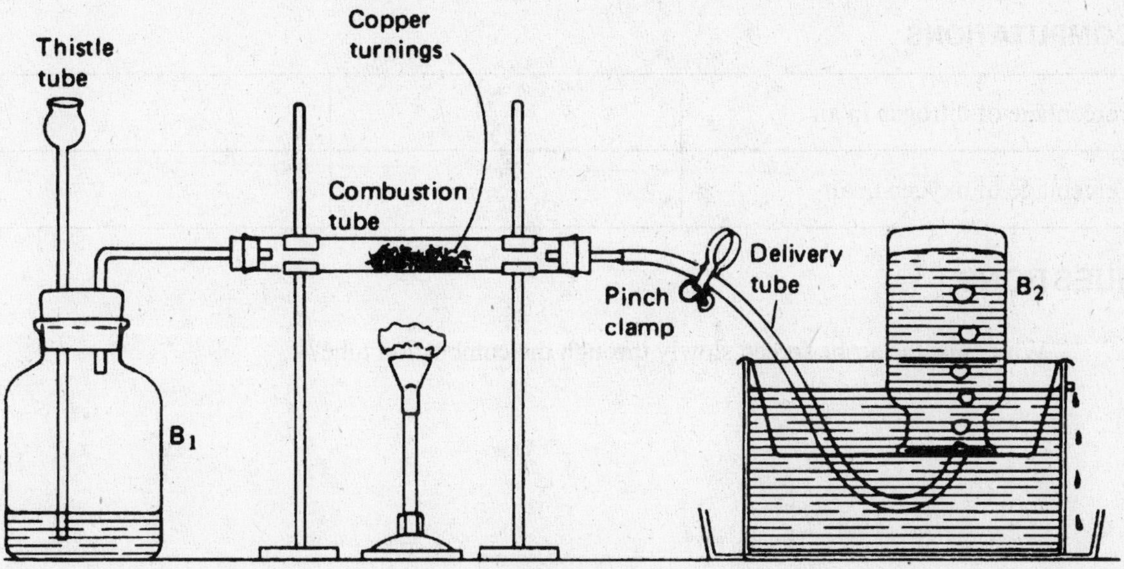

Figure 31.1

Without applying any heat to the combustion tube, slowly pour about 100 mL of water into the thistle tube. As the water drips into the bottle B_1, air will be forced very slowly through the combustion tube and will bubble from the end of the delivery tube. If it does not, check for leaks. Tighten carefully all clamps, connections, and stoppers. Adjust the bubbling rate by means of the pinch clamp to about two bubbles per second.

When the water has completely gone through the thistle tube, put the flame under the combustion tube and begin heating the tube. A few bubbles due to expansion of air in the combustion tube will come out of the end of the delivery tube. When these have ceased to appear, put the end of the delivery tube under bottle B_2. Accurately measure out in a graduated cylinder 300 mL of water. Pour this into the thistle tube a little at a time (do not let the thistle tube top become empty at any time) until all 300 mL are in the bottle. Note that this second batch of water will force 300 mL of air from the bottle B_1 through the combustion tube. The nitrogen residue from this 300 mL of air will be collected in bottle B_2.

When all the water has drained from the thistle tube, and the bubbles cease appearing in bottle B_2, remove the delivery tube from the trough and the burner from under the combustion tube. Slip a cover glass carefully over the mouth of bottle B_2, remove the bottle from the trough and set it upright on the table. Empty the water from B_1 and make a second trial.

2. To obtain the percentage of nitrogen in the air, fill a graduated cylinder with water to its highest reading. Pour some of this water into bottle B_2, which holds the nitrogen gas. *Fill* B_2 *up to the brim.* Record the number of mL of water required. This is the volume of nitrogen in 300 mL of air.

DATA TABLE 31.1

	First Trial	Second Trial
Number of mL of nitrogen		
Number of mL of air forced through		

COMPUTATIONS

Percentage of nitrogen in air		
Percentage of oxygen in air		

QUESTIONS

1. Why must the air be forced slowly through the combustion tube?

2. If part of the 300 mL of water were lost by an overflow of the thistle tube, how would the results of the experiment be affected?

3. If an air leak developed at the right end of the combustion tube, how would the results of the experiment be affected?

Experiment 32

An Exothermic Chemical Reaction

INTRODUCTION

A chemical reaction produces a chemical change that yields one or more new substances. During a chemical reaction, energy may be absorbed or released. When heat energy is absorbed, the reaction is called an *endothermic chemical reaction.* When energy is released the reaction is called an *exothermic chemical reaction.* The energy released may be in the form of heat, light, or electricity. An example of heat released is the burning of a match.

Heat is a form of energy (energy in transit) that is difficult to measure directly. In this experiment, an indirect measurement of the amount of heat released during the reaction of hydrochloric acid (HCl) and sodium hydroxide (NaOH, a base) will be obtained by measuring the temperature change that occurs when the two substances are mixed. The reaction is represented empirically by the following equation:

$$HCl + NaOH \rightarrow NaCl + H_2O + Heat$$

The heat energy released during this exothermic reaction is directly proportional to the concentration of the reacting substances. Another way of stating this is: *The energy released is directly proportional to the number of molecules taking part in the chemical reaction.* The amount of heat released during this chemical reaction is also a function of the nature of the substances taking part in the reaction.

A 1 molar (1.0 M) solution is one mole of the substance per liter of solution. One mole of anything contains Avogadro's number (6.023×10^{23}) of particles. A 2.0 molar solution contains two moles per liter of solution and contains 12.046×10^{23} particles. In this experiment 2.0 M solutions of NaOH and HCl will be used. The 2.0 M solutions will be referred to as the *standard solutions.*

In this experiment, everything is held constant except the strength (number of molecules per cubic centimeter) of the hydrochloric acid and the sodium hydroxide. As the strength of these two solutions is varied by diluting with water, the temperature changes will be measured and recorded. The temperature changes during the mixing of the two solutions of different concentrations will give an indication of how much heat is released during the chemical reaction. This is true because the heat released is proportional to the change in the temperature when the volumes are kept constant.

LEARNING OBJECTIVES

After completing this experiment, you should be able to do the following:

- Define an *exothermic chemical reaction,* and give an example.
- State the relationship between the energy released in a chemical reaction and the concentration of the reacting solutions.

APPARATUS

Calorimeter, thermometer, two 50 mL graduated cylinders, one 25 mL beaker, distilled water, 2.0 M sodium hydroxide solution, 2.0 M hydrochloric acid solution.

Safety glasses or goggles are recommended. Be careful when handling or pouring acid. The HCl used here is dilute, but any contaminated skin surface should be thoroughly flushed with clear water. Use caution when handling chemicals.

PROCEDURE

The purpose of the procedure is to collect data to determine the relationship between the concentration of the reacting solutions (sodium hydroxide and hydrochloric acid) and the amount of heat released during the chemical reaction that occurs when the two solutions are mixed.

The volumes of the two solutions will be kept constant at 20 mL, and the concentration of the two solutions varied. The original temperature of the sodium hydroxide and the hydrochloric acid will be measured and held constant, preferably at room temperature. When the two solutions are mixed, the temperature change will be measured and recorded.

Step 1 Measure 20 mL of the standard solution of sodium hydroxide,* and place it in the calorimeter. Measure and record the temperature of the 20 mL solution; it should be at room temperature. This is the original temperature T_0.

Step 2 Measure 20 mL of the standard solution of hydrochloric acid,* and transfer it to the 25-mL beaker. Measure and record the temperature of the HCl; it should also be at room temperature.

Step 3 Transfer and mix the HCl solution with the NaOH in the calorimeter. Stir the mixture with the thermometer as the HC1 is added. Measure and record the highest temperature of the mixture. This will be recorded as T_h in Data Table 32.1. The highest temperature reading will occur almost immediately upon mixing the two solutions. The temperature reading will be slightly lower than the true temperature, because (1) some heat is lost to the surroundings and (2) it takes time for the mercury to rise in the thermometer.

Step 4 Reduce the concentration of the NaOH and HC1 solutions. Place 15 mL of the standard solution of sodium hydroxide in the graduated cylinder. Add 5 mL of distilled water and transfer the 20 mL (75% solution) to the calorimeter. Place 15 mL of the standard solution of HCl in the other graduated cylinder. Add 5 mL of distilled water and transfer the 20 mL (75% solution) to the 25-mL beaker.

Step 5 Repeat Step 3.

Step 6 Reduce the concentration of the NaOH and HC1 solutions. Place 10 mL of the standard NaOH solution in the graduated cylinder. Add 10 mL of distilled water and transfer the 20 mL (50% solution) to the calorimeter. Place 10 mL of the standard HCl solution in the other graduated cylinder. Add 10 mL of distilled water and transfer the 20 mL (50% solution) to the 25-mL beaker.

Step 7 Repeat Step 3.

* Use separate graduated cylinders to measure the volume of sodium hydroxide and hydrochloric acid.

EXPERIMENT 32 NAME (print) ______________________ DATE ______________
LAST FIRST

LABORATORY SECTION ______ PARTNER(S) ______________________

Step 8 Reduce the concentration of the NaOH and HCl solutions. Place 5 mL of the standard NaOH solution in the graduated cylinder. Add 15 mL of distilled water and transfer the 20 mL (25% solution) to the calorimeter. Place 5 mL of the standard HCl solution in the other graduated cylinder. Add 15 mL of distilled water and transfer the 20 mL (25% solution) to the 25-mL beaker.

Step 9 Repeat Step 3.

Step 10 Reduce the concentration of the NaOH and HCl solutions. Place 2.5 mL of the standard NaOH solution in the graduated cylinder. Add 17.5 mL of distilled water, and transfer the 20 mL (12.5% solution) to the calorimeter. Place 2.5 mL of the standard HC1 solution in the other graduated cylinder. Add 17.5 mL of distilled water, and transfer the 20 mL (12.5% solution) to the 25-mL beaker.

Step 11 Repeat Step 3.

Step 12 Analyze the data by means of a graph. Plot the temperature changes on the *y* axis and the percent concentration on the *x* axis. Refer to Experiment 1 for information on how to correctly plot and label a graph. Should the intersection of the *x* and *y* axes be a point on your graph? (*Hint:* What would be the temperature change when the percent concentration is zero?)

DATA TABLE 32.1

Concentration of Solutions	T_0	T_h	ΔT
100%			
75%			
50%			
25%			
12.5%			

T_0 = original temperature of the NaOH and HC1 solutions
T_h = highest temperature of the mixed solution
ΔT = change in temperature $(T_h - T_0)$

QUESTIONS

1. What does the graph indicate concerning the relationship between percent concentration of solution and the changes in temperature?

EXPERIMENT 32 NAME (print) ______________________________ DATE ______________
LAST FIRST

LABORATORY SECTION ______ PARTNER(S) ______________________________

2. Explain the relationship between the heat released by the chemical reaction and the change in temperature.

3. Explain the slope of your graph.

4. How would the slope of your graph change if the volume of the solutions were doubled? Explain your answer.

5. Assume that the chemical reaction between the NaOH and the HC1 is completed. (a) With the 100% standard solution, how many sodium ions react with how many chlorine ions? (b) With the 25% solution, how many water molecules were formed?

Experiment 33

Avogadro's Number

INTRODUCTION

The elements that make up our environment vary greatly in their physical and chemical properties and in the types of compounds they form. Elements with similar properties can be classified and placed in groups or arranged in order relative to some physical or chemical property. The periodic table is one method used to group the elements. This grouping is based on the *periodic law,* which states that the physical and chemical properties of the elements are periodic functions of their atomic number. That is, the properties of elements appear in a certain order, and elements with similar properties occur at particular intervals when arranged in order of increasing atomic number.

The periodic table also gives the relative weight of each element. The relative weight is given with respect to the carbon-12 atom, which was established as the reference standard by the International Union of Pure and Applied Chemistry in 1961. The relative weights are proportional to the actual weights of the atoms.

The relative weight of a molecule can be obtained by adding the relative atomic weights of the atoms that combine to form the molecule.

In work requiring weights of elements, it is convenient to use grams to express relative weight. The concept *gram atomic weight* is defined as the amount of the element, expressed in grams, that is numerically equal to the relative atomic weight. For example, the gram atomic weight of hydrogen is 1.0080 g and of oxygen, 15.9994 g. Similarly, the *gram molecular weight* is the mass of a molecular substance in grams equal to its molecular weight. For example, the gram molecular weight of water (H_2O) is 18.0154.

One gram atomic weight of any element contains the same number of atoms. This number is called *one mole of atoms* and has the value of 6.022×10^{23}. One mole of atoms is defined as the number of atoms in exactly 12 grams of carbon-12.

The number 6.022×10^{23} is known as *Avogadro's number.* Amedeo Avogadro (1776–1856), an Italian physicist, announced in 1811 a hypothesis to explain the behavior of gas, which stated that under conditions of equal pressure and temperature, equal volumes of gases contain equal numbers of molecules.

Several methods can be used to determine Avogadro's number (symbol N_0). In this experiment, the method of electrolysis will be used.

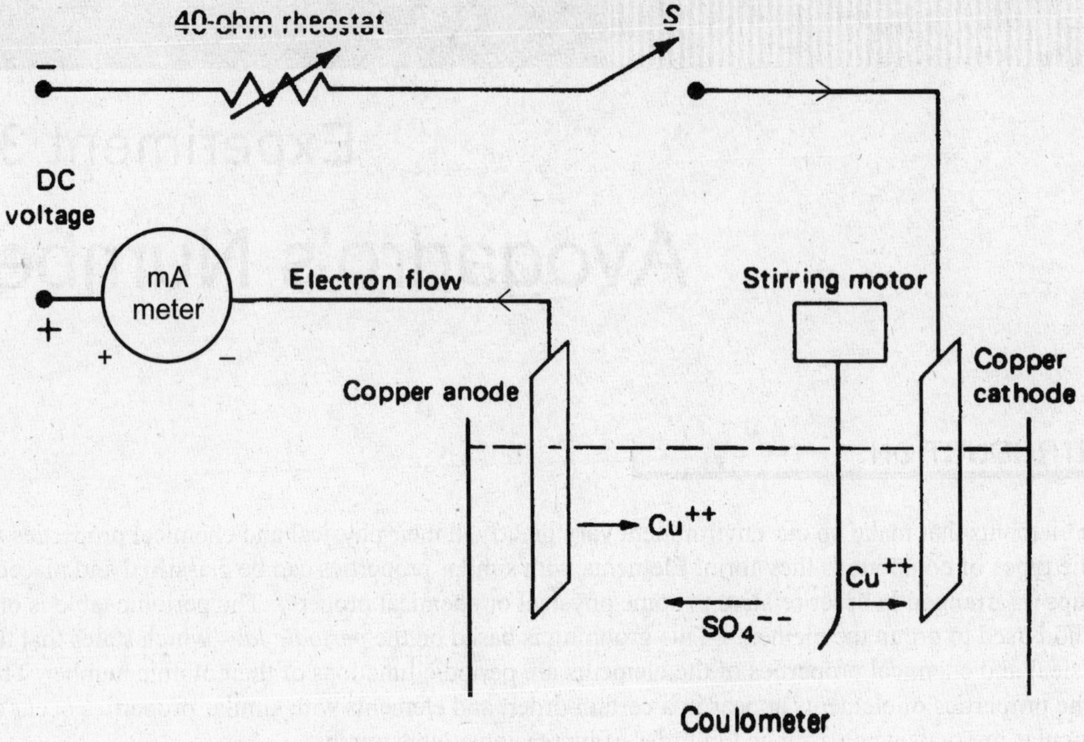

Figure 33.1

A typical example of electrolysis is shown in Fig. 33.1, which uses a solution of copper sulfate containing Cu^{++} ions and SO_4^{--} ions. When the switch *S* is closed, an electric field is placed between the copper electrodes; the Cu^{++} ions are drawn toward the negatively charged cathode, and the SO_4^{--} ions are drawn toward the positively charged anode. If the action is allowed to continue for a period of time, we should find that a mass *m* of metallic copper has been deposited on the cathode and an equal mass *m* of metallic copper has been removed from the anode. The concentration of the solution remains the same.

LEARNING OBJECTIVES

After completing this experiment, you should be able to do the following:

- Define *gram atomic weight, gram molecular weight, mole,* and *Avogadro's number.*
- Determine experimentally the numerical value of Avogadro's number.

APPARATUS

Coulometer, good milliammeter (range 0–500 or 0–1000 mA), copper electrodes, 40-ohm rheostat, copper sulfate solution containing 100 g copper sulfate, 18 mL concentrated H_2SO_4, 34 mL ethyl alcohol in 1000 mL distilled water (*Note*: Filter through glass wool), one single-pole/single-throw switch, stirring motor, glass stirrer.

EXPERIMENT 33 NAME (print) ______________________ DATE ______________
LAST FIRST

LABORATORY SECTION ______ PARTNER(S) ______________________

PROCEDURE

1. Connect the circuit as shown in Fig. 33.1.
2. If copper sulfate solution is not in the coulometer, obtain it from the instructor, and fill the glass container.
3. Close switch *S* and adjust the rheostat *R* until the milliammeter reads 400 to 800 mA; then open the switch. The circuit is now adjusted for operation.
4. Remove the anode and cathode from the coulometer, dry them with paper toweling, polish them with steel wool, wipe them clean with soft paper, and weigh each on a balance to ±0.01 g.
5. Position the anode and cathode in the coulometer, close the switch *S*, start the timer (your watch will do) and the stirring motor, adjust the rheostat for 400 to 800 mA, and *keep the current constant* for 1 hour.
6. At the end of 1 hour, open the switch *S* and quickly remove the anode and cathode, wash them in running water, rinse them with acetone (or ethyl alcohol), and air dry.
7. Weigh the anode and cathode to ±0.01 g.
8. Record data in Data Table 33.1.

DATA TABLE 33.1		
1. Current	$I =$	A
2. Time	$t =$	s
3. Loss of mass of anode (original mass – final mass)	$m_{lost} =$	g
4. Gain of mass of cathode (final mass – original mass)	$m_{gained} =$	g
5. Average mass transferred $m_{average} = \dfrac{m_{lost} + m_{gained}}{2}$	$m_{average} =$	g

CALCULATIONS

1. Determine the amount of flow of charge q.

$$q = It$$

$q =$ ________ coulombs

2. Determine the number of electrons n_e flowing through the solution, where

$$n_e = \frac{q}{1.6 \times 10^{-19} \text{ coulomb}}$$

$n_e =$ ________ electrons

3. Determine the number of copper atoms that were transferred from the anode to the cathode. The charge carried by each copper atom is +2.

$$N_c = \text{no. of copper atoms transferred} = \frac{n_e}{2}$$

$N_c =$ ________ electrons

This is the number of copper atoms in the amount of mass transferred between the anode and cathode ($m_{average}$). Using this, we can set up a ratio of this mass to the mass of one gram mole of copper (63.54 g) and set this equal to the ratio of copper atoms transferred to Avogadro's number:

$$\frac{m_{average}}{m_{1\ mole}} = \frac{N_c}{N_0}$$

4. Determine the number of copper atoms N_0 in 1 mole of copper. This will be equal to Avogadro's number

$$N_0 = \frac{m_{1\ mole}}{m_{average}} N_c$$

N_0 = ________

5. Determine your percent error using $N_0 = 6.022 \times 10^{23}$ atoms/mole as the accepted value.

________ %

QUESTIONS

1. Define *one mole,* and give an example.

2. State the numerical value of Avogadro's number.

3. How many sodium ions are there in one mole of sodium chloride (NaCl)? How many chloride ions?

4. How many molecules are there in 47.8 grams of $CuSO_4$?

5. What do you think is the greatest source of error in this experiment? Why?

Experiment 34

Molecular Structure

INTRODUCTION

A covalent bond is formed between two atoms when they share electrons. Covalently bonded molecules are formed by these atoms because each atom tends toward a stable configuration, and the greatest stability is achieved when the outer shell of electrons is closed, or has its full complement of electrons orbiting the nucleus. In the case of hydrogen, one electron orbits the nucleus of the atom, but two electrons are required to complete the shell and render stability to the molecule. Two hydrogen atoms tend to share electrons to form a molecule of hydrogen gas.

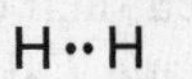

Dots between atoms indicate that the electrons are shared. Each hydrogen atom in a molecule has two electrons going around it.

If we concentrate on the lighter elements, ignoring the *d* and *f* electrons, we find that it takes eight electrons to make a closed shell (except for hydrogen and helium, in which two electrons form a closed shell). In the electron dot notation, we draw a dot for every electron outside closed shells. Thus, carbon, nitrogen, and chlorine, which have four, five, and seven electrons outside closed shells, are represented as

For a carbon atom to make a stable molecule, it must obtain four electrons from other atoms. Nitrogen needs three, and chlorine only one. Simple arrangements of atoms of carbon, nitrogen, and chlorine each united with the necessary number of hydrogen atoms are CH_4, NH_3, and HCl. They are expressed in the electron dot notation as:

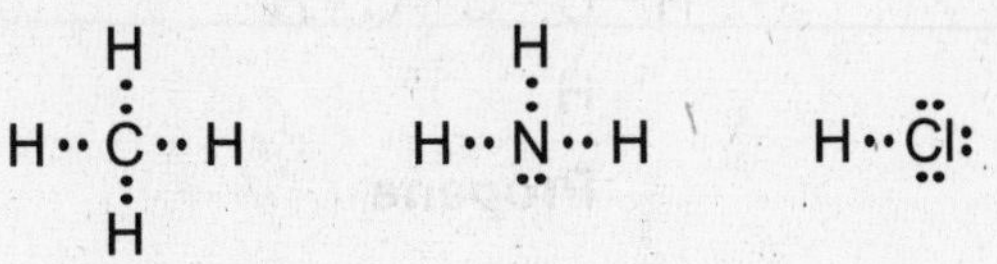

Note that each atom in these compounds has a closed shell of electrons.

Because it has four unpaired electrons, carbon forms an unusually large number of different compounds. Carbon compounds are called *organic compounds.* Many of the organic compounds are quite complicated. To portray them, a line is frequently used to indicate two electrons. For example, methane, CH_4, and ethane, C_2H_6, are drawn as in the accompanying structural formulas.

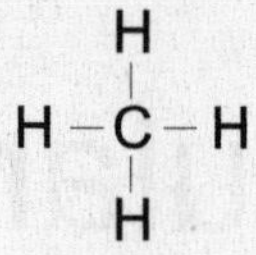

Methane

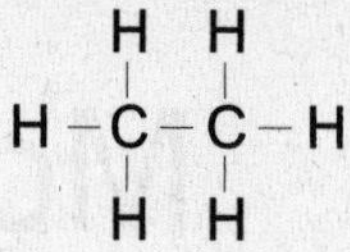

Ethane

Double and triple bonds are formed when four or six electrons are shared among atoms. Ethylene, C_2H_4, and acetylene, C_2H_2, are examples of simple molecules with double or triple bonds. Their structural formulas are shown below.

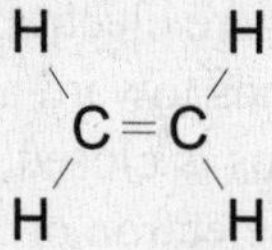

Ethylene

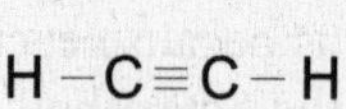

Ethyne (Acetylene)

Note that each carbon and hydrogen has a filled shell of electrons, as it must.

Carbon compounds such as cyclopropane, C_3H_6 also can be formed in ring structures.

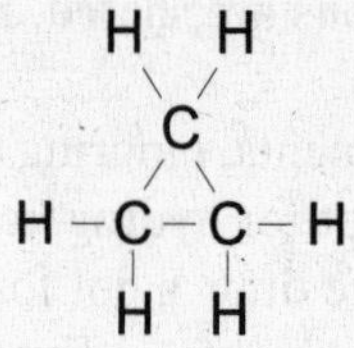

Cyclopropane

More complicated compounds of carbon have oxygen, nitrogen, the halogens, and other elements combined in complicated configurations.

When two compounds have exactly the same chemical constituents but different structures, they are called *isomers.* Two isomers of C_3H_6 are cyclopropane and propene, whose structures are shown.

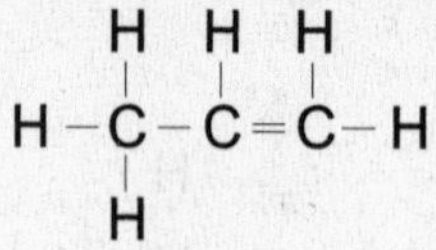

Propene

In the study of isomers, it is necessary to understand that two atoms joined by a single bond are each free to rotate independently about the axis formed by the bond.

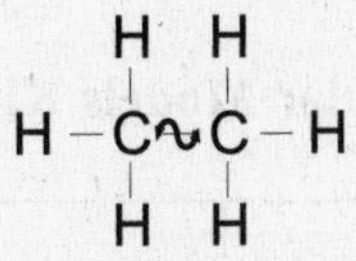

Free rotation about C–C bond

With a double or triple bond, the rotation is restricted so there is no actual rotation.

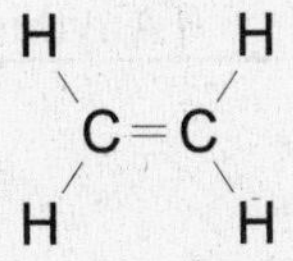

No rotation about C = C bond

In order to study molecular structure effectively, we must work with three-dimensional molecules. Hence we need models in three dimensions. For example, CH_4 in three dimensions has the hydrogens at the corners of a tetrahedron with the carbon in the middle.

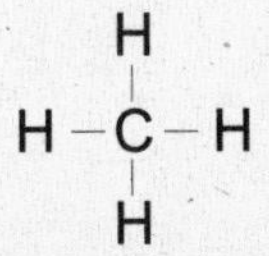

Two-dimensional representation

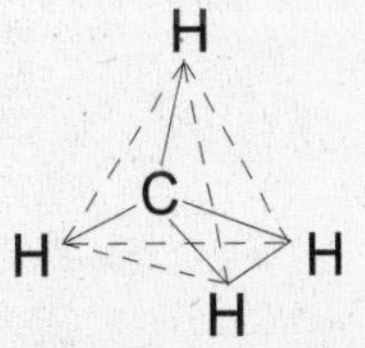

Three-dimensional representation

LEARNING OBJECTIVES

After completing this experiment, you should be able to do the following:

- ▼ Define the terms *covalent bond* and *isomers.*
- ▼ Construct models of molecules using a molecular model kit.

APPARATUS

A molecular model kit consisting of beads with holes inserted, spring connectors, and a protractor.

Table 34.1 gives the color code information on the molecular model kit. Each different-colored bead can be used to represent a different atom. Each hole in a bead represents an unpaired electron.

Each spring represents a covalent bond of two electrons. A single bond is represented by one spring, a double bond by two springs. Use the short spring for C–H bonds and the longer springs for other bonds.

The models should be assembled by insertion of the springs with a partial clockwise turn. In order to avoid damage to the springs, the models should be dismantled by withdrawing the springs with the same clockwise turn.

Table of Color Code Information, Molecular Models Kit

Color of Bead	Number of Holes	Element Represented
Black	4 holes	Carbon
White	1 hole	Hydrogen
Red	2 holes	Oxygen
Green	1 hole	A halogen (fluorine, chlorine, bromine, or iodine)
Orange or yellow	4 holes	Anything

EXPERIMENT 34 NAME (print) ______________________________ DATE ______________________
LAST FIRST

LABORATORY SECTION ______ PARTNER(S) ____

PROCEDURE 1

1. Make a model of CH_4. Measure the angle between the two bonds with your protractor. Record it.

   Bond angle = __________

2. Make a model of CH_3Cl. Are there two isomers with configurations as drawn below? Or are these the same?

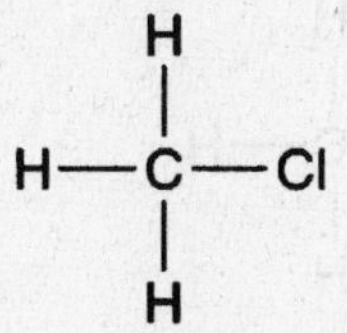

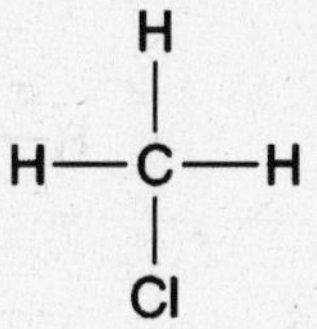

3. Make models of CH_2ClBr. Are there two isomers as shown below? Or are these the same?

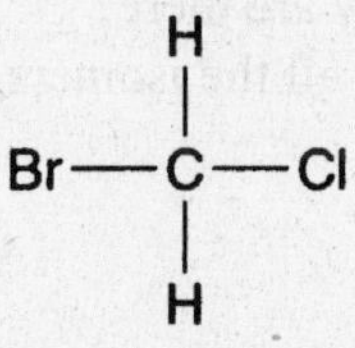

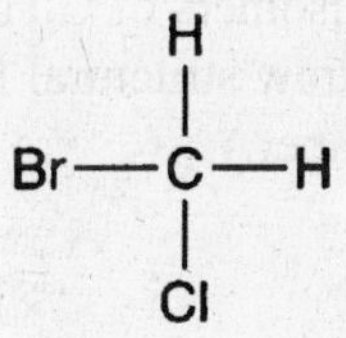

4. Are the models in Steps 2 and 3 planar? __________

5. Make a model of C_2H_2Cl.

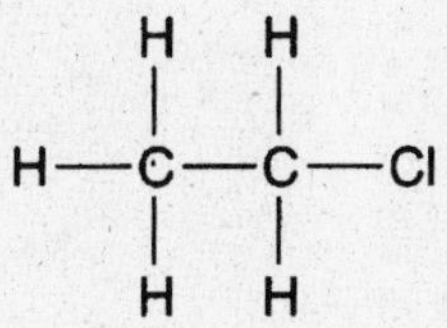

   Is this molecule planar? __________

   How many isomers are there? __________

6. Make a model of C_2H_3Cl.

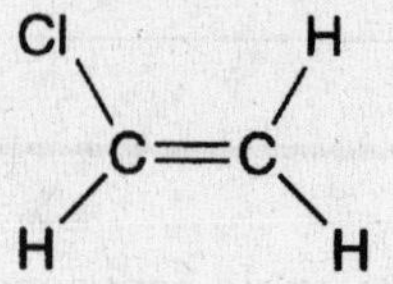

Is this molecule planar? ___________

How many isomers are there? ___________

7. Make as many isomers of C_3H_8 as you can. One of them is

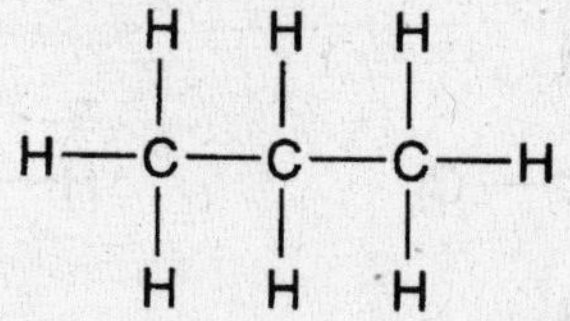

How many are there? ___________

8. Make models for all isomers of C_3H_6. How many are there? ___________
In the space below, draw structural formulas for all the isomers. (If you need more space, use a separate piece of paper.)

9. Make models for all the isomers of C_3H_5Cl. How many are there? ___________
In the space below, draw structural formulas for all the isomers. (If you need more space, use a separate piece of paper.)

EXPERIMENT 34 NAME (print) ________________________ DATE ____________
LAST FIRST

LABORATORY SECTION ______ PARTNER(S) ________________________

10. Make models for all the isomers of CHBrClF. How many are there? __________
An example of one isomer is:

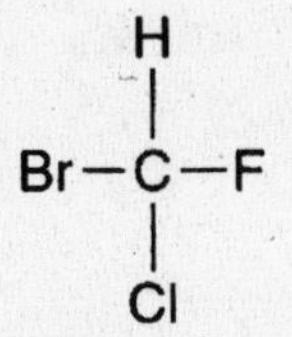

11. Make models for all the isomers of C_4H_{10}. How many are there?

12. Make models for all the isomers of C_4H_9Cl. How many are there? __________

13. Make models for all the isomers of C_5H_{12}. How many are there? __________

14. Make a model of ethyl alcohol. Its structure is

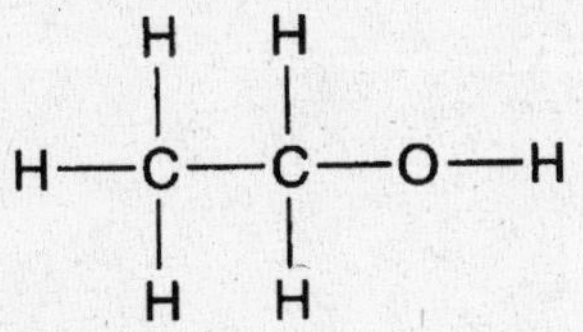

15. Make a benzene ring model. Its structure is

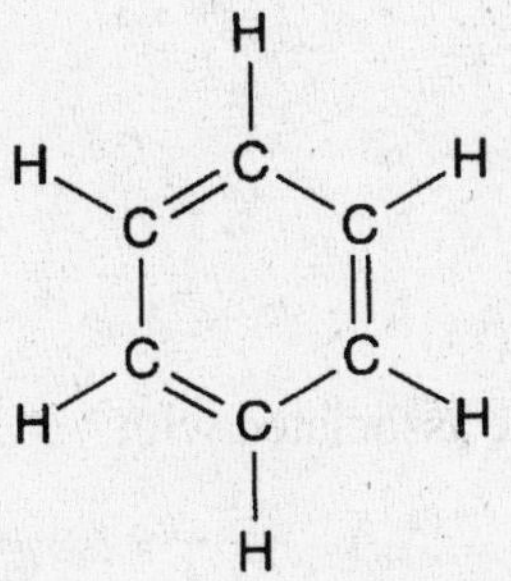

Have the instructor inspect it and check here. __________

16. Make a long-chain molecule of $C_{10}H_{22}$. Could you make an indefinitely long chain?

17. Make a model of nicotine. Its structure is

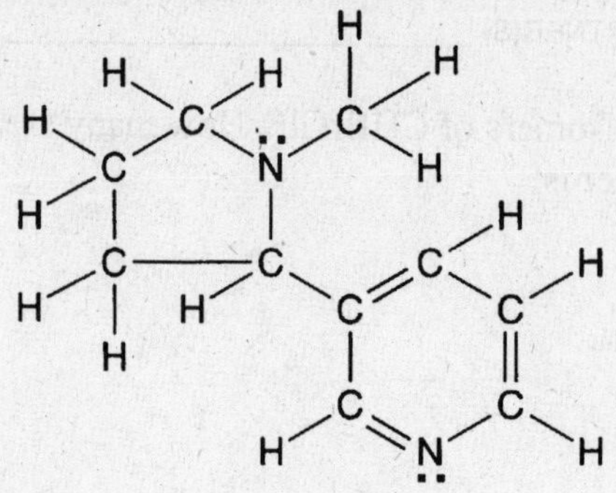

Have the instructor inspect your model and check here. ________

QUESTIONS

1. What do the "dots" around and between the chemical symbols in the drawings in this experiment represent?

2. What do the single, double, and triple "bars" between the chemical symbols in the drawing represent?

3. How many valence electrons are associated with:

a hydrogen atom ________

a carbon atom ________

a nitrogen atom ________

a chlorine atom ________

Experiment 35

Solutions and Solubility

INTRODUCTION

A *mixture* is a sample of matter composed of two or more substances in varying amounts that are not chemically combined. Mixtures can be homogeneous or heterogeneous. A *solution* is a homogeneous mixture of two or more substances. The substance present in greatest quantity in a solution is called the *solvent.* The substance dissolved is the *solute.* Solutions in which water is the solvent are called *aqueous solutions.* A solution in which the solute is present in only a small amount is called a *dilute solution.* If the solute is present in a large amount, the solution is a *concentrated solution.* When the maximum amount of solute possible is dissolved in the solvent, the solution is called a *saturated solution.* The concentration is frequently expressed in terms of the number of grams of solute dissolved in a given quantity of solvent. Example: a solution of salt in water might contain 5 g of salt for each 100 g of water.

The *solubility* of a given solute is the amount of solute that will dissolve in a specified volume of solvent (at a given temperature) to produce a saturated solution. The solubility depends on the temperature of the solution. If the temperature increases, the solubility of the solute will almost always increase.

LEARNING OBJECTIVES

After completing this experiment, you should be able to do the following:

- Define *mixture, solvent, solute, solution,* and *solubility.*
- Determine the concentration of a salt in water solution.

APPARATUS

Salt in water solution of unknown concentration, salt, distilled water, evaporating dish, glass beaker, ring stand with holder to support beaker, thermometer, Bunsen burner, lighter, and beam balance.

Safety glasses or goggles are recommended. Open flame can be dangerous. Do not leave a lighted burner unattended or reach across flame. Do not leave the gas turned on with an unlighted burner, and light with care.

PROCEDURE

1. The concentration of a solution can be determined by separating the solute from the solvent. Find the weight of each; then calculate the ratio of grams of solute to 100 g of solvent.
   (a) Weigh a clean, dry evaporating dish. Record in Data Table 35.1.
   (b) Obtain from the instructor about 4 cm^3 of a salt solution of unknown concentration. Pour the solution into the evaporating dish.
   (c) Weigh the dish plus solution. Record in Data Table 35.1.
   (d) Place the evaporating dish on a beaker of gently boiling water, and let the water evaporate from the salt.
   (e) When the water has completely evaporated from the dish, remove the dish from the beaker, cool, and then remove any moisture from the bottom of the dish. Weigh the dish and residue. Record in the data table.
   (f) Determine from your data the weight of the solute and the weight of the solvent in your sample.
   (g) Determine the concentration as grams of solute per 100 g solvent.

2. The solubility of a substance as a function of temperature can best be done in a limited time period by using one substance and having each group of students at a laboratory table obtain the solubility at a specific temperature and share the results. The following temperatures are suggested: room temperature (which will usually be about 22°C), 30°C, 40°C, 50°C, 60°C, and 70°C. If there are more or less than six groups, duplicate a temperature or let a group do two temperatures. Use two sets of apparatus to save time if a group is to do two temperatures.
   (a) Place about 20 mL distilled water in a clean beaker and bring to the correct temperature. Keep the temperature constant and add small amounts of salt while stirring continuously. Add salt until it fails to dissolve in the water.
   (b) Keep the temperature constant and allow the excess salt to settle to the bottom of the beaker. Weigh a clean, dry evaporating dish and record its weight in Data Table 35.1, then pour about 4 mL of the solution, still at the constant temperature, into the dish.
   (c) Weigh the dish plus the solution. Record in Data Table 35.1.
   (d) Place the evaporating dish on a beaker of gently boiling water, and let the water evaporate from the salt.
   (e) When the water has completely evaporated from the dish, remove the dish from the beaker, cool, and then remove any moisture from the bottom of the dish. Weigh the dish and residue. Record in Data Table 35.1.
   (f) Determine from your data the weight of the solute and the weight of the solvent in your sample.
   (g) Determine the concentration as grams of solute per 100 g of solvent.
   (h) Plot a graph of solubility as a function of temperature. Plot solubility on the y axis and temperature on the x axis.

EXPERIMENT 35 NAME (print) ______________________________ DATE ____________________
LAST FIRST

LABORATORY SECTION ______ PARTNER(S) ______________________________

DATA TABLE 35.1 (Part 1)	
Weight of clean, dry evaporating dish	g
Weight of evaporating dish plus solution	g
Weight of evaporating dish plus dry residue	g
Weight of solvent (computed)	g
Weight of solute (computed)	g
Grams of solute per 100 g of solvent (computed)	g

DATA TABLE 35.1 (Part 2)

Temperature	Room Temp. ________°C	30°C	40°C	50°C	60°C	70°C
Weight of clean, dry dish (g)						
Weight of dish plus solution (g)						
Weight of dish plus dry residue (g)						

COMPUTATIONS

Temperature	Room Temp. ________°C	30°C	40°C	50°C	60°C	70°C
Weight of solute (g)						
Weight of solvent (g)						
Grams of solute per 100 g of solvent (g)						

QUESTIONS

1. Define the term *solution.* Give an example.

2. What is a *saturated solution?*

3. Distinguish between *solvent* and *solute.*

4. Define *solubility.* Give an example.

5. Is solubility a function of temperature? Explain.

Experiment 36

Pressure-Volume Relationship of Gases

INTRODUCTION

A *gas* is a substance that has no definite shape or volume. If we investigate a sample of any gas, we discover that it fills the entire volume of the container it occupies. Close examination reveals the gas molecules to be in rapid motion, undergoing collisions with one another and with the walls of the container.

The behavior of gases is understandable if we contemplate an ideal gas. Such a gas is composed of molecules that are small in size compared with the total volume of a container and where molecules have no attraction for one another. This description can be used as an approximation for the behavior of most gases found in our environment.

The physical states used to describe the gases are volume, pressure, and temperature. The relationships of these states and the laws governing such relationships were formulated by Robert Boyle in the sixteenth century, and Jacques Charles and Joseph Gay-Lussac in the seventeenth century.

The *absolute temperature* of a gas can be defined as a measure of the average kinetic energy of the gas molecules:

$$T \propto \frac{mv^2}{2}$$

where T = absolute temperature (in degrees Kelvin, K)
m = mass
v = velocity

Pressure can be defined as the force per unit area:

$$P = \frac{F}{A}$$

The *general gas law,* which is a combination of Boyle's and Charles's laws, deals with the expansion and compression of a gas at different temperatures. The law states that the pressure P of a given amount of gas is proportional to the absolute temperature T and inversely proportional to the volume V, or

$$\frac{PV}{T} = k$$

where the constant k has a different value for different samples. When samples of the same gas are taken that have different total masses $m,$ we find experimentally that

$$\frac{PV}{T} \propto m$$

When equal masses of gases are taken that have different molecular weights M_w, we find that

$$\frac{PV}{T} \propto \frac{1}{M_w} \quad \text{thus} \quad \frac{PV}{T} \propto \frac{m}{M_w}$$

But $\frac{m}{M_w}$ is the number of moles n in the sample (constant throughout this experiment). Therefore,

$$\frac{PV}{T} \propto n \quad \text{or} \quad \frac{PV}{T} = Rn$$

where R is known as the universal gas constant.

The numerical value of the constant can be evaluated by considering a gas at standard temperature (273 K) and standard pressure [1 atmosphere = 14.7 lb/in^2 = 76 cm of mercury (Hg) = 1.013×10^5 newtons/meter^2]. The measured volume of any gas at these conditions is approximately $n \times 22.4$ L, where n is the number of moles, and 1 L = 1×10^{-3} m^3. Thus we obtain

$$\frac{PV}{T} = \frac{P_0 V_0}{T_0}$$

where the zero subscript indicates standard conditions.

Substituting known values into the equation yields

$$\frac{PV}{T} = \frac{\left(1.013 \times 10^5 \ \text{N/m}^2\right) \times \left(n \times 22.4 \times 10^{-3} \ \text{m}^3\right)}{273 \ \text{K}}$$

or $$\frac{PV}{T} = n \times 8.31 \ \text{N}^2/\text{K}$$

Thus, if $n = 1$ mole $$R = \frac{PV}{T} = 8.31 \ \text{N}^2/\text{K}$$

LEARNING OBJECTIVES

After completing this experiment, you should be able to do the following:

- Define the terms *gas, pressure,* and *absolute temperature.*
- State in words the general gas law and Boyle's law.
- Determine experimentally the relationship that exists between pressure and volume of a gas when the temperature is held constant.

APPARATUS

One elasticity of gases apparatus (see Fig. 36.1), seven mass units, 1 kg each.

Caution Safety Tips

When placing weights on the top of the plunger top board, keep the entire stack balanced, and do not leave stacked up weights unattended. Do not get hands or feet below weights because they might fall.

EXPERIMENT 36 NAME (print) ______________________________ DATE ______________________
LAST FIRST

LABORATORY SECTION ______ PARTNER(S) ____

PROCEDURE 1

1. Remove the plunger from the syringe (refer to Fig. 36.1), and carefully measure the diameter (D) of the syringe tube and record this value above Data Table 36.1.

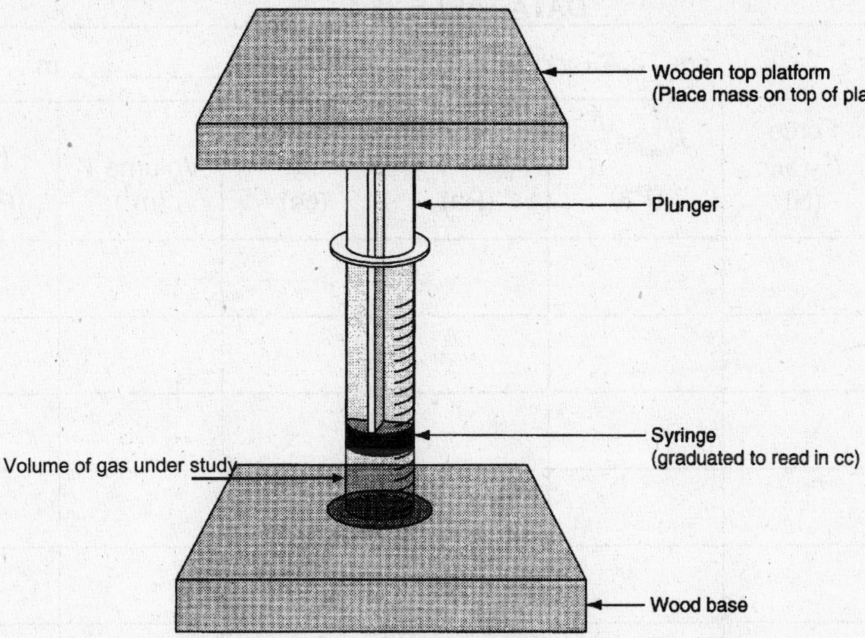

Figure 36.1

2. Remove the end cap of the syringe, insert the plunger, and adjust the position of the plunger to indicate about 20 to 25 cc. Replace the end cap snugly.
3. Set syringe assembly in wooden holder, and place the wooden top platform in position.
4. Read the initial volume V_{cc}, and record this value in Data Table 36.1.
5. Place about 1 kg of mass on the wooden top platform, taking care to balance it carefully. Record the mass M, and the new volume V_{cc}, in Data Table 36.1.
6. Repeat Step 5 for 6 to 8 readings by placing additional mass, in increments of about 1 kg, on the wooden top platform. Always be careful to balance the masses and keep the apparatus from falling over. Record the total mass M on the platform and the new volume V_{cc} for each reading in Data Table 36.1.
7. Determine the current atmospheric pressure P_{atm} in the room, and record its value below Data Table 36.1. If you cannot find the exact current atmospheric pressure, you may use 1.01×10^5 Pa for this value with very little loss of accuracy in the experiment.
8. Perform all necessary calculations to complete Data Table 36.1 using the following relationships:

$$F = mg \qquad \text{(in Newtons)}$$

$$P_{meas} = \frac{F}{A} \qquad \text{where } A = \pi r^2 \qquad \text{(in m}^2\text{)}$$

$$P_{abs} = P_{meas} + P_{atm} \qquad \text{(in N/m}^2 \text{ or Pa)}$$

Also convert the volume V_{cc} from cubic centimeters into SI units of cubic meters; then calculate the value of V (in cubic meters), times P_{abs} in pascals (Pa) and record these values in Data Table 36.1. One pascal (Pa) is equal to one newton per meter squared.

9. Determine the percentage difference for the values of P and V calculated in the last column of Data Table 36.1.

10. Plot a graph of P as a function of $1/V$. *Note:* You will have to calculate values for $1/V$ so make a listing of these values.

DATA TABLE 36.1

D = ______ cm $A = \pi r^2$ = ______ cm² = ______ m²

Mass M (kg)	Force $F = mg$ (N)	$P_{meas} = \frac{F}{A}$ (Pa)	$P_{abs} = P_{meas} + P_{atm}$ (Pa)	Volume V_{cc} (cc)	Volume V (m³)	$P_{abs} \times V$ (Pa·m³)

P_{atm} = ______ Average value of $P_{abs} \times V$ ______

Percent difference for all values in the last column of Data Table 36.1 ______ %

(show work below)

EXPERIMENT 36 NAME (print) ______________________________ DATE ____________________
LAST FIRST

LABORATORY SECTION ______ PARTNER(S) ____

QUESTIONS

1. Explain the shape of the curve you have plotted.

2. State the relationship between the volume and pressure of a gas when the temperature is held constant (Boyle's law).

3. Do the data taken in the experiment support the idea that the pressure times the volume for an ideal gas is constant when the temperature remains at the same temperature?

4. The temperature of a gas is increased while the volume is held constant. Does the pressure decrease, increase, or remain the same? Explain.

5. A volume of gas is decreased by half while the temperature remains constant. How much does the gas pressure change?

Experiment 37

Chemical Qualitative Analysis

INTRODUCTION

The systematic effort to determine what elements are present in a sample of unknown constituency is called *qualitative analysis.* Such elements can be determined by using the knowledge we have about their chemical characteristics. The procedure may be illustrated with a test for the metals *lead, silver,* and *mercury,* commonly known as *Group I,* with the common characteristic that their chlorides are largely insoluble in water. In a solution of their compounds, these elements can be detected by adding hydrogen chloride to the solution; the metals will be precipitated as chlorides and may be filtered out and thereby separated from any other metals present.

Lead chloride is very soluble in hot water and may be separated from the other precipitated chlorides by washing the mixture with hot water. The presence of silver can then be detected by adding ammonium hydroxide to the mixture. The silver ion will combine with ammonia to form a complex silver ammonium ion, whose chloride is soluble; the silver is thereby separated and put into solution as a complex ion. The mercury chloride that is left can be separated by filtration.

The presence of each of these elements in solution can be confirmed as follows: If potassium chromate is added to the hot water solution, a yellow precipitate will confirm the presence of lead. The presence of the silver in the filtrate obtained after the ammonium hydroxide is added may be confirmed by adding to the filtrate some nitric acid. This neutralizes the ammonium hydroxide and silver chloride is precipitated. The presence of mercury is confirmed immediately when ammonium hydroxide is first added to the mixture, because a black precipitate, indicating the presence of mercury, will remain on the filter while the silver ammonium compound passes through the filter as a solution. A flow diagram summarizing the above information is provided as a part of these instructions.

LEARNING OBJECTIVES

After completing this experiment, you should be able to do the following:

- Define *qualitative analysis.*
- Test experimentally by quantitative analysis for lead, silver, and mercury ions.

APPARATUS

1.0 M solution hydrochloric acid, 1.0 M solution ammonium hydroxide, 1.0 M solution nitric acid, potassium chromate, test solution containing Ag^+, Pb^{++}, and Hg^+ ions, ring stand, funnel, filter paper, one 250-mL beaker, three test tubes, hot water.

Safety glasses or goggles are recommended. Be careful when handling or pouring acid. Acids used here are dilute, but any contaminated skin surface should be thoroughly flushed with clear water. Do not get in your eyes.

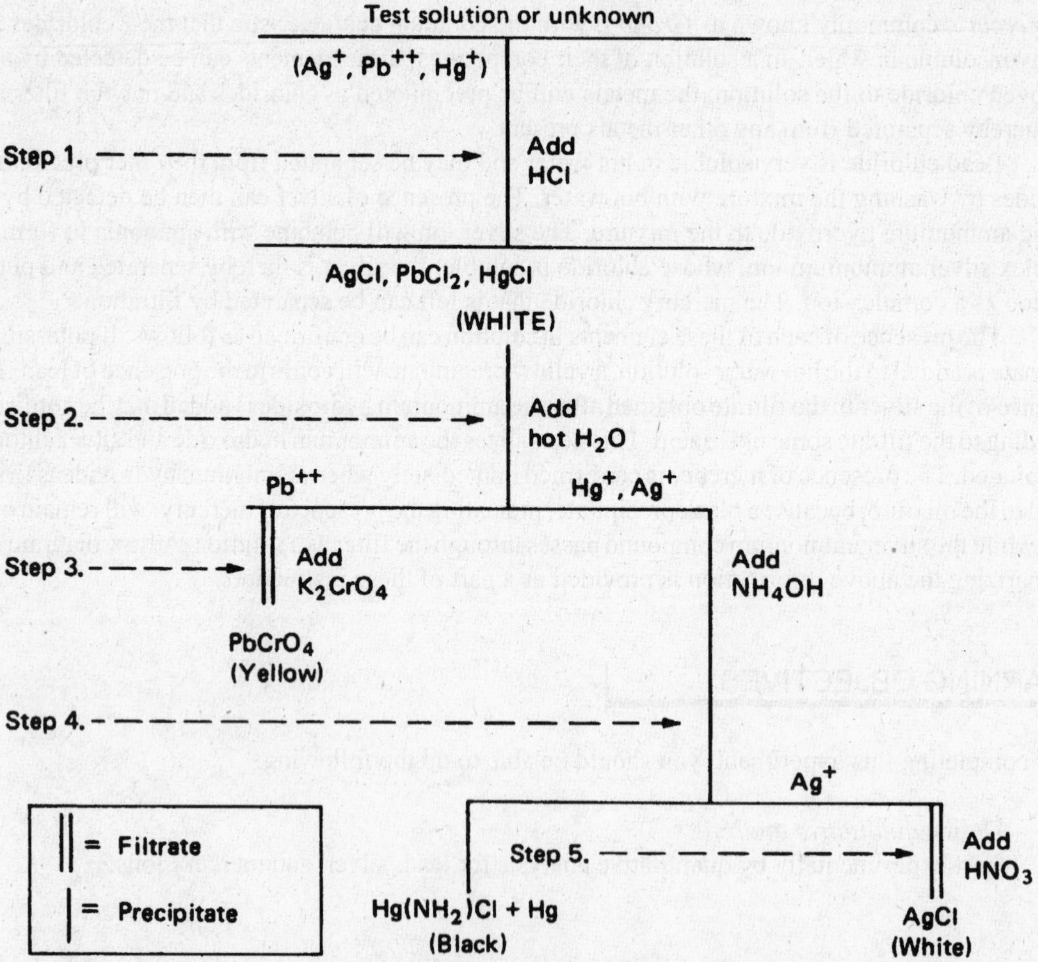

Figure 37.1 Flow diagram. Separation of Ag^+, Pb^{++}, and Hg^+ in a known or unknown solution.

EXPERIMENT 37 NAME (print) ______________________ DATE ______________
LAST FIRST

LABORATORY SECTION ______ PARTNER(S) ______________________________

PROCEDURE

Use the flow diagram (Fig. 37.1) and proceed as follows:

1. Pour about 5 mL of the test solution (which contains lead, silver, and mercurous ions) into a clean, dry test tube.
2. Add four or five drops of hydrochloric acid.
3. Filter out the precipitate, and discard the filtrate (the solution that goes through the filter paper).
4. Without removing the precipitates from the filter paper or funnel, wash thoroughly with hot water, using about half a test-tube full, and collect the filtrate in a clean test tube.
5. Check for lead in the filtrate by adding potassium chromate. If lead is present, a yellow precipitate of lead chromate will be formed. *Note:* The yellow color will always be present. The solution will turn cloudy if lead is present.
6. After placing another test tube under the funnel, add ammonium hydroxide to the precipitate remaining on the filter paper. If mercury is present, the precipitate on the filter paper will turn black.
7. Check the filtrate for silver by acidifying it with nitric acid. The appearance of a precipitate (look carefully for it; the solution will show only a slight cloudiness) confirms the presence of silver.
8. Obtain from the instructor an unknown sample and, using the procedure given above, determine which of the above elements are present.

Number of unknown __________

Indicate the elements present or not present.

Pb __________

Hg __________

Ag __________

QUESTIONS

1. What is the purpose of qualitative analysis?

2. How is any lead chloride in the sample removed in step 2?

3. What are the final indicators that show the presence of the following elements.

Pb:

Hg:

Ag:

Experiment 38

Chemical Quantitative Analysis (Volumetric)

INTRODUCTION

Determination of the exact amounts of elements or substances present in an unknown sample is called *quantitative analysis.* The two most common methods of doing this are the *gravimetric method* and the *volumetric method.* The latter will be used in this experiment.

In the volumetric method, the solutions used in the reactions are of known strength. The volumes of these solutions needed to complete reactions are measured, and the quantities involved are calculated from the known strengths of the solutions. The completion of a reaction is made visible by the use of an indicator, a material added to the solutions. It shows by its color when the reaction is complete. The solutions are usually measured out of long, graduated tubes that have stopcocks at the end so that small quantities, even to a fraction of a drop, may be obtained as desired. Such tubes, called burettes, are commonly graduated to 0.10 ml.

All solutions used for testing in these experiments will be made up on the molar basis. *Thus a one molar solution will contain one gram molecular weight of a substance per liter of solution. A liter is 1000 mL.* A 0.50 molar solution will contain 0.50 g molecular weight of the substance per liter of solution, and so forth.

LEARNING OBJECTIVES

After completing this experiment, you should be able to do the following:

- Define *quantitative analysis, molarity,* and *one molar solution.*
- Calculate the percentage of acetic acid in vinegar.
- Calculate the molarity of a solution.
- Determine experimentally by quantitative analysis (volumetric method) the exact amount of a substance in an unknown sample.

APPARATUS

Two burettes, two Erlenmeyer flasks, phenolphthalein, 1.0 M hydrochloric acid, 1.0 M sodium hydroxide, vinegar, two 250-mL beakers.

Safety glasses or goggles are recommended. Be careful when handling or pouring acid. The HCl used here is dilute, but any contaminated skin surface should be thoroughly flushed with a large quantity of clean water. Use caution when handling chemicals.

EXPERIMENT 38 NAME (print) ______________________ DATE ______________
LAST FIRST

LABORATORY SECTION ______ PARTNER(S) ______________________

PROCEDURE

1. Fifteen milliliters of vinegar (which includes acetic acid) will be placed in your Erlenmeyer flask together with two drops of phenolphthalein (an indicator that is colorless in acid solutions but turns red in basic solutions). About 50 mL of one molar sodium hydroxide will be put in your burette. Record the level before you start. Now from this burette slowly titrate sodium hydroxide into the flask containing the vinegar, keeping the solution well stirred by swirling it gently in the flask, until the drop most recently added causes the solution to remain a pale pink color. The appearance of this color indicates that the acid has just been neutralized by the base and the last drop has made the whole solution slightly basic. Read the burette carefully and record the volume of sodium hydroxide used in Data Table 38.1.

Because a one molar solution contains one gram molecular weight per liter, each milliliter contains one-thousandth of a gram molecular weight. If, for example, a quantity of 17 mL of one molar sodium hydroxide is required in the experiment above, then $17 \times 0.001 = 0.17$ (or 17/1000) of a gram molecular weight of sodium hydroxide is required. From the equation

$$NaOH + HC_2H_3O_2 \rightarrow H_2O + NaC_2H_3O_2$$

we know that one molecule of sodium hydroxide neutralizes one molecule of acetic acid, or in other words, one gram molecular weight of sodium hydroxide neutralizes one gram molecular weight of acetic acid.

If, as in the above example, 17/1000 of a gram molecular weight of sodium hydroxide is required, then 17/1000 of a gram molecular weight is in the vinegar. The molecular weight of acetic acid is 60; therefore, the weight of acetic acid present in the vinegar sample is 17/1000 of 60 g, or 1.02 g.

Calculate the percentage of acetic acid present in the vinegar sample that is provided for the experiment. Federal law requires a minimum of 4%.

2. The instructor will provide you with a known volume of hydrochloric acid. Slowly titrate sodium hydroxide to the volume of hydrochloric acid, and record the amount added in Data Table 38.2.

Calculate the percentage of sodium hydroxide present in the solution supplied for you in the laboratory. The sodium hydroxide will be neutralized by a solution of hydrochloric acid whose strength will be given to you in the laboratory. The equation is

$$NaOH + HCl \rightarrow H_2O + NaCl$$

DATA TABLE 38.1		
Volume of vinegar	15 mL	
Volume of sodium hydroxide	________	mL

COMPUTATIONS

Weight of volume of vinegar	15 g	
Weight of acetic acid present	________	g
Percentage of acetic acid	________	%

DATA TABLE 38.2		
Volume of hydrochloric acid	15 mL	
Volume of sodium hydroxide solution	________	mL

COMPUTATIONS

Weight of volume of NaOH solution	15 g	
Weight of NaOH present	________	g
Percentage of NaOH	________	%

QUESTIONS

1. Find the number of grams of KCl in 500 mL of a 3 molar solution.

2. What would be the molarity of an NaCl solution if 250 g of salt is used to make 5 L of solution?

Experiment 39

Kepler's Laws of Planetary Motion

INTRODUCTION

In this experiment we investigate the effects of a central force on the motion of an object. A *central force* may be defined as one that is always directed toward some particular point. The gravitational attraction between Earth and the Sun is an example of an inverse-square central force (Fig. 39.1). The path resulting is an ellipse. All central forces produce elliptical paths. In the seventeenth century, Kepler discovered several properties of these ellipses. We investigate one of these, the law of equal areas, which may be formally stated as follows: *When an object is traveling in an ellipse under the action of a central force, a line drawn from the force center to the object sweeps out equal areas in equal times.*

The drawing in Fig. 39.2 illustrates the path of an object traveling in an ellipse under the action of a central force. The time interval between the dots shown on the diagram is 0.10 second.

LEARNING OBJECTIVES

After completing this experiment, you should be able to do the following:

- ▼ State Kepler's law of equal areas.
- ▼ Determine experimentally the validity of the law of equal areas.

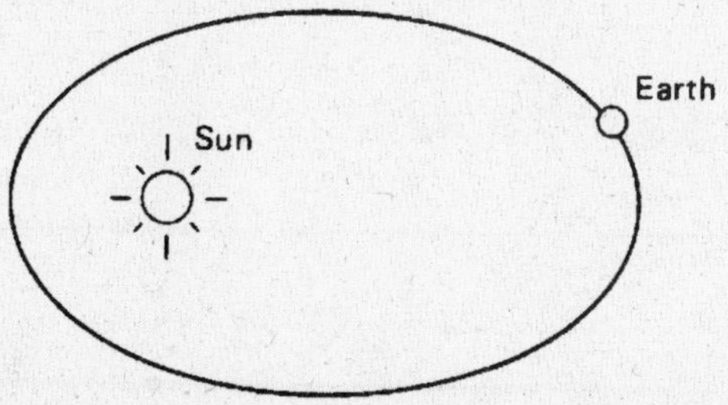

Figure 39.1 Earth revolves around the Sun in an elliptical path, with the Sun at one focus.

APPARATUS

Reproduction of the path points for a stroboscopic photograph of the path of a pendulum bob moving under the influence of a central force; rulers, protractors, soft board or sheet of corrugated cardboard, string, straight pins.

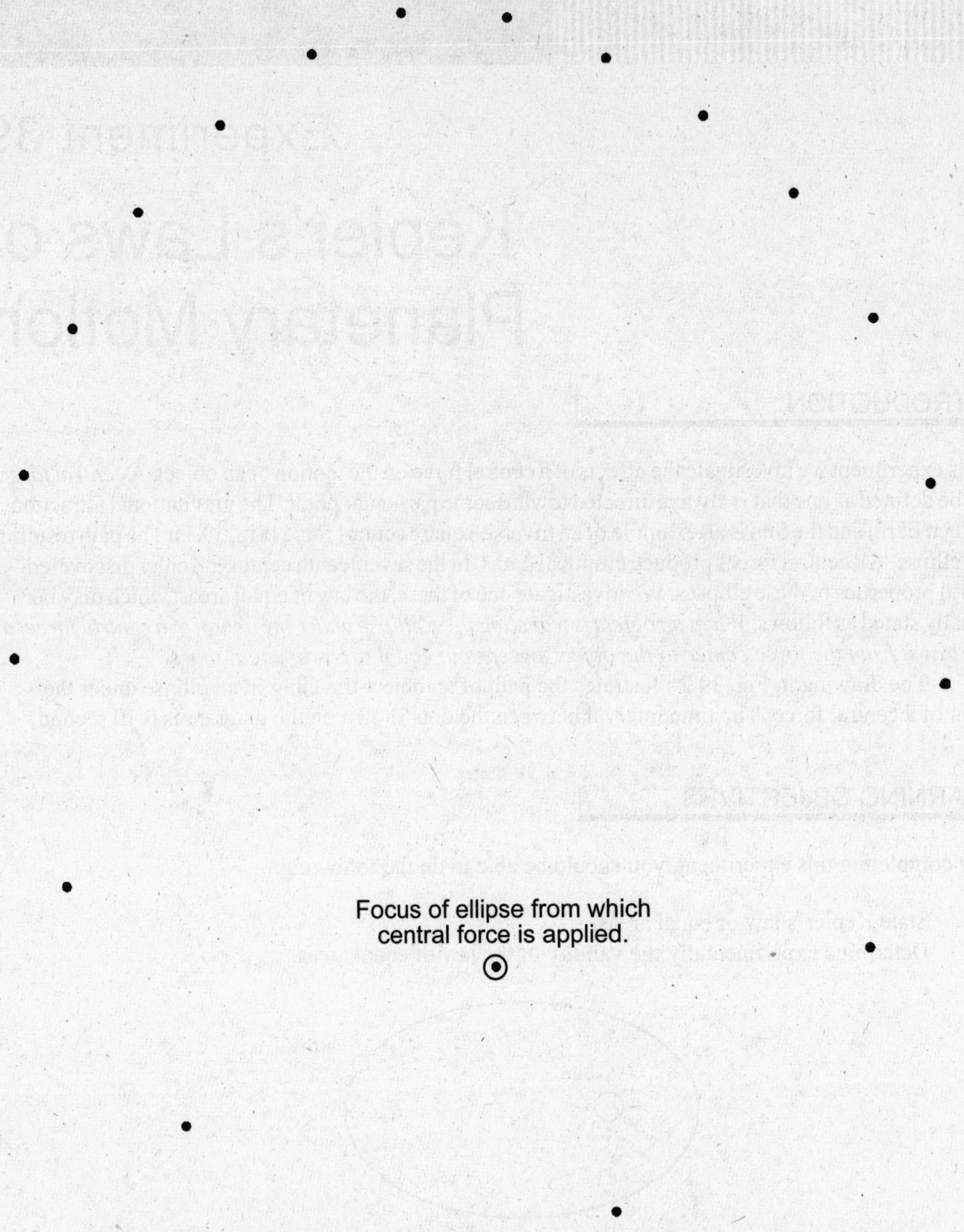

Figure 39.2 Reproduction showing the elliptical path of a revolving body around a central force.

EXPERIMENT 39 NAME (print) ______________________________ DATE ______________________
LAST FIRST

LABORATORY SECTION ______ PARTNER(S) ____

PROCEDURE 1

Planetary Orbits

Place a clean sheet of paper lengthwise on the soft-surface board provided, and pin it down at the corners with straight pins. Draw a small circle slightly to the left of the center of the page and label this "Sun," and place a straight pin at the center of this small circle. Now choose a point slightly to the right of the center of the page, and stick a second pin into the board at this point. Take a piece of string about 1 foot long and tie it into a large loop. Place the loop around the two pins and also around the tip of your pencil while the pencil is touching the paper. Move the pencil point outward until the string is snug. Move the pencil point in a roughly circular motion keeping the string taut so that it guides the pencil around the two pins. The path drawn on the paper with the pencil point is an ellipse and could represent the path followed by Earth (or any other planet) as it orbits around the Sun.

Experiment with different sized loops of string and with the pins at different separation distances to see how this affects the size and shape of the ellipse formed. Do six different ellipse drawings of various shapes and sizes.

PROCEDURE 2

Law of Equal Areas

A planet follows an elliptical path around the Sun because the restoring force (gravity in this case) is always toward the center. According to Kepler's second law, the area of triangles formed by connecting adjacent positions of the planet to the central focus and to each other should be equal for each equal time interval. We can determine the area of these triangles by multiplying the width of the triangle by the height of the triangle and dividing this product by 2. See Fig. 39.3.

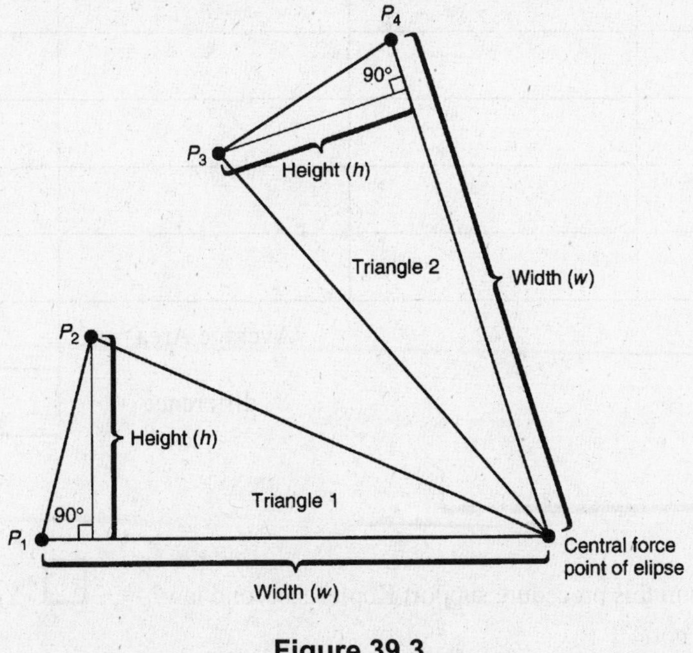

Figure 39.3

$$\text{Area} = \frac{\text{width} \times \text{height}}{2}$$

Note: The height must be at a *right angle* to the width for this equation to give the correct area. The width is the *longer* of the two sides of the triangle that go the central force point.

To complete the experiment, follow the step-by-step procedure outlined below using Fig. 39.2.

1. Draw lines from the central force point of the ellipse to any two adjacent points. (P_1 and P_2).
2. Draw a line connecting the two points chosen. (This will give you a triangle.)
3. Construct a perpendicular from the *longest* side of the triangle to the dot opposite it. Measure the lengths between both P_1 and P_2 and the central force point. These will be considered the sides of the triangle. In the example shown in Fig. 39.3, the distance from P_1 to the central force point is longest, so this side is chosen as the width of this triangle. (Your laboratory instructor can show you how to use a protractor to draw this perpendicular line.)
4. Carefully measure the length of the perpendicular line and the length of the long side W. Be as accurate as possible in these measurements. Record the data in Data Table 39.1.
5. Calculate the area of the first triangle using the formula for the area given above.
6. Repeat this procedure with five other triangles formed in a similar manner on the ellipse provided.
7. Calculate the average area for all triangles measured.
8. Calculate the percent difference for the six area values found in Data Table 39.1.

DATA TABLE 39.1

Triangle	Long side Width W (cm)	Perpendicular Height h (cm)	Area of Triangle $\frac{(W+h)}{2}(\text{cm}^2)$
1			
2			
3			
4			
5			
6			
		Average Area =	
		% difference =	

CONCLUSION

Do your calculations in this procedure support Kepler's second law? ☐ Yes ☐ No

Explain why or why not.

EXPERIMENT 39 NAME (print) ______________________________ DATE ____________________
LAST FIRST

LABORATORY SECTION ______ PARTNER(S) ______________________________

QUESTIONS

1. What would happen if both pins in Procedure 1 were located at the same point, that is, right next to each other actually touching? (Try it if you need to, so that you can see what the result would be.)

2. In Figure 39.2 the time interval between the dots is 0.10 seconds. What is the overall period for one swing of the pendulum bob that was photographed to make this figure?

3. What is the period (time for one complete revolution) for the planet Earth to travel around the Sun on its elliptical path?

4. If Fig. 39.2 represents a planet with the same period as Earth, how long does it take for the planet to move from one dot to the next?

5. Remembering that each set of dots is separated by the same time interval, where on its elliptical path will the planet be traveling fastest?

6. If Earth is closest to the Sun in January of each year, what can we say about the speed of Earth in its orbit around the Sun at this time of year?

Experiment 40

Stars and Their Apparent Motions

INTRODUCTION

The best way to study the stars and their motions would be to observe them night after night for a year or more. Because weather conditions would not permit this, and the astronomy laboratory for this course is only two hours long, other methods must be used. The next best method would be to use a planetarium, where the observer is placed at the center of a large half-sphere or dome. The stars and their apparent motions are projected by optical means on the dome. However, the method we shall use in this experiment will be that of the celestial sphere. A *celestial sphere* is a small model sphere (they are made in several sizes) that may be set to portray the sky as seen from any latitude on Earth, on any day and for anytime of day.

LEARNING OBJECTIVES

After completing this experiment, you should be able to do the following:

- ▼ Define the terms *celestial latitude, celestial longitude, altitude, declination, ecliptic, vernal equinox, zenith, perpetual apparition,* and *perpetual occultation.*

With the aid of the celestial sphere:

- ▼ Determine the declination of the Sun on any day of the year.
- ▼ Determine the time of sunrise and sunset for any latitude on any day of the year.
- ▼ Determine the approximate number of hours of daylight and darkness from any latitude on any day.
- ▼ Determine the altitude of the Sun at 12 noon local solar time for any latitude on any day of the year.
- ▼ Identify and give the location and apparent motion across the sky of the brighter stars and constellations.

DISCUSSION

The apparent position of a star changes in the sky as the observer changes latitude. For example, Polaris, the North Star, appears directly overhead for an observer at 90°N. When the observer is at 0° latitude, the North Star will appear on the northern horizon. If the observer travels northward 1°, the North Star will appear 1° above the northern horizon. At 40°N, the North star will appear 40° above the northern horizon. If the observer travels south of the equator 1°, then the North star will be 1°

below the northern horizon and will not be visible to the observer. Thus a change in latitude by the observer makes an obvious change in the positions of the stars with respect to the horizon.

To understand how the celestial sphere is used to view the positions of the stars, place the celestial sphere in front of you as you read this discussion, and locate each item on the model as it is discussed. Refer also to Figs. 40.1 and 40.2.

The celestial sphere represents the sky, not the Earth. Earth is an imaginary point located at the center of the sphere. As viewed from above the North Pole, Earth turns counterclockwise or eastward. The eastward rotation of Earth about its axis produces the apparent westward motion of the celestial sphere. Thus, when you rotate the celestial sphere westward the rotation represents an apparent, not actual, motion. For example, we observe the Sun rises on the eastern horizon, travel westward across the sky, and set in the west. This apparent westward motion of the Sun is due to Earth's eastward rotation about its axis.

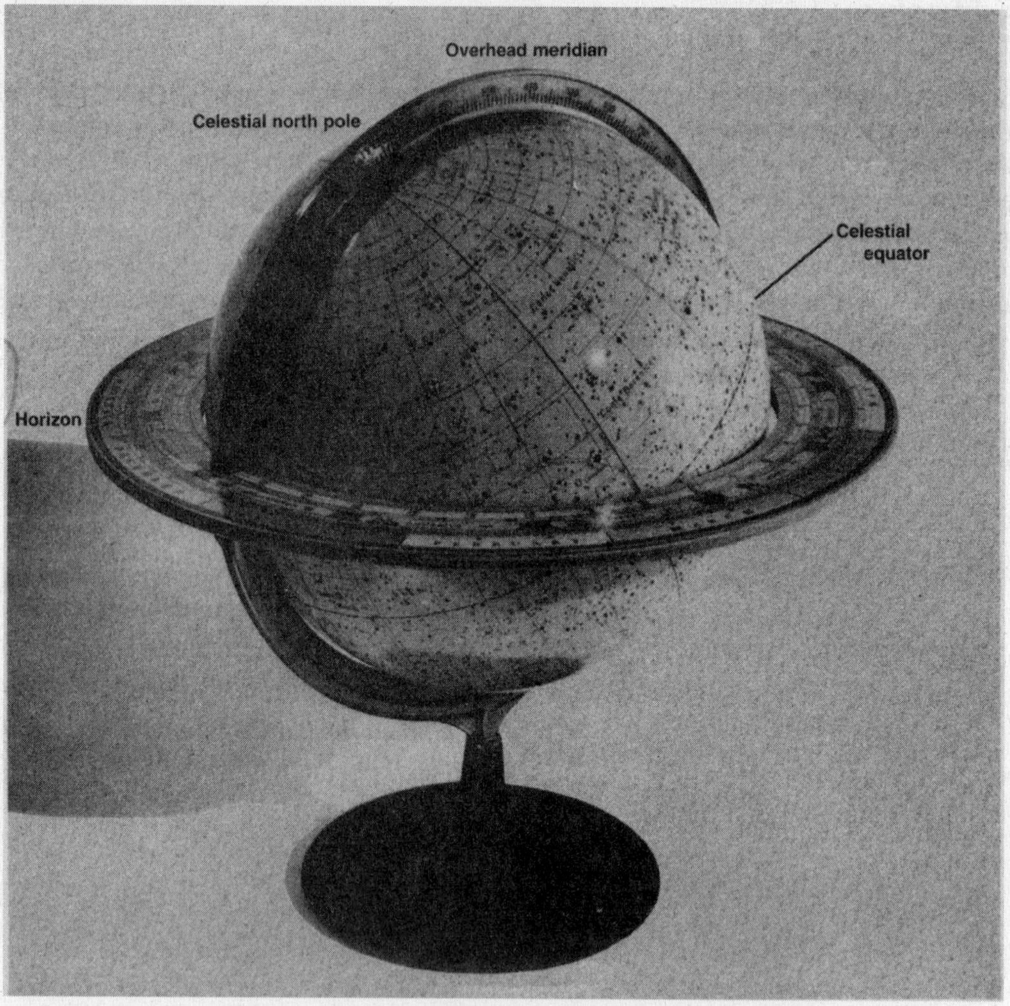

Figure 40.1 Celestial sphere, a model sphere portraying the sky. The celestial equator is the line circling the sphere midway between the north and south celestial poles. The ecliptic plane is at an angle of 23½° to the celestial equator. Earth is located at an imaginary point in the center of the sphere. (Courtesy of Dennoyer-Geppert Company)

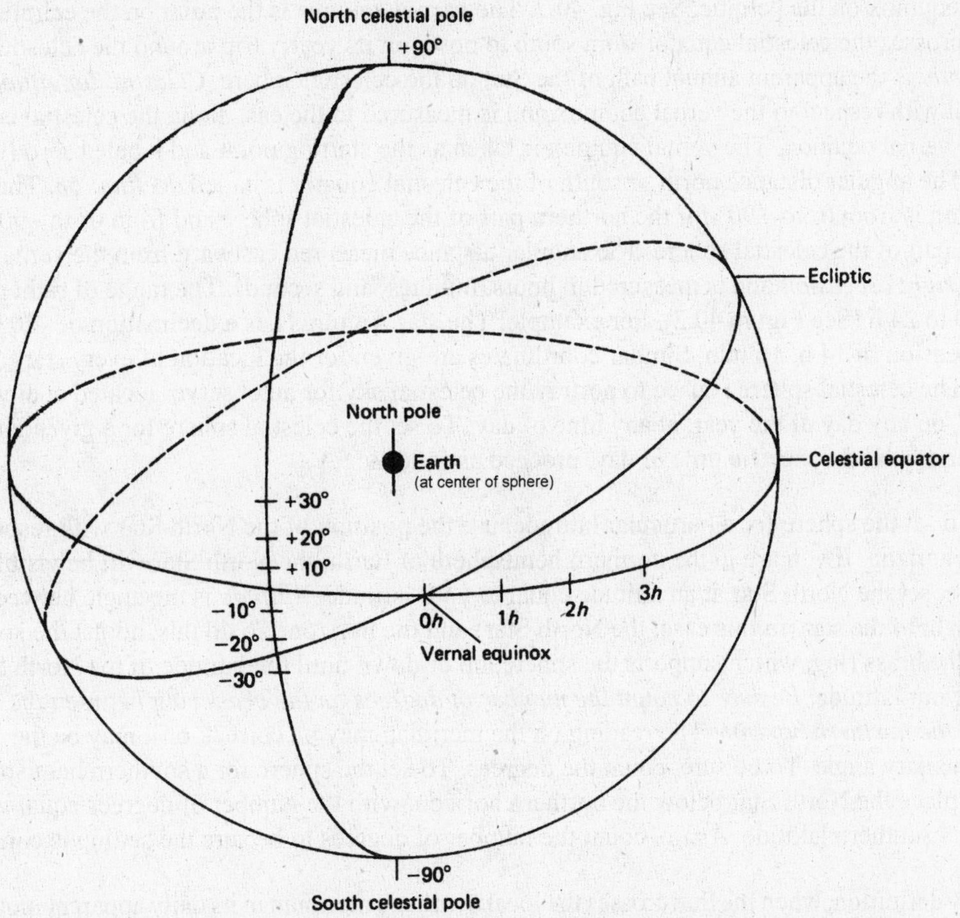

Figure 40.2 The Celestial Sphere. Diagram illustrating declination and right ascension. Declination is measured from 0° to +90° for the northern half of the celestial sphere, and from 0° to −90° for the southern half. Earth is represented by the small imaginary dot at the center of the celestial sphere.

The large wooden ring constructed around the sphere, parallel to the floor, represents the observer's horizon. If you are outdoors, the horizon appears as the dividing line between Earth and the sky. The horizon appears 90° down from directly overhead (the observer's zenith), and extends completely around the observer for a full 360°. The *zenith* is the point on the celestial sphere opposite the direction in which gravity pulls, that is straight downward. Only those stars above the horizon (the wooden ring) can be seen by the observer. The metal ring positioned at 90° to the horizon, and circling vertically over the top of the sphere, represents the observer's overhead meridian.

The position of a star on the celestial sphere is given with respect to two reference frames. The first is the *celestial equator,* an imaginary line circling the celestial sphere halfway between the north and south celestial poles. If Earth's equator were extended outward, it would pass through the celestial equator. That is, Earth's equator and the celestial equator are in the same plane. *Celestial latitude* is measured with respect to the celestial equator and is similar to latitude measurements for the surface of Earth, except that degrees toward the North celestial pole are labeled plus (+) and degrees toward the South celestial pole are negative (–). The second reference frame is the *celestial prime meridian.* This is an imaginary line running from the North celestial pole to the South celestial pole perpendicular to the equator, and intersecting the celestial equator at the point of the vernal (spring) equinox on the ecliptic. See Fig. 40.2. The *vernal equinox* is the point on the ecliptic where the Sun crosses the celestial equator from south to north on its yearly trip around the celestial sphere. The *ecliptic* is the apparent annual path of the Sun on the celestial sphere. *Celestial longitude* is measured with respect to the vernal equinox and is measured to the east along the celestial equator from the vernal equinox. The vernal equinox is taken as the starting point and labeled zero (0) hour.

The angular distance north or south of the celestial equator is called *declination.* The range of declination is from 0° to +90° for the northern part of the celestial sphere and from 0° to –90° for the southern part of the celestial sphere. The angular distance measured eastward from the vernal equinox is called *right ascension* and is measured in hours, minutes, and seconds. The range of right ascension is from 0 to 24 h (See Figure 40.2). For example: The star Arcturus has a declination of –20° and a right ascension of 14 h, 16 min. Similar coordinates are given for the location of every star.

The celestial sphere is used to portray the celestial sky for an observer located at any latitude on Earth, on any day of the year, at any time of day. To set the celestial sphere for a given latitude, a particular day, and a specific time of day, proceed as follows:

Step 1 To set the sphere for a particular latitude, use the position of the North Star with respect to the northern horizon. If you are in the northern hemisphere of Earth, the North Star will be visible. Therefore, set the North Star at an altitude equal to your latitude. *Altitude* is the angle between the line of sight to the star (in this case, the North Star) and the horizon. To do this, adjust the sphere by moving the brass ring, which supports the sphere, up or down until the latitude of the North Star is equal to your latitude. *Be sure to count the number of degrees on the brass ring between the North Star and the northern horizon.* The reading on the meridian may be correct, or it may be the complementary angle. To be sure, count the degrees. To set the sphere for a southern hemisphere latitude, place the North Star below the northern horizon, with the number of degrees equal to the observer's southern latitude. Again, count the number of degrees to be sure the setting is correct.

Step 2 By definition, when the Sun crosses the local overhead meridian in its daily apparent motion across the sky, it is 12 noon, local solar time. Hence, to portray the sky as seen at 12 noon, it is necessary to locate the Sun on the ecliptic and then turn the sphere until the Sun is at the overhead meridian. To do this, find the day of the year as marked on the celestial equator. Place this date under the overhead meridian. This automatically places the Sun under the overhead meridian. Remember, the Sun is on the ecliptic, not the celestial equator. The Sun will cross the celestial equator twice each year, once apparently going northward (vernal equinox) and again apparently going southward (autumnal equinox).

Step 3 Once the position of the sphere is set for 12 noon on a particular day, it is easy to set the sphere to correspond to any other hour of the same day by rotating the sphere westward 15° for each hour past noon, or eastward 15° for each hour before noon. Thus, any hour of the day can be established on the sphere. Because there are 60 min. in 1 h and 15° represents 1 h on the celestial sphere, each degree represents 4 min. of time. (Sixty min. divided by 15° = 4 min./degree.) Thus the celestial sphere can be set for any time of day to within a few minutes.

Once the celestial sphere has been set for a particular latitude, day, and time of day, everything above the horizon (the wooden ring) is visible to the observer. If it is nighttime, the stars can be seen. If it is daytime, the stars are there but the brightness of the sunlight prevents them from being seen.

APPARATUS

Celestial sphere and one small, ¼-in, circular disk, cut from masking tape (yellow preferred) to be used to represent the Sun's position on the ecliptic.

EXPERIMENT 40 NAME (print) ______________________ DATE ______________
LAST FIRST

LABORATORY SECTION ______ PARTNER(S) ______________________________

PROCEDURE 1

Place the celestial sphere on the laboratory table so that the overhead meridian is in a north–south direction, with the north pole of the sphere in the general direction of geographic north. The word *north* may appear on the north wall of the laboratory. Record the date of this experiment.

Today's date ____________

1. Set the celestial sphere to portray the sky as seen by an observer located at Earth's equator (0° latitude). Place Polaris on the northern horizon. Rotate the globe until the date you recorded above appears under the overhead meridian. This sets the celestial sphere for 0° latitude and 12 noon local solar time on this date. Stick the ¼-in circular disk of masking tape on the ecliptic where it passes under the overhead meridian. This represents the Sun and its position on the ecliptic.

   Determine the declination of the Sun.
   Declination of the Sun ____________

   Determine the right ascension of the Sun.
   Right ascension of the Sun ____________

   Determine the altitude of the Sun at 12 noon local time on this date.
   Altitude of the Sun above the horizon ____________

   Determine the altitude of the North Star (Polaris) at this time and date.
   Altitude of Polaris above the horizon ____________

2. Set the celestial sphere for 4 P.M. local time for the latitude (0°) and date given above. This is done by rotating the sphere westward 4 h or 60° from the 12 noon position. *Note:* Westward is to your left, if you are facing north. Look for the sign (north) on the wall of the laboratory. Also, the word *west* or W may be printed on the horizon ring.

   Determine the approximate altitude of the Sun at this time. State your answer in degrees above or below the western or eastern horizon. Example: 16° above western horizon.

   Altitude of the Sun at 4 P.M. local time ____________

   Altitude of Polaris at this time ____________

3. Set the celestial sphere to portray the sky as seen by an observer located at the North Pole (90° N) on today's date. Rotate the globe until the date recorded above appears under the overhead meridian. This sets the celestial sphere for 90°N and 12 noon local solar time on this date. Stick the ¼-in circular disk of masking tape on the ecliptic where the ecliptic passes under the overhead meridian. This represents the Sun and its position on the ecliptic.

Determine the altitude of the Sun and Polaris at 12 noon local time on today's date.

Altitude of the Sun at 12 noon local time ____________

Altitude of Polaris at this time ____________

4. Rotate the sphere westward until 8 P.M. is represented by the celestial sphere.

Altitude of the Sun at 8 P.M. local time ____________

Altitude of Polaris at this time ____________

5. Set the sphere to portray the sky as seen from Washington, D.C. (39°N), on the date of this experiment or use the latitude of your local city or town. Determine the altitude of the Sun at 12 noon local time. Use the small circular disk to represent the Sun.

Altitude of the Sun at 12 noon local time ____________

Altitude of Polaris at this time ____________

Determine the local time of sunrise and sunset. Sunrise occurs when the Sun is on the eastern horizon. Sunset occurs when the Sun is on the western horizon. Proceed as follows to determine the time of sunset. Place your forefinger on the Sun, which should be under the overhead meridian. The time is 12 noon local mean solar time. Rotate the sphere westward counting the hours (the hours are marked on the celestial equator) as they pass under the overhead meridian. Rotate the sphere until your finger touches the western horizon. The time of sunset is the number of hours you counted plus any minutes past 12 noon. To determine the time of sunrise proceed as follows: Place your forefinger on the Sun, which should be under the overhead meridian. Rotate the sphere eastward counting the hours as they pass under the overhead meridian. Rotate the sphere until your finger touches the eastern horizon. The time of sunrise is the number of hours counted plus any minutes subtracted from 12 noon.

Time of sunrise ____________

Time of sunset ____________

How many hours of daylight did Washington, D.C., or your home city have on the date of this experiment? Hours of daylight ____________

How many hours of darkness? Hours of darkness ____________

How many hours of daylight did the equator (0° latitude) have on this date? (Determine the time of sunrise and sunset.)

Sunrise ____________ Sunset ____________ Hours of daylight ____________

How many hours of daylight did the North Pole have on this date? ____________

How many hours of daylight did the South Pole have on this date? ____________

EXPERIMENT 40 NAME (print) ______________________________ DATE ______________________
LAST FIRST

LABORATORY SECTION ______ PARTNER(S) ____

State the regions of perpetual apparition and perpetual occultation. That is, state the regions on the celestial sphere that always appear (regions that never go below the horizon to an observer at 39°N latitude), and the regions that never appear (regions that never come above the horizon to an observer at this latitude). Rotate the sphere 360° and observe the regions that stay above and below the horizon. The answers will be in degrees. Example: 52° north to 90° north.

Area of Apparition ______________________

Area of Occultation ______________________

6. On June 21 and December 21 the Sun will be at its maximum declination north and maximum declination south, respectively. Set the celestial sphere to portray the sky from 39°N. Determine the declination of the Sun on June 21 and December 21. Determine the altitude of the Sun on these dates and the degrees north or south of due east where the Sun rises on these two dates, and the degrees north or south of due west where the Sun sets on these two dates.

Declination of the Sun (June 21) ______________

Altitude of the Sun on June 21 ______________

Degrees north or south of due east for sunrise on June 21 ______________

Degrees north or south of due west for sunset on June 21 ______________

Declination of the Sun (December 21) ______________

Altitude of the Sun on December 21 ______________

Degrees north or south of due east for sunrise on December 21 ______________

Degrees north or south of due west for sunset on December 21 ______________

PROCEDURE 2

1. Set the celestial sphere to portray the sky as seen from Washington, D.C., or your local city on the date of this experiment at 9:00 P.M., local solar time. Show the location of the Little Dipper, Big Dipper, and Cassiopeia by drawing them on the diagram, Figure 40.3 in this experiment. Remember, the observer is located at the center of the celestial sphere looking up and northward to see these constellations. Check your drawing by observing these constellations in the actual sky at 9:00 P.M., local solar time, on the date of this experiment if you can.
2. Determine the local solar time at which the Great Nebula in Andromeda (designated M31 on most celestial spheres) is on the overhead meridian. If your model does not show this notation, pick one of the brightest stars in this constellation.

____

3. State the declination and right ascension of the Galaxy M3 1 in Andromeda.

Declination ____________

Right ascension ____________

Note: M31 can be seen with the unaided eye on a clear night. Check the celestial sphere for the position of M31, then observe on a clear night in the actual sky if you can.

4. The brightest star in the sky is Sirius. At what local solar time will Sirius rise on this day at Washington, D.C., or your local city? ____________

5. At what local solar time will Sirius be on the overhead meridian? ____________

6. What is the altitude of Sirius at the local solar time of crossing the overhead meridian? ____________

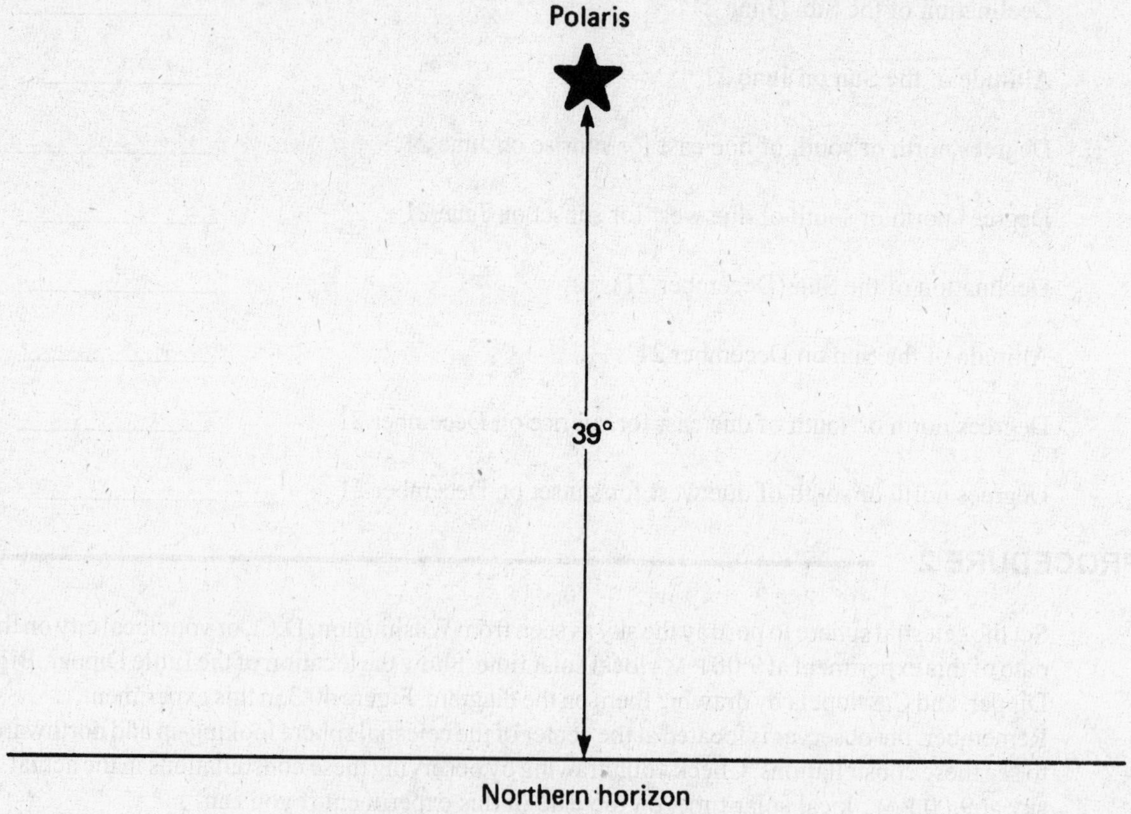

Figure 40.3 Student Diagram

Use the above space to draw the constellations given in Procedure 2, Step 1.

EXPERIMENT 40 NAME (print) ____________________ LAST ____________________ FIRST DATE ____________________

LABORATORY SECTION ______ PARTNER(S) ____________________

QUESTIONS

1. How does the declination of the Sun vary over 1 year?

2. Where on the ecliptic is the vernal equinox located?

3. How does the altitude of the North Star vary for an observer located at the equator? At 39°N?

4. Why was the globe rotated westward to position the celestial sphere to portray a P.M. local time?

5. Does the Sun ever set north of due west for an observer located at Washington, D.C. (39°N)? What time of year would this occur?

Experiment 41

Locating Stars in the Night Sky

INTRODUCTION

This experiment is a combination of in-lab work that will teach you about the stars in the night sky that can be observed during this season of the year and an actual observation of the sky that can be made at a "star party" held by your instructor or as an individual homework assignment. Experiment 40 is also about the stars. It explains the use of a celestial sphere. You can refer to Experiment 40 if you are not familiar with how to use a celestial sphere when you do Step 2 of this experiment. If a celestial sphere is not available, you can get the information needed for Data Tables 41.1 and 41.2 from a star atlas or by using handouts available from your instructor.

LEARNING OBJECTIVES

After completing this experiment, you should be able to do the following:

- List at least six constellations that are visible at this season of the year.
- Find constellations and stars on a celestial sphere and determine their right ascension, declination, and zenith angles.
- Explain how to use a star atlas to find information about the stars.
- Locate the prominent constellations and stars in the sky at this time of year and identify them.

APPARATUS

Star Charts 41.1–41.4, blank Star Charts 41.5–41.8, star atlas or handouts from instructor, celestial sphere.

PROCEDURE

1. Select the star chart provided that represents the current season of the year. Study it for about 10 minutes to get to know the location of the constellations that are now present in the night sky.
2. Take one of the blank star charts and fill in as much information about the night sky as you can remember from your previous 10-minute study. Do not refer back to the "full" star chart. You will get to correct any mistakes or omissions in the next step. Be as neat and as accurate as you can.
3. Now using the "full" star chart, fill in all of the details that you missed in Step 2. This procedure is designed to prepare you to successfully locate the principle stars in the night sky when you get outside under the stars.
4. Fill in the information in Data Table 41.1 for the stars listed in Table of Stars that appear in the current night sky. Use the celestial sphere to find this data. Set the celestial sphere for 9:00 P.M. local solar time on today's date and a location corresponding to your latitude. Use 40°N if you do not know your latitude. (If you do not have a celestial sphere, skip this step and go on to Step 5. You can fill in the data in Data Table 41.1 from a star atlas or from handouts passed out by your instructor.)

Table of Stars (Seasons of the Year)

This table lists the stars to use in Steps 4 and 5 for the various seasons of the year. You will be able to see those indicated for the current season when you actually perform sky observations later in this experiment (Step 6).

Fall	Winter	Spring	Summer
Aldeberan	Aldeberan	Arcturus	Altair
Algol	Betelgeuse	Capella	Antares
Altair	Capella	Castor	Arcturus
Capella	Castor	Polaris	Deneb
Deneb	Polaris	Pollux	Polaris
Formalhaut	Pollux	Procyon	Regulus
Polaris	Regulus	Regulus	Spica
Vega	Rigel	Sirius	Vega

EXPERIMENT 41 NAME (print) ______________________ DATE ______________
LAST FIRST

LABORATORY SECTION ______ PARTNER(S) ______________________________

DATA TABLE 41.1

Name of Star	Constellation	Right Ascension	Declination	Altitude	Zenith Angle	Compass Direction
Aldeberan						
Algol						
Altair						
Capella						
Deneb						
Formalhaut						
Polaris						
Vega						

Note: Altitude + zenith angle = 90°

5. Using one of the star atlases, look up the stars used in Data Table 41.1. Check the right ascension and declination values to make sure the ones you found on the celestial sphere are correct, and fill in the other data in Data Table 41.2.

DATA TABLE 41.2

Name of Star	Right Ascension	Declination	Apparent Magnitude	Luminosity Compared to Sun	Spectral Class

6. Fill in the names of the stars and the constellations in Data Table 41.3 using Table of Stars and the appropriate seasonal star chart. Then go outside and look at the stars in the night sky some evening when the weather is clear. Fill in the remainder of the data in Data Table 41.3 from your own personal observations. You will not need a telescope; you can see all of these stars with the unaided eye on a clear night.

DATA TABLE 41.3

Name of Star	Constellation	Approximate Zenith Angle	Approximate Compass Direction	Apparent Color of Star	Apparent Magnitude of Star

7. Your instructor will tell you if any planets are in position to be observed at this time. If there are any planets visible, you can observe them as you study the stars. Answer the following questions.

1. What planet(s) are visible tonight? ____________________

2. In which constellation(s)? ____________________

3. Your estimate of planet(s) brightness: ____________________

4. Apparent color of planet(s): ____________________

EXPERIMENT 41 NAME (print) ______________________________ DATE ______________
LAST FIRST

LABORATORY SECTION _______ PARTNER(S) __________________________________

Star Chart 41.1 (Full)
Sky as Seen from 40°N Latitude

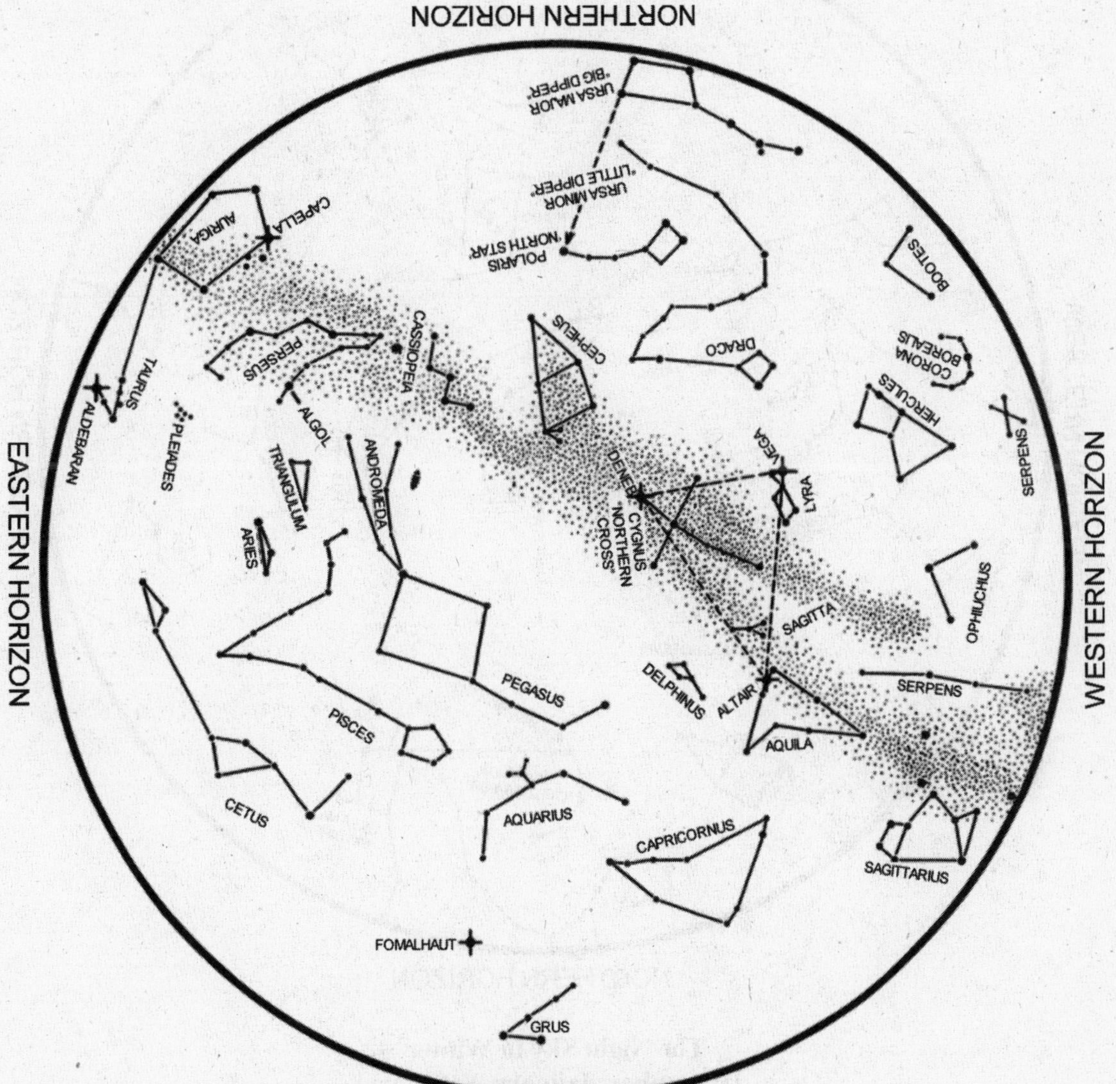

The Night Sky in Fall
(September, October, November)
11 P.M. 9 P.M. 7 P.M.

Star Chart 41.2 (Full)
Sky as Seen from 40°N Latitude

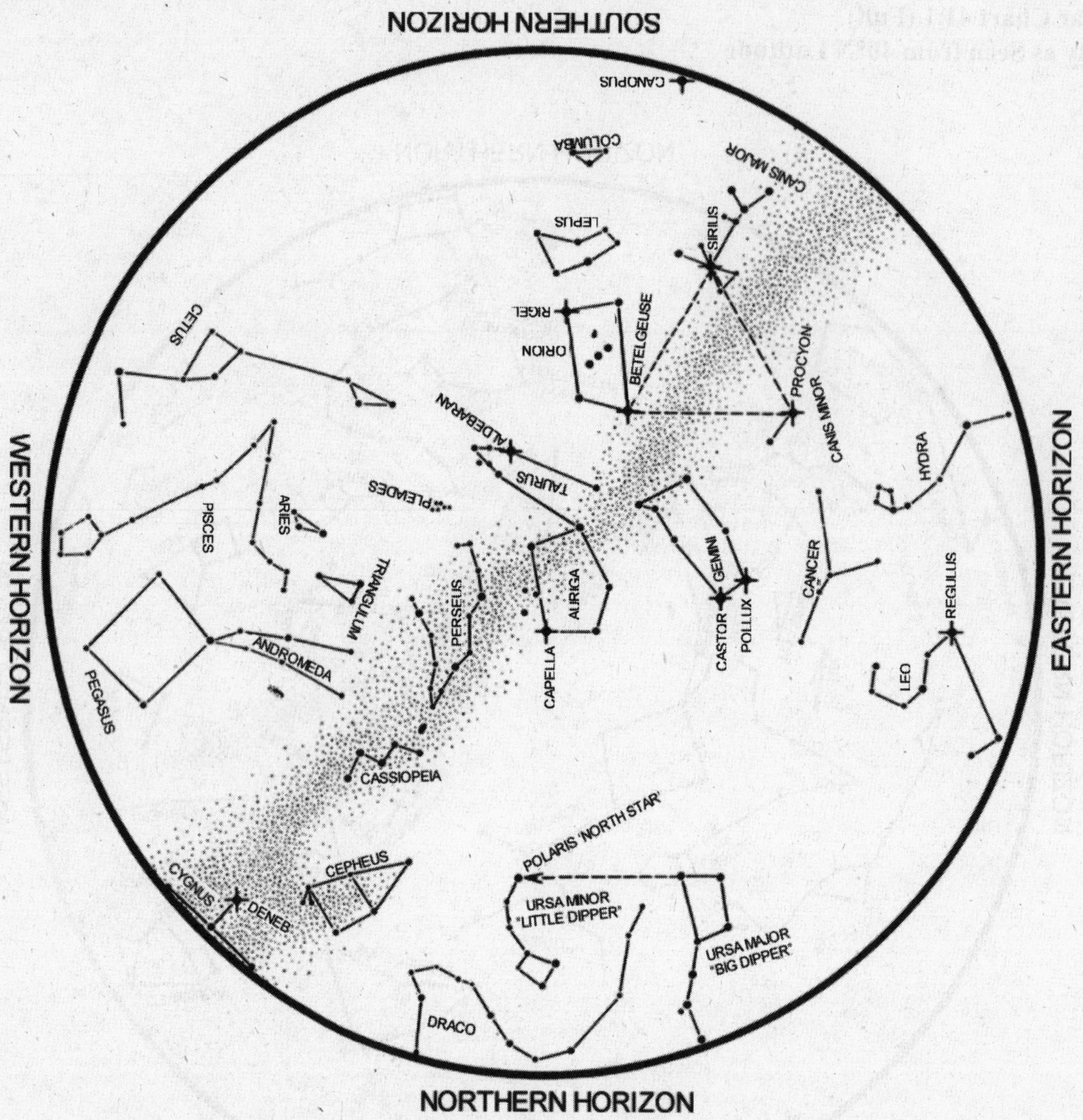

The Night Sky in Winter
(December, January, February)
11 P.M. 9 P.M. 7 P.M.

EXPERIMENT 41 NAME (print) ______________________________ DATE ______________
LAST FIRST

LABORATORY SECTION _______ PARTNER(S) ______________________________

Star Chart 41.3 (Full)
Sky as Seen from 40°N Latitude

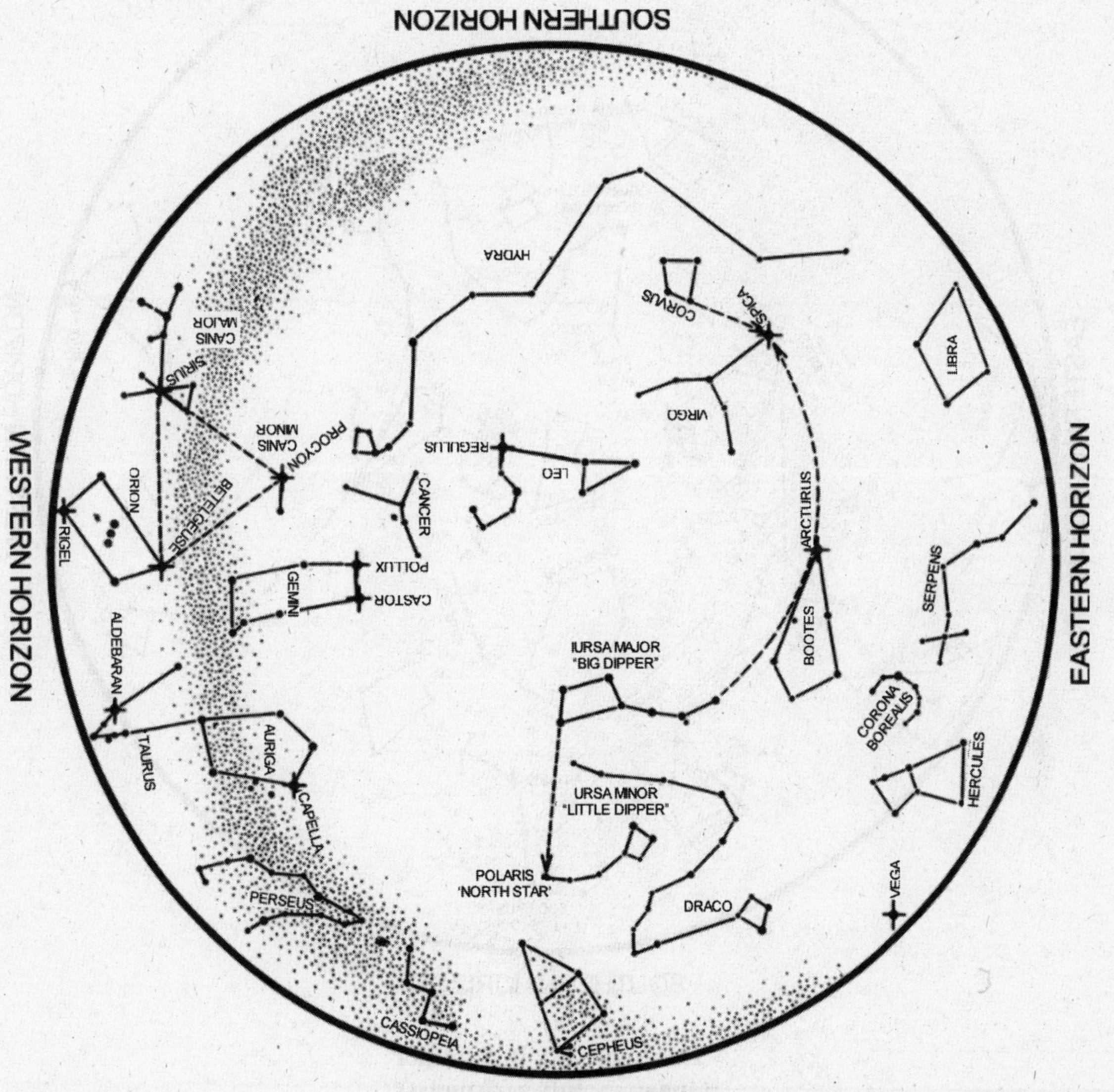

The Night Sky in Spring
(March, April, May)
11 P.M. 9 P.M. 7 P.M.

Star Chart 41.4 (Full)
Sky as Seen from 40°N Latitude

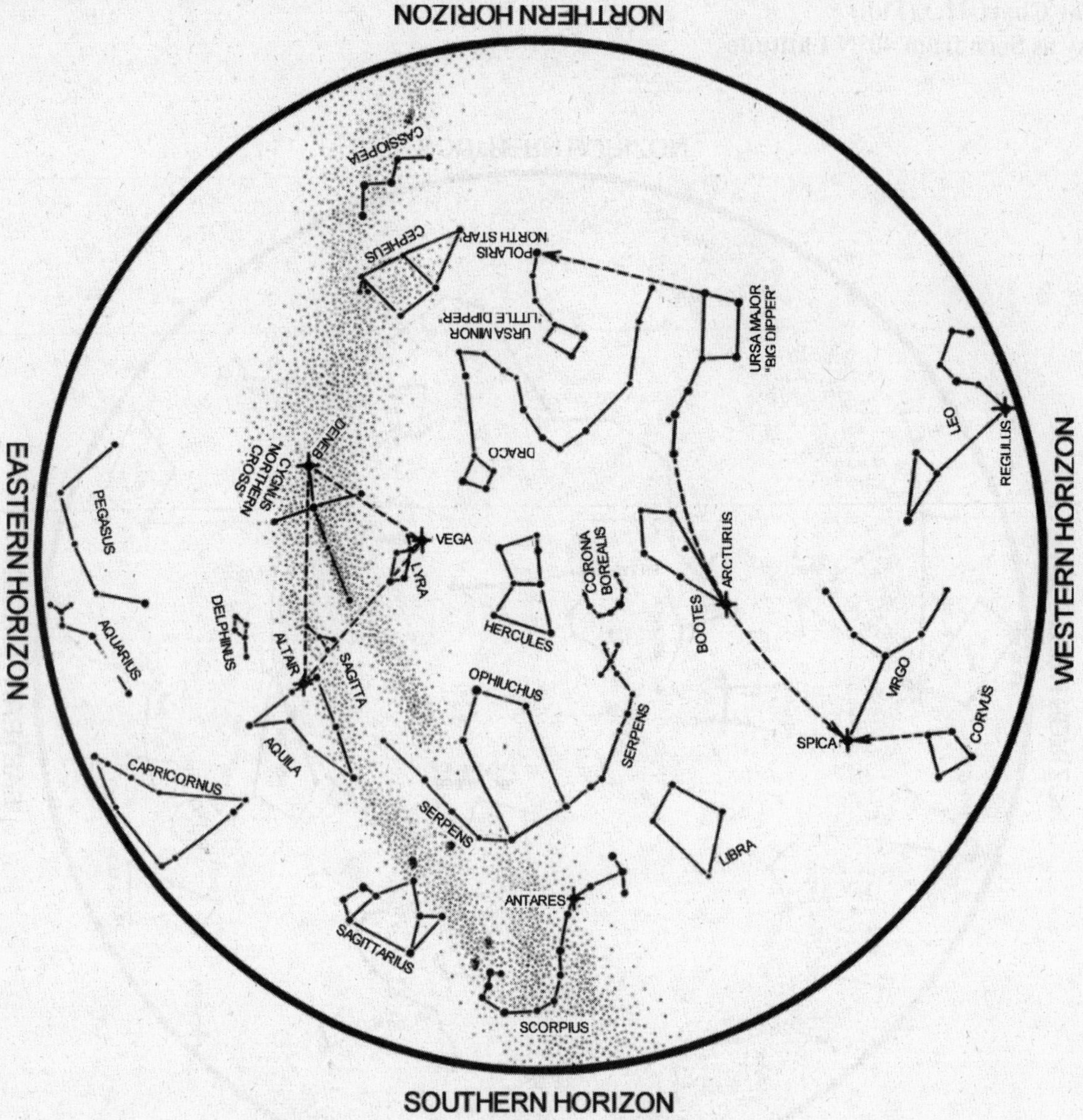

The Night Sky in Summer
(June, July, August)
11 P.M. 9 P.M. 7 P.M.

EXPERIMENT 41 NAME (print) ______________________ DATE ______________
LAST FIRST

LABORATORY SECTION _______ PARTNER(S) ______________________________

Star Chart 41.5 (Blank)
Sky as Seen from 40°N Latitude

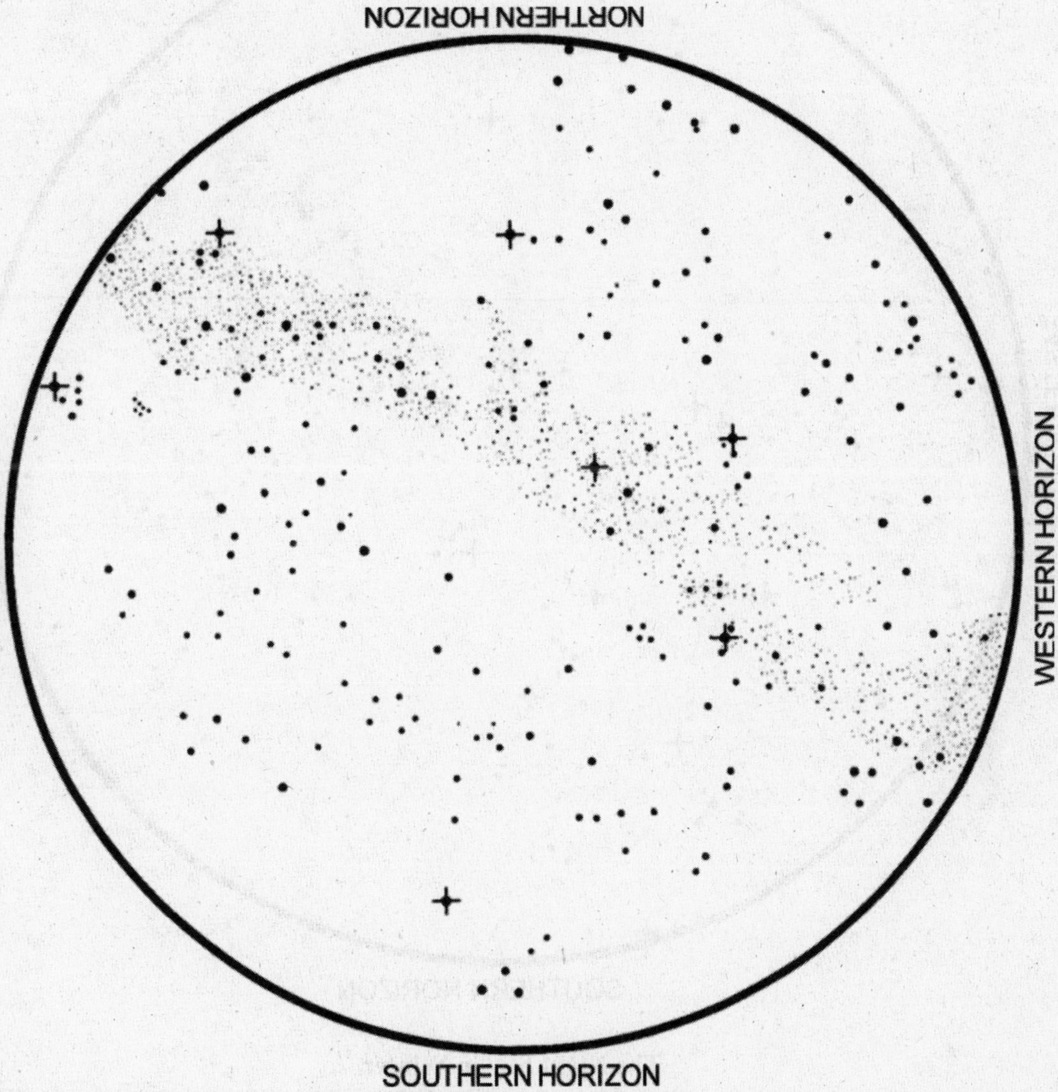

The Night Sky in Fall
(September, October, November)
11 P.M. 9 P.M. 7 P.M.

Star Chart 41.6 (Blank)
Sky as Seen from 40°N Latitude

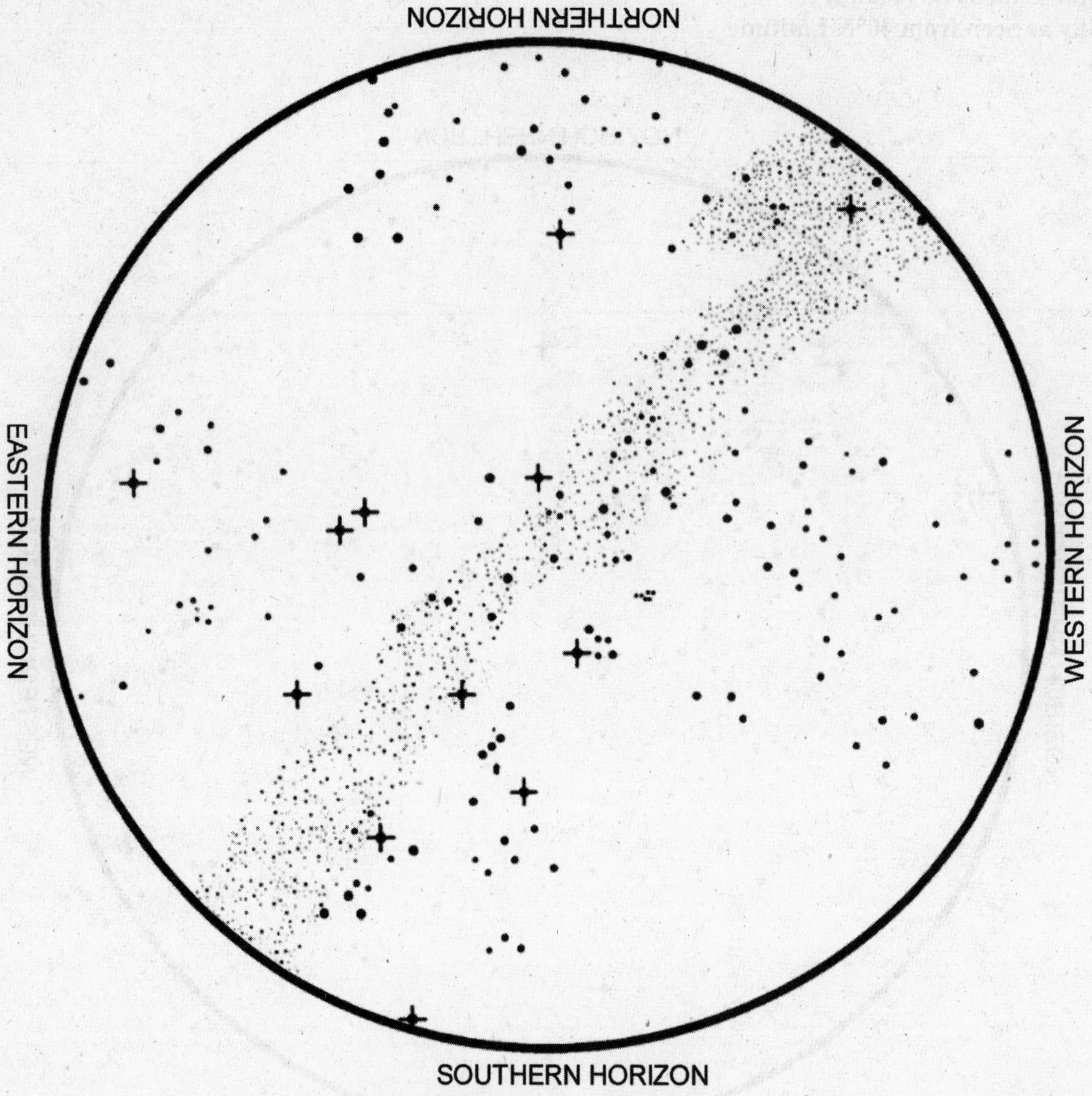

The Night Sky in Winter
(December, January, February)
11 P.M. 9 P.M. 7 P.M.

EXPERIMENT 41 NAME (print) ______________________ DATE ______________
LAST FIRST

LABORATORY SECTION _______ PARTNER(S) ______________________________

Star Chart 41.7 (Blank)
Sky as Seen from 40°N Latitude

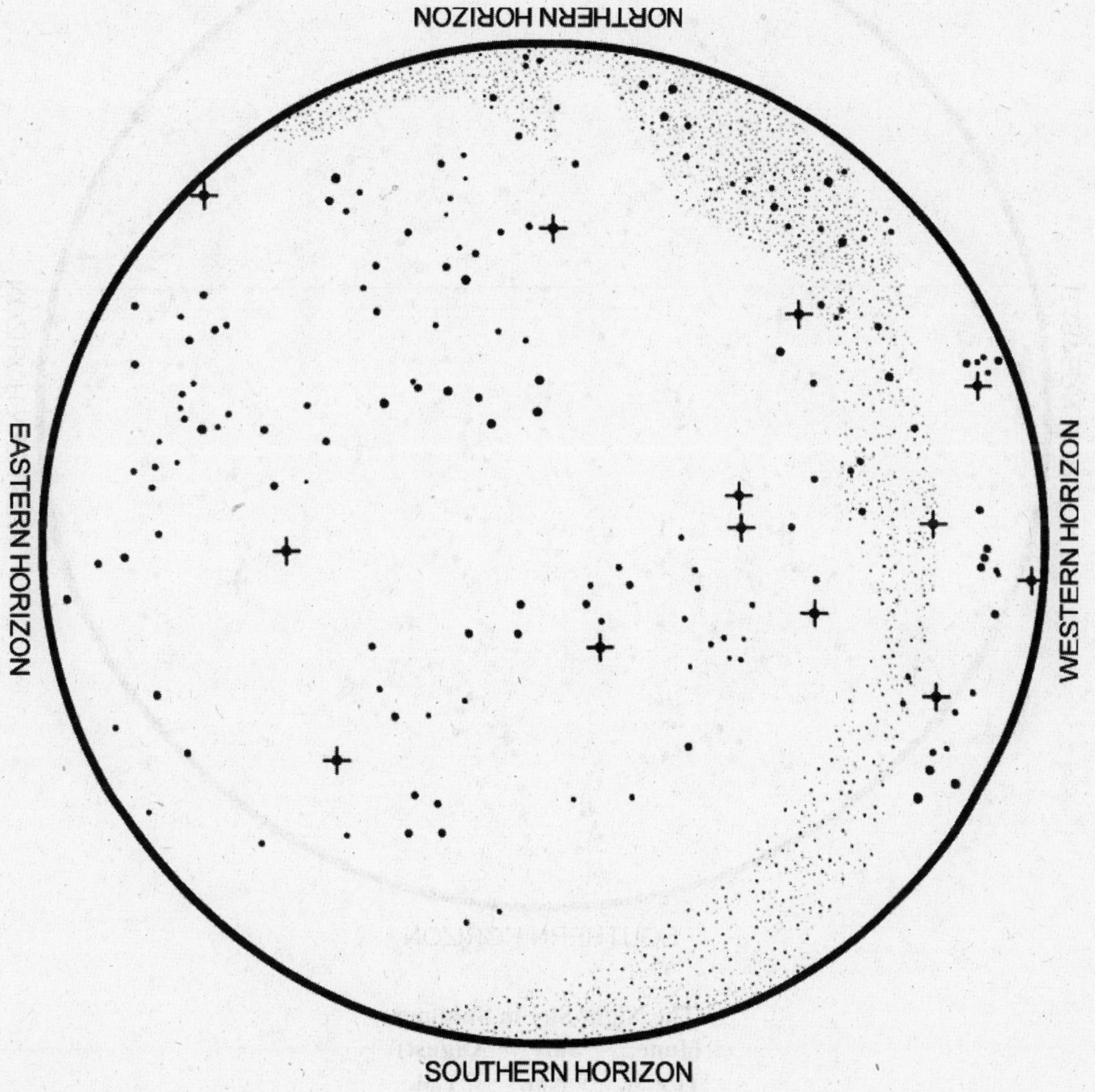

The Night Sky in Spring
(March, April, May)
11 P.M. 9 P.M. 7 P.M.

Star Chart 41.8 (Blank)
Sky as Seen from 40°N Latitude

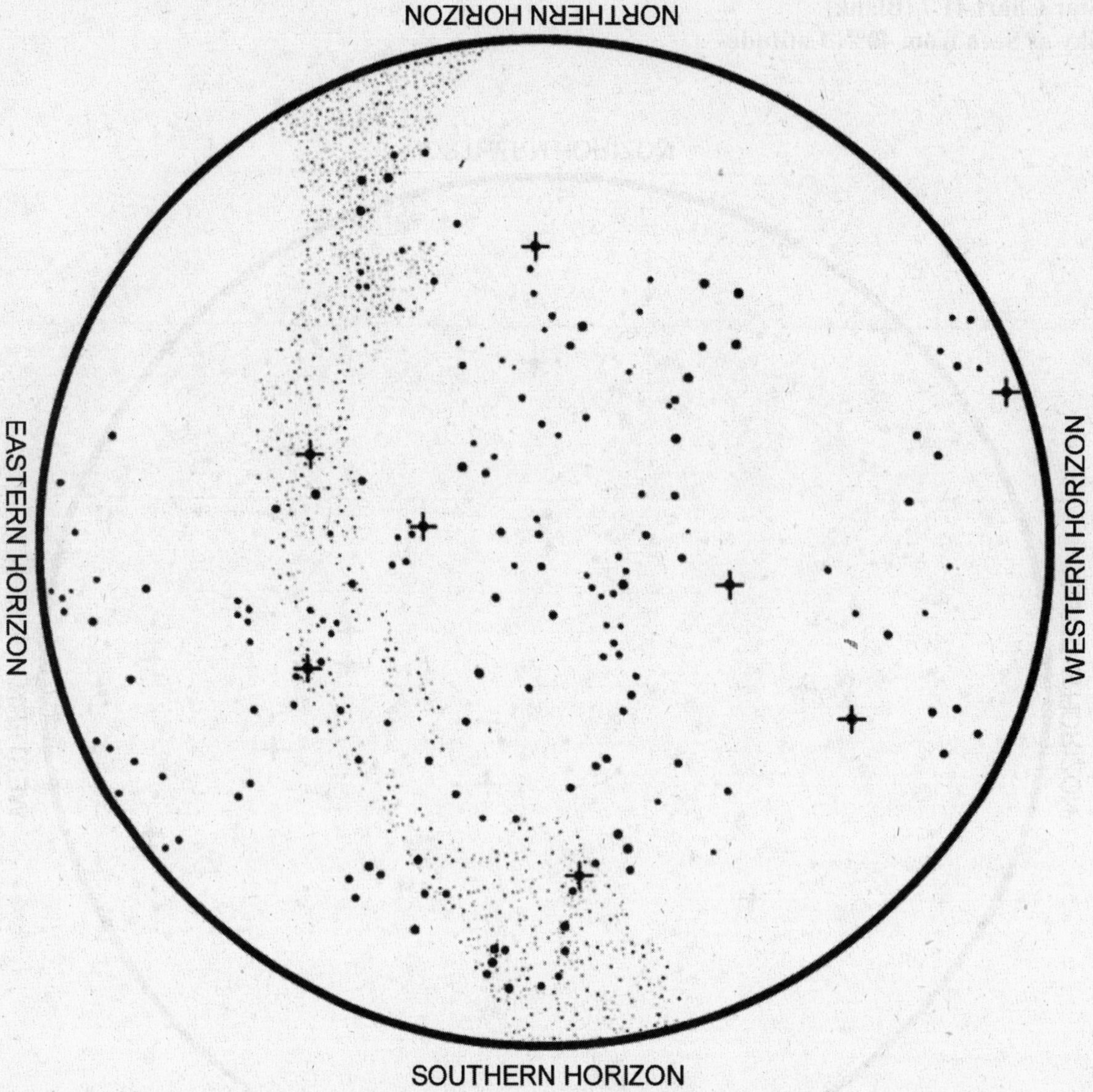

The Night Sky in Summer
(June, July, August)
11 P.M. 9 P.M. 7 P.M.

Experiment 42

Motions and Phases of the Moon

INTRODUCTION

The Moon is a satellite of Earth that makes one revolution eastward around Earth every 29½ solar days. The orbit is an ellipse inclined 5° to the ecliptic and crossing the ecliptic twice each revolution.

By definition, when the Moon is on the same meridian with the Sun, it is in new phase. Since the dark side of the Moon is toward Earth at this time, the new moon cannot be seen by the observer on Earth (see Fig. 42.1). When the Moon is 90° east of the Sun, the Moon is in the first-quarter phase. At this time it will appear on the meridian at 6:00 P.M., local solar time, in the form of one-half of a fully lighted disk. As the Moon revolves eastward, it appears larger in face size until at 180° east of the Sun, the Moon is in full phase, and appears on the meridian at 12 midnight, local solar time, as a fully lighted disk. Another 90° eastward will locate the Moon 270° east of the Sun, or 90° west of the Sun. The Moon now appears on the meridian at 6:00 A.M., local solar time, as a quarter moon. This phase is known as third-quarter or last-quarter moon.

Although the moon's orbit is inclined at an angle of 5° with the ecliptic plane, for simplicity in this experiment we shall assume that the Moon travels in the ecliptic plane.

LEARNING OBJECTIVES

After completing this experiment, you should be able to do the following:

- ▼ State the different phases of the Moon and tell the local times when they will be seen on the overhead meridian.
- ▼ With the use of the celestial sphere determine the phase of the Moon, and its approximate altitude, when the time of the meridian crossing is known.

APPARATUS

Celestial sphere and a small circular disk cut from masking tape as called for in Experiment 40 to represent the Sun plus a half-circular disk of the same diameter for representing the Moon's position on the ecliptic.

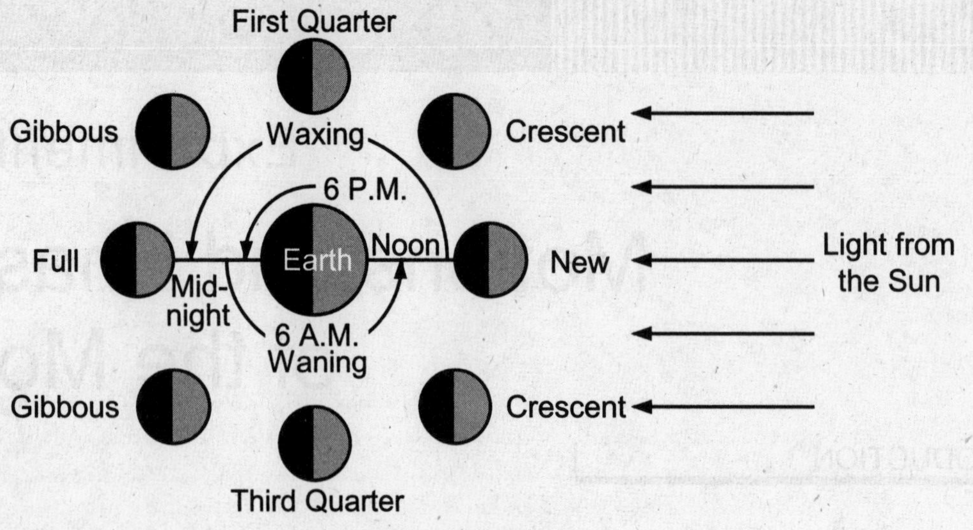

Figure 42.1 Phases of the Moon as observed from a position in space above the north pole of Earth.

EXPERIMENT 42 NAME (print) ______________________ DATE ______________
LAST FIRST

LABORATORY SECTION ______ PARTNER(S) ______________________________

PROCEDURE 1

1. Set the celestial sphere to portray the sky as seen from Washington, D.C. (39°N), or from your local city on December 22. Use the two pieces of masking tape to show the positions of the Sun and the Moon in the following procedure:

   (a) Using the celestial sphere, determine the altitude of the new moon as observed from Washington or your local city if the Moon was in the new phase on December 22.

   Altitude when the new Moon is on the overhead meridian ____________

   (b) Seven and three-eighths days later, the Moon would be in the first-quarter phase. For simplicity, assume that the Sun remains at the same declination (23½ S) and determine the altitude, rising time, and setting time of the first-quarter moon.

   Altitude when the first quarter Moon is on the overhead meridian ____________

   Local solar time the first quarter Moon rises ____________

   Local solar time the first quarter Moon sets ____________

   (c) At the end of another 7⅜ days, the Moon will be in the full phase. Again, in order to make solving the problem easier, assume that the Sun remains at the same declination. Determine the altitude, rising time, and setting time of the full moon.

   Altitude when the full Moon is on the overhead meridian ____________

   Local solar time the full Moon rises ____________

   Local solar time the full Moon sets ____________

   (d) As the Moon revolves eastward, it will be in the third-quarter or last-quarter phase another 7⅜ days later. Assuming that the Sun remains at 23½°S, find the altitude, rising time, and setting time of the last-quarter moon.

   Altitude when the third quarter Moon is on the overhead meridian ____________

   Local solar time the third quarter Moon rises ____________

   Local solar time the third quarter Moon sets ____________

2. Obtain from the instructor in charge of the laboratory the date in the current month when the new moon occurs. (Use this date throughout this question.)

   Date of the new moon ____________

(a) Using the celestial sphere, determine the maximum altitude of the new moon as observed from Washington or your local city on this date.

Altitude when the new Moon is on the overhead meridian ____________

Note: For this series of problems, we shall *not assume* that the Sun remains at the same declination, as we did in Procedure 1. The Sun appears to move about one degree per day. Allow for this movement as you solve the following problems. To find the declination of a particular phase of the Moon, remember its position in respect to the Sun, as given in the introduction to this experiment.

(b) Seven and three-eighths days after the date of the new moon, the Moon will be in the first-quarter phase. Solve for the following information using previously learned facts and the celestial sphere.

Date of first-quarter moon ____________

Local solar time first-quarter moon is on the overhead meridian ____________

Standard time first-quarter moon is on the overhead meridian ____________

Altitude of first-quarter moon when on the overhead meridian ____________

Rising time (local solar) of first-quarter moon ____________

Rising time (standard) of first-quarter moon ____________

Setting time (local solar) of first-quarter moon ____________

Setting time (standard) of first-quarter moon ____________

(c) With the passing of another 7⅜ days, the Moon will be in full phase. Solve for the following information using previously learned facts and the celestial sphere.

Date of full moon ____________

Local solar time full moon is on the overhead meridian ____________

Standard time full moon is on the overhead meridian ____________

Altitude of full moon when on overhead meridian ____________

Rising time (local solar) of full moon ____________

Rising time (standard) of full moon ____________

Setting time (local solar) of full moon ____________

Setting time (standard) of full moon ____________

EXPERIMENT 42 NAME (print) ______________________ DATE ______________
LAST FIRST

LABORATORY SECTION ______ PARTNER(S) ______________________________

Degrees north or south of due east at which full moon rises ____________

Degrees north or south of due west at which full moon sets ____________

(d) When the Moon has revolved to 270° east of the Sun, it will be in the last-quarter phase. Solve for the following information using previously learned facts and the celestial sphere.

Date of last-quarter moon ____________

Local solar time last-quarter moon is on the meridian ____________

Standard time last-quarter moon is on the meridian ____________

Altitude of last-quarter moon when on overhead meridian ____________

Rising time (local solar) of last-quarter moon ____________

Rising time (standard) of last-quarter moon ____________

Setting time (local solar) for last-quarter moon ____________

Setting time (standard) for last-quarter moon ____________

QUESTIONS

1. During what month will the full moon achieve its maximum altitude, as observed from Washington, D.C. (39°N)? When will it have its minimum altitude?

2. Is it possible to see a waxing crescent moon on the overhead meridian at 10 A.M. local time? How about at 4 P.M. local time? Explain your answers.

3. In this experiment, what is the maximum possible error in degrees that could be made in determining the Moon's altitude by assuming the Moon traveled in the same plane as the ecliptic?

4. Assuming the Moon actually travels in the *same plane* as the ecliptic, how many total lunar eclipses would occur each year? How many total solar eclipses?

Experiment 43

Observing the Phases of the Moon

INTRODUCTION

This experiment is a long-term project in which you will observe the phases of the Moon and keep a log of its position in the sky along with the date and time of your observations. To best study the complete phase cycle of the Moon, your observations should be made over at least four weeks, but the work can be done in as short a period as two weeks if time is limited.

Experiment 42 also deals with the phases of the Moon, but uses a celestial sphere for gathering the data. You may be assigned to do Experiment 42 or at least look it over before you complete your work on this experiment. The two experiments complement each other nicely; this one is based on actual observation of the Moon in the sky, while the previous one is an in-class study of the same general material.

The Moon revolves around Earth once every 27.3 sidereal days, but because Earth is also moving in its orbit around the Sun, the cycle of the phases of the Moon as seen from Earth is about two days longer than this. The lunar phases follow a cycle called the *synodic period,* which is 29.5 days. The Moon's orbit is inclined five degrees to the plane of the ecliptic so that even though the Moon's path crosses the ecliptic plane twice each revolution, eclipses of the Moon and the Sun do not occur every month.

The Moon's surface is illuminated by sunlight and one-half of the lunar surface is illuminated at all times. The fractional portion of this lighted hemisphere that can be seen from Earth depends on the relative positions of the Sun, Earth, and the Moon when the observation is being made. As this fractional portion of the Moon's lighted surface changes over time, an observer on Earth sees the monthly cycle of lunar phases.

When the Moon is on your overhead meridian at the same time as the Sun, the moon is in the *new phase.* This occurs at 12 o'clock noon local solar time (LST). Because the dark side of the Moon is toward Earth and the Sun is at the same location in the sky at that time, the Moon is quite difficult to observe.

Over the next 7⅜ days, the position of the Sun and the Moon will shift relative to Earth in such a way that an increasing amount of the lighted portion of the lunar surface will become visible from Earth. At first only a small crescent of the lighted surface can be seen, but this steadily increases until half of the surface facing Earth is lighted. During this week, the Moon is said to be in the *waxing crescent phase.*

When the Moon is 90 degrees east of the Sun, one-half of its illuminated surface is visible, and the Moon is then said to be in *the first-quarter phase.* At this time, it will appear on an observer's overhead meridian at 6:00 P.M. LST. After the first-quarter phase, the lighted portion of the lunar surface that is visible from Earth continues to increase over the next 7⅜ days and the Moon is said to be in the *waxing gibbous phase.*

When the entire hemisphere of the Moon facing Earth is lighted, the observer will see a *full phase.* This happens when the moon reaches a position 180 degrees east of the Sun and appears on your overhead meridian at 12 o'clock midnight LST as a fully illuminated disk. After full phase, the portion of lighted lunar surface as seen from Earth decreases. This portion of the monthly lunar cycle is called the *waning gibbous phase.*

At 270 degrees east of the Sun (or 90 degrees west of the Sun), the Moon appears on the meridian at 6:00 A.M. LST, and has reached the *third-quarter* or *last-quarter phase.* At this time, one-half of its lighted surface is visible again from Earth. As the Moon continues its eastward motion, it will eventually be on the overhead meridian at the same time as the Sun and thus return to the *new moon phase.* During the period between the third-quarter and new phase, the lighted portion of the Moon visible from Earth is still decreasing, and this portion of the cycle is called the *waning crescent phase.*

The Moon is only in its new, first-quarter, full, and last-quarter phases for an instant. These are well defined positions relative to the Sun and Earth. Therefore, the Moon is in its waxing crescent, waxing gibbous, waning gibbous, or waning crescent phase most of the time during any month. Each of the crescent or gibbous phases will last about one week. Figure 43.1 shows the relative positions of Earth, the Sun, and the Moon and the local solar times at which the Moon crosses an observer's overhead meridian.

LEARNING OBJECTIVES

After completing this experiment, you should be able to do the following:

- List the phases of the Moon and tell the approximate local solar time when they will be on your overhead meridian.
- Locate the Moon in the sky at various times of day depending on its phase.
- Identify the phase of the Moon that you are observing.
- Tell the approximate rising and setting times for the various phases of the Moon.

APPARATUS

Notebook to record observational data, small plumb bob and protractor to aid in measuring angles like altitude, zenith angle, etc.

EXPERIMENT 43 NAME (print) ______________________________ DATE ______________
LAST FIRST

LABORATORY SECTION ______ PARTNER(S) ____

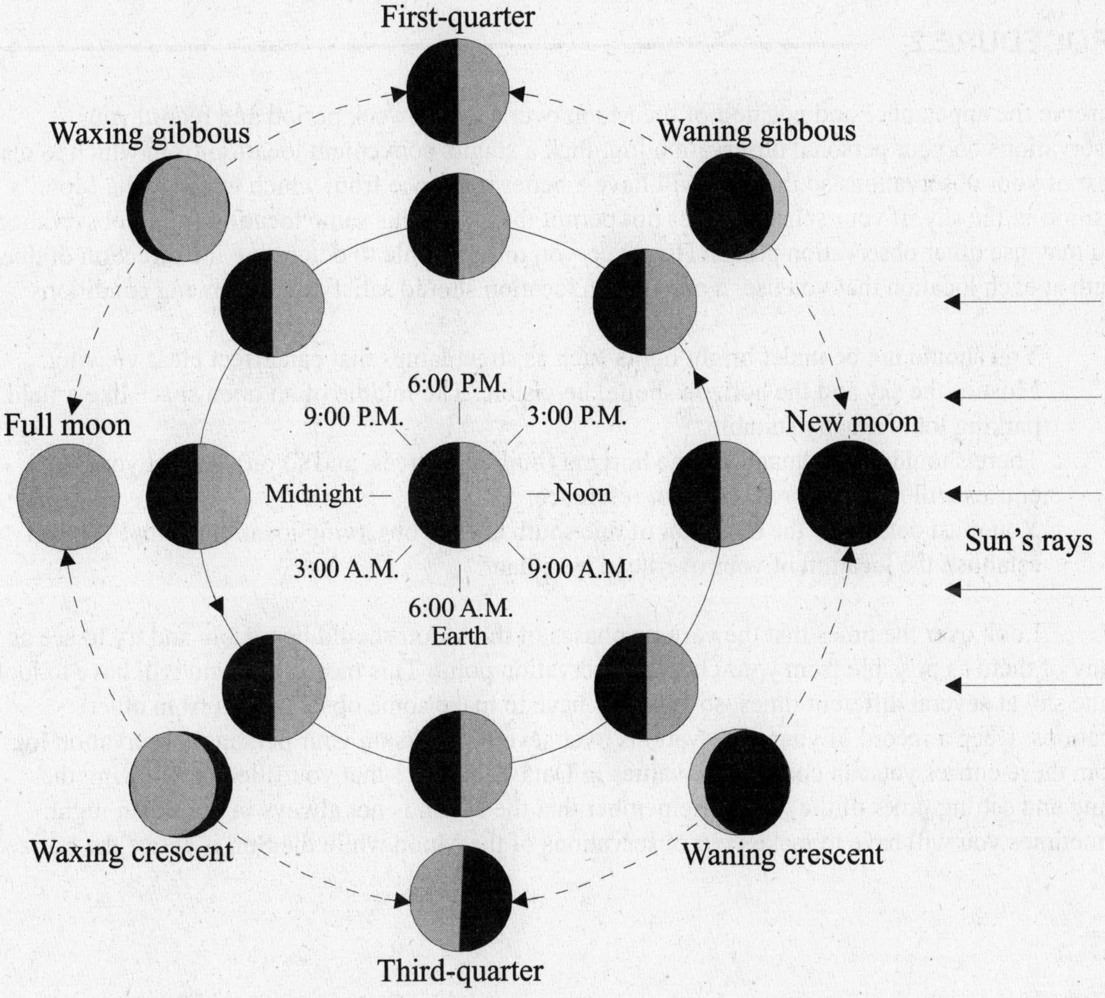

Figure 43.1 Phases of the Moon. The inner circle of spheres in the diagram shows the relative positions of the Sun (off the right hand edge of the page), Earth, and the Moon. The outer spheres show the Moon as seen from the Earth's surface. The Moon is on the meridian at the local solar times indicated.

PROCEDURE 1

Use Fig. 43.1 and complete Data Table 43.1, which can be done anytime after the experiment has been assigned.

PROCEDURE 2

Observe the appearance and position of the Moon over a 2 to 6-week period and record your observations on your personal observation log. Pick a single, convenient location from which to make most of your observations so that you will have a better reference from which to judge the Moon's position in the sky. If your schedule does not permit the use of this same location for all observations, you may use other observation points. However, you must be able to determine the direction of due south at each location that you use. Your chosen location should satisfy the following conditions:

1. You should not be under bright lights such as street lamps that can affect clear viewing.
2. Most of the sky and the horizon should be visible. The middle of an open space like a field or parking lot is usually suitable.
3. There should be landmarks on the horizon (buildings, trees, and so on), so that your log entries will have a more consistent reference.
4. You must determine the direction of due south at each observing location so that you can establish the location of your overhead meridian.

Look over the times that the various phases of the Moon should be visible and try to see as many of them as possible from your chosen observation point. This means that you will have to look at the sky at several different times, so you may have to make some observations from other locations. Keep a record of your observations over several weeks on your personal observation log. From these entries you can confirm the values in Data Table 43.1 that you filled in, showing the rising and setting times of the Moon. Remember that the Moon is not always in the sky at night. Sometimes you will have to make your observations of the Moon while the Sun is above the horizon.

EXPERIMENT 43 NAME (print) ________________________________ DATE ____________________
LAST FIRST

LABORATORY SECTION ______ PARTNER(S) ____

Assume a 12-hour period for the Moon to be above the horizon. This is, of course, not strictly true for all seasons of the year, but it will give you a rough guide as to when to expect the moon to rise or set in its various phases. Note: For the waxing and waning phases, the times in the table will be a RANGE of time, not a single hour.

DATA TABLE 43.1 Approximate Local Solar Times of Moonrise and Moonset

Phase	Moon Rises	Moon on Overhead Meridian	Moon Sets
New moon	6:00 A.M.	Noon	6:00 P.M.
Waxing crescent	6:00 A.M. to Noon		
First quarter			
Waxing gibbous		6:00 P.M. to Midnight	
Full moon			
Waning gibbous			
Third quarter			
Waning crescent			6:00 P.M. to Midnight

QUESTIONS

1. Why does an observer on Earth always see the same side of the Moon's surface?

2. During which phase of the Moon can a lunar eclipse occur? During which phase can a solar eclipse occur?

3. Make a drawing illustrating the positions of the Sun, the Moon, and Earth during a total solar eclipse. Show the shadows of both the Moon and Earth. (It is not necessary to make the drawing to scale.)

4. The term *terminator* refers to the boundary line dividing day and night on the surface of a planet or moon. In this experiment, it is the line between the bright and dark side of the Moon. State the phase of the Moon when the terminator is the (a) sunrise line, (b) sunset line.

Experiment 44

Hubble's Law and the Expanding Universe

INTRODUCTION

Most astronomers presently support the theory of an expanding universe. That is, the galaxies or clusters of galaxies that make up the universe are moving away from one another. One reason for believing this theory is the shift in their spectral lines toward the red end (longer wavelengths) of the electromagnetic spectrum. This red shift is known as the *cosmological red shift.*

In 1929, Edwin P. Hubble, American astronomer, published a paper in the *Proceedings of the National Academy of Sciences* indicating that the radial velocities of some observed galaxies are roughly proportional to their distances. Hubble's discovery, now known as *Hubble's law,* can be written as

$$v = Hd \qquad \text{Eq.(44.1)}$$

where v = recessional velocity of the galaxy
d = distance of galaxy from the observer
H = a constant called the *Hubble constant*

The Hubble constant is believed to have a value of 45 to 90 kilometers per second per million parsecs. For example: If H is 50 km/s/mpc, the observed galaxy is moving away from the observer 50 km/s for every 1 million parsecs (megaparsecs) the galaxy is from the observer. Hubble's law is extremely important because it gives astronomers vital information about the structure of the universe.

LEARNING OBJECTIVES

After completing this experiment, you should be able to do the following:

- State Hubble's law.
- Determine an experimental value for the Hubble constant.
- Calculate the approximate age of the universe.

APPARATUS

Hand calculator, distance versus red shift diagrams (Fig. 44.1).

Table of Useful Conversion Factors
(Additional conversion factors are inside the Manual's back cover.)

1 light-year (ly)	$= 9.46 \times 10^{12}$ kilometers (km)
1 parsec (pc)	$= 3.26$ ly
1 parsec (pc)	$= 3.09 \times 10^{13}$ km
1 megaparsec (mpc)	$= 3.09 \times 10^{19}$ km
1 megaparsec (mpc)	$= 1.00 \times 10^{6}$ parsec (pc)
1 day	$= 86{,}400$ seconds
1 year	$= 365$ days

EXPERIMENT 44 NAME (print) ______________________________ DATE ______________________
LAST FIRST

LABORATORY SECTION ________ PARTNER(S) ______________________________________

PROCEDURE 1

Figure 44.1 is a collection of photographs of five galaxies, their spectra, plus two bright-line spectra for comparing wavelengths. From top to bottom at the left, the photographs show the galaxies at increasing distance from the observer. On the right side are spectra showing H and K calcium lines. They are the two dark vertical lines in the spectrum. The spectrum of each galaxy is bordered, top and bottom, by two bright-line spectra for wavelength reference. The photographs also give the distance to each galaxy in light-years and the recessional velocity in kilometers per second.

In this experiment one objective is to determine the value of Hubble's constant in kilometers per second per megaparsec. To accomplish this, we must convert the distance presented in light-years in Fig. 44.1 to megaparsecs. The Table of Useful Conversion Factors located under the apparatus section may be used here.

Step 1 Data Table 44.1 gives the distances to the five galaxies in light-years. These distances were obtained from Fig. 44.1. Convert these distances to megaparsecs and record in Data Table 44.1. Example:

$$7.8\times10^{7}\ \text{ly} = 7.8\times10^{7}\ \cancel{\text{ly}}\times\frac{1\ \cancel{\text{pc}}}{3.26\ \cancel{\text{ly}}}\times\frac{1\ \text{mpc}}{10^{6}\ \cancel{\text{pc}}} = 24\ \text{mpc}$$

Step 2 Record the recessional velocities, given in Fig. 44.1, in the data table.

DATA TABLE 44.1

	Distance in Light-Years	Distance in Megaparsecs	Recessional Velocity km/s	Hubble's Constant km/s/mpc
Virgo	7.80×10^{7} =			
Ursa Major	1.00×10^{9} =			
Corona Borealis	1.40×10^{9} =			
Bootes	2.90×10^{9} =			
Hydra	3.96×10^{9} =			
			Average value for Hubble's constant	

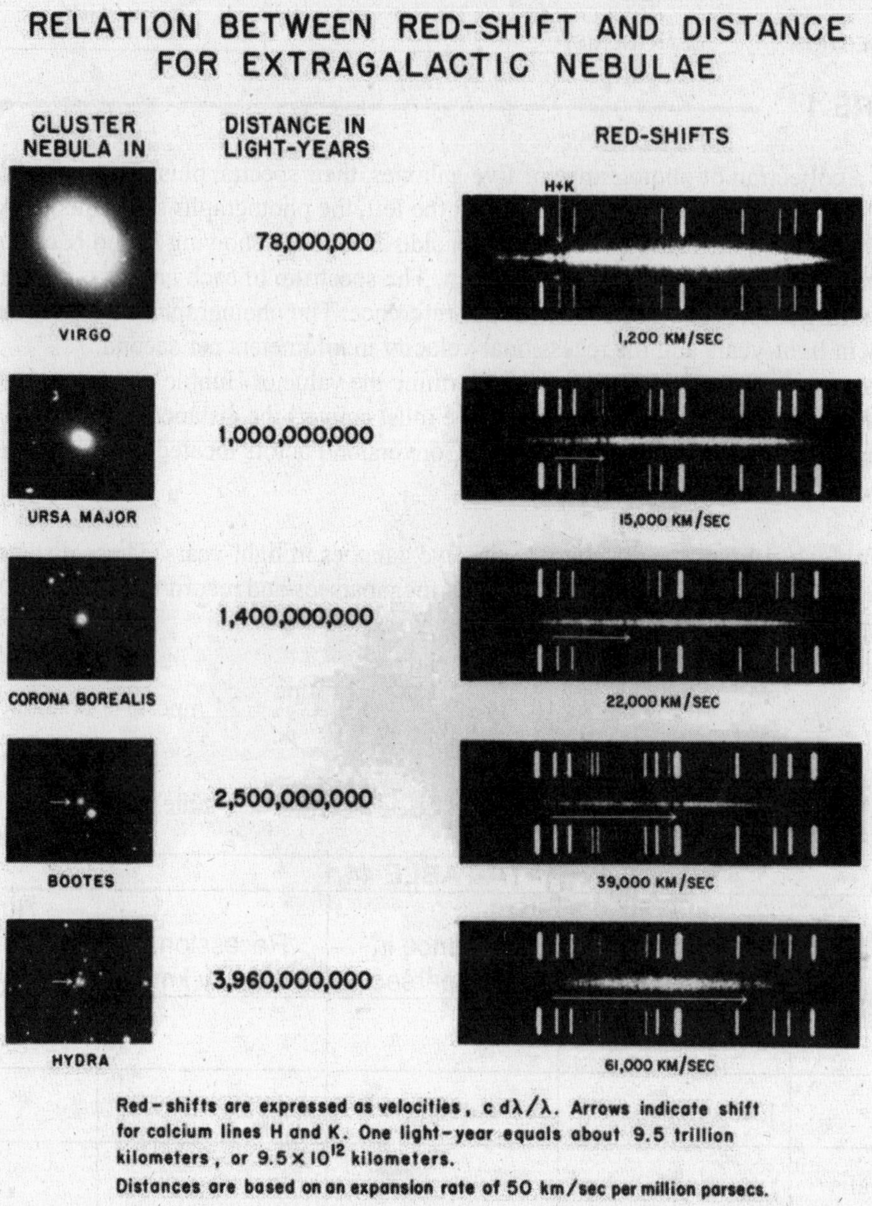

Figure 44.1 Relationship between red shift and distance for extragalactic nebulae. On the left are photographs of five individual elliptical galaxies. From top to bottom the photographs show the galaxies at increasing distance from the observer. On the right, the spectrum (the broad white band) of each galaxy is shown between an upper and lower comparison spectrum. The H and K lines of ionized calcium are the two dark vertical lines in the galaxy's spectrum. The arrows indicate the shift in the calcium H and K lines. The red shifts are expressed as velocities. (Palomar Observatory Photograph).

EXPERIMENT 44 NAME (print) ______________________________ DATE ______________
LAST FIRST

LABORATORY SECTION ______ PARTNER(S) ______________________________

Step 3 Calculate the value of Hubble's constant in kilometers per second per megaparsec for each galaxy and record the value in the data table. Use Eq. 44.1. Determine the average value and record in the data table.

Step 4 Plot a graph with recessional velocity on the *y* axis and distance on the *x* axis. Determine the slope of the curve. Refer to Experiment 1 for help in plotting a graph.

PROCEDURE 2

Step 1 An interesting aspect of the Hubble constant is that if you convert both distance units in the constant to the same value, the resulting constant turns into a quantity whose units become one over time. This means that the reciprocal of the Hubble constant becomes a direct estimate of the age of the universe if we consider the present expansion rate of observable objects in space to have been uniform from the Big Bang until today.

To calculate this age we must first change the units in the Hubble constant so that the distance units will cancel each other and the time is expressed in years. To accomplish this we will use conversion factors to change the units from kilometers per second per megaparsec to kilometers per year per kilometer. Take the average experimental value for the Hubble constant as recorded in Data Table 44.1 and apply the proper conversion factors from the Table of Useful Conversion Factors given earlier in this experiment, until you have converted megaparsecs to kilometers and seconds to years. Use the space below and show your work.

__________ km/s/mpc (required conversion factors) = __________ km/year/km

Step 2 With the Hubble constant in this form we can obtain a rough estimate of the age of the universe. The elapsed time since the Big Bang, our estimate of the age of the universe, is the time of separation that the galaxies have been receding from on another. This time can be found by taking the distance measured and dividing it by the velocity at which this recession has been occurring.

$t = \frac{d}{v}$ but from Hubble's Law $v = Hd$ so this equation becomes

$t = \frac{d}{Hd}$ and once the *d* is canceled

$t = \frac{1}{H}$ or time is equal to the reciprocal of the Hubble constant.

Calculate the value of t, in years, using the value for the Hubble constant that you calculated in Part 2, Step 1. Show your work. Now we finally have our estimate of the age of the universe.

Estimated age of the universe = ___________ years

QUESTIONS AND CALCULATIONS

1. How does the slope of the curve plotted in Procedure 1, Step 4 compare to the average value of the Hubble constant calculated in Procedure 1, Step 3?

2. Would our estimate of the age of the universe increase, older estimate, or decrease, younger estimate, if the value of the Hubble constant increases?

3. How would our estimate of the age of the universe change if we considered that the force of gravity between galaxies has been gradually decreasing the separation velocities of the galaxies ever since the Big Bang?

4. Recent date from astronomical studies seems to indicate that there is a very large amount of **dark matter** throughout the universe that is distributed in such a way that the overall effect of gravity on all of the parts of the universe is to make separation velocities between galaxies faster over time instead of slower. How would this change our estimate of the age of the universe using the Hubble constant method?

Experiment 45

Measuring the Radius of the Observable Universe

INTRODUCTION

The universe is the totality of all energy, mass, space, and time. The constituents of the universe and their proportions are: 4 percent normal matter, 22 percent dark matter, and 74 percent dark energy. The universe is a flat universe (that is, one with a negative curvature) and will expand forever. The expansion is accelerating due to a force supplied by dark energy. Presently, the expansion rate is believed to be 77 km/sec/Mpc plus or minus 5 percent. The position of the universe that we can actually detect or make measurements on, is called our observable universe, or our Hubble Volume, or our Horizon Volume.

The normal matter of the universe is what we observe. From the smallest entities (quarks) to the largest entities (voids, the relatively empty regions of space) we observe structure. Quarks are the building blocks of protons and neutrons, which in turn compose atoms. Atoms compose the larger structures — planets and stars. The stars are arranged in clusters and galaxies. And the galaxies are also arranged in clusters and superclusters. Larger structures also exist such as the broad string of galaxies known as the Great Wall.*

Why is the universe so big? Light waves (electromagnetic radiation) from galaxies observed at the farthest distance from Earth, left their origin and traveling at 300,000 km/s (186,000 mi/s) arrive at planet Earth some 13 billion years later. This distance is the radius of our Hubble Volume. Try and visualize and comprehend traveling 13 billion years at the speed of light. I doubt that any human mind can. But astronomers can photograph (see color photo on the back of the Laboratory Guide) out to 13 billion light years. And they can measure this distance. This experiment will show how.

LEARNING OBJECTIVES

After completing this experiment, you should be able to do the following:

- ▼ Distinguish between the different methods used to calculate the distance to stars, galaxies, and other observable celestial objects.
- ▼ Calculate the distance to any observable object beyond Earth when sufficient data is given.

* See textbook pages 525-526.

CALCULATION

Determine the radius (in kilometers or miles) of our Hubble Volume.

APPARATUS

Hand calculator, writing paper, and class textbook.
The following units for distance are used in this experiment.

AU 1 Astronomical Unit
The distance from Earth to the Sun $= 9.3 \times 10^7$ miles $= 1.496 \times 10^8$ kilometers $\left(1.5 \times 10^8 \text{ km}\right)$.

ly 1 light year
The distance light travels in 1 year $= 9.46 \times 10^{12}$ km $= 5.88 \times 10^{12}$ miles $= 0.307$ pc.

pc 1 parsec $= 3.09 \times 10^{13}$ km $= 3.26$ ly $= 2.06 \times 10^5$ AU.

Mpc 1 megaparsec $= 1.00 \times 10^6$ pc.

The word parsec comes from the first three letters of parallax* plus the first three letters of the angular measuring unit the second. Parallax is the apparent motion, or shift, that occurs between two fixed objects when the observer changes position.

* See textbook page 444, Fig. 16.10.

EXPERIMENT 45 NAME (print) ______________________________ DATE ______________________
LAST FIRST

LABORATORY SECTION ______ PARTNER(S) ____

PROCEDURE 1

Step 1 Radar ranging In the past astronomers have used triangulation based on Euclidean geometry to measure the distance to the moon and to the planets. Today they use radar to measure these distances, and these measurements constitute the base for obtaining an absolute scale for the solar system.

The first step in learning the methods to measure the radius of the observable universe begins with measuring the distance to the moon using radar ranging. Radar is an acronym for radio detection and ranging. To determine the distance to the moon, a beam of electromagnetic energy, at radio frequency, is transmitted to the moon where it is reflected back to Earth. The speed of the radiation is 3.00×10^5 km/s (186,000 mi/s). The time for the radiation to go to the moon and back is measured. Knowing the time and the speed, the distance can be computed. The results give a value within 8.5 centimeters of the exact distance.

QUESTIONS

1. Calculate the distance to the planet Venus at its closest approach to Earth. The time for radio waves transmitted to Venus, at closest approach, and returned to Earth is 299.2 seconds rounded off to four significant figures.

2. Using your answer to question one and knowing that Venus is 0.300 AU from Earth at its closest approach, calculate the length of one astronomical unit.

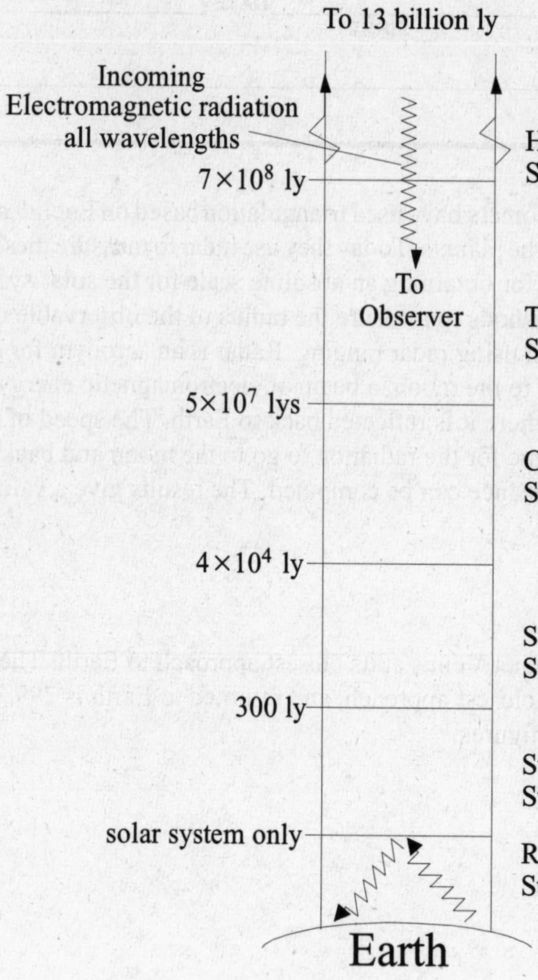

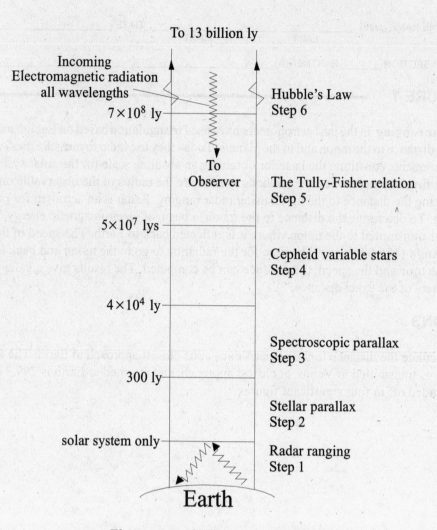

Electromagnetic Distance—Ladder
Extended to the Observable Limit of the Universe

Figure 45.1 Electromagnetic steps used to measure successively more distant stars and galaxies.

EXPERIMENT 45 NAME (print) ______________________________ DATE ______________
LAST FIRST

LABORATORY SECTION ______ PARTNER(S) ______________________________

PROCEDURE 2

Step 2 Stellar parallax The terms parallax and parsec, which were introduced and defined at the end of the list of units, are the fundamental concepts for the study of stellar parallax; measurement of distance to stars. Refer to Fig. 15.11 in the textbook, which illustrates stellar parallax.

The distance to a star with parallax p is given by the following equation.

$$\text{distance } d = \frac{1}{\text{the parallax in seconds of arc}}$$

A star's parallactic angle p is defined to be half its apparent displacement relative to the background stars as the observer moves from one side of Earth's orbit to the other. The parallax is measured in seconds of arc, and the distance is in parsecs. One parsec is defined as the distance to a star with a parallax of one second of arc. Astronomers conceived the term because knowing the parallax in seconds of arc, the distance is given directly in parsecs by dividing the value into one.

A star's parallactic angle p is very difficult to measure, and the greater the distance away the smaller the angle. If you held this sheet of paper about 75 centimeters (arm's length) in front of you and view it edge on, the thickness of the paper displays and angle of about 30 arc seconds.

QUESTIONS

1. The nearest star to Earth (excluding the Sun) is Proxima Centauri, which has a parallax of 0.772 arc seconds. Calculate the star's distance from Earth.

2. Our Earth's observatories cannot measure parallax smaller than 0.01 seconds of arc. Calculate the maximum distance, in parsecs, that we can measure to a star using stellar parallax.

3. The star Altair, in the constellation Aquila, has a parallax of 0.200 arc seconds. Calculate the distance, in light years, to the star.

4. The maximum base line for measuring parallax of stars is the major axis of Earth's orbit around the Sun. With our modern technology, suggest how astronomers can increase the base line.

EXPERIMENT 45 NAME (print) ________________________________ DATE ______________
LAST FIRST

LABORATORY SECTION ______ PARTNER(S) ________________________________

PROCEDURE 3

Step 3 Spectroscopic parallax Spectroscopic parallax is the third method used for measuring the distance of the observable universe. The method is based on the stellar properties (spectra and color) of stars out to thousands of parsecs. When a star's apparent brightness and absolute brightness are known the distance can be calculated using the inverse-square law. Although the term parallax is included in the method's name, parallax is there in name only. Parallax of a star cannot be measured at these great distances.

Apparent brightness refers to the amount of energy received on the eye or on a unit area of a light-sensitive surface. Absolute brightness is the apparent brightness a star would have if it were placed 10 parsecs (a standard distance) from the observer. The absolute brightness can be obtained from the H-R (Hurtzsprung-Russell) diagram.* From a photograph of the spectrum of a star, an astronomer can determine its spectral class, which gives its location on the horizontal axis. The astronomer can also determine the star's luminosity (the total energy at all wave lengths per second) class by examining the widths of its spectral lines. This gives the star's location on the vertical axis. Once a plot of this point on the H-R diagram, the absolute brightness can be read from the vertical axis. This method assumes that distant stars are similar to stars closer to Earth. Also, the main sequence on the H-R-diagram is not a narrow line but has width. This will cause some error in placing the star's position on the main sequence.

The magnitude-distance equation provides another way to determine the distance to a star using the values for apparent and absolute magnitudes. The equation is as follows:

$$m - M = -5 + 5\log_{10}(d)$$

where *m* is the apparent magnitude†, and *M* is the absolute magnitude. Solving for *d* (in parsecs)

$$d = 10^{\frac{m - M + 5}{5}}$$

QUESTIONS

1. Calculate the distance to a star with an apparent magnitude of 7 and an absolute magnitude of 2.

* See textbook page 510, Fig 18.13.

† Brightness and magnitude are terms that have the same meaning.

2. Calculate the distance (in kilometers or miles) that light travels in one year.

3. Calculate the radius (in parsecs) of the Hubble Volume.

PROCEDURE 4

Step 4 Pulsating stars; Cepheids A small group of stars known as Cepheid pulsating variables take center stage in the next step to increase radial distance measurements of the Hubble volume. Cepheid intrinsic variable stars are small in number, but they have unique physical properties that provide astronomers information to measure radial space distance out to about 15 million parsecs. The uniqueness comes from the fact that the radiation from the Cepheids varies in intensity over time with large changes in brightness in a periodic and distinctive manner.

The distance to a Cepheid star can be determined from its period (the time of one cycle) and its average apparent brightness as observed from Earth. Observing the variations in brightness over time, astronomers can obtain its period and apparent brightness.

The pulsating period correlates closely with the star's brightness. When astronomers view a large number of Cepheid variables relatively close to Earth, they can measure distance using the method given in Procedure 3. From this data they can calculate the Cepheids absolute brightness. When a graph is made with the star's pulsation period on the horizontal axis and the absolute brightness on the vertical axis, a straight line is the result. This graph can now be used to obtain the absolute brightness (luminosity) of a more distant Cepheid when its pulsation period is determined. Astronomers assume a good correlation between near and more distant Cepheids.

The following proportionality shows the relation between apparent brightness, luminosity, and distance; a typical inverse square relation.

$$\text{apparent brightness} \propto \frac{\text{luminosity}}{\text{distance}^2}$$

EXPERIMENT 45 NAME (print) ______________________ DATE ______________
LAST FIRST

LABORATORY SECTION _______ PARTNER(S) ______________________________

Cepheid stars can be detected out to about 15 million parsecs, but beyond this distance the pulsation period is difficult to measure.

QUESTIONS

1. On a clear night sky the galaxy Andromeda can be seen with unaided eyes. Was astronomer Hubble able to observe Cepheids in Andromeda? Justify your answer.

2. Can the magnitude-distance equation be used to find the distance to pulsating Cepheids? Justify your answer.

3. Name the two parameters that are directly measured by observing pulsating Cepheids.

4. If the observations were made from the surface of the moon, could the distance to Cepheids be increased beyond 15 million parsecs? Justify your answer.

PROCEDURE 5

Step 5 The Tully-Fisher relation The Tully-Fisher relation is named for Brent Tully and Richard Fisher, astronomers, who in the 1970s discovered that the width of the hydrogen 21-cm emission line of a spiral galaxy is related to the galaxy's absolute brightness. The relation or correlation is a very close one. The wider the line, the brighter the galaxy. The discovery is one of the best methods for measuring deep space distance to galaxies out to 200 million parsecs. At greater distances the spectral line broadens so much it becomes difficult to measure.

Knowing the absolute brightness of the galaxy, astronomers compare it with apparent brightness and determine its distance as was done in procedure 3 using the magnitude-distance equation.

QUESTIONS

1. What is the significance of the Tully-Fisher relation?

EXPERIMENT 45 NAME (print) ______________________________ DATE ____________________
LAST FIRST

LABORATORY SECTION ______ PARTNER(S) ______________________________________

2. **What type of galaxy was observed by Tully and Fisher?**

3. **What is the relation between galaxy brightness and the width of the emission lines?**

4. **How are the Tully-Fisher calculations for the distance to a galaxy different from those used for spectroscopic parallax?**

PROCEDURE 6

Step 6 Hubble's law In 1912 astronomers discovered that most spiral galaxies observed were moving away, in all directions, from Earth. This was shown by the shift of their spectral lines toward the red end of their spectrum. It indicated the galaxies are actually moving away from one another and the universe was expanding. In the 1920s, Edwin Hubble and Milton Humason photographed the spectra of several galaxies with the 100-inch telescope on Mount Wilson, California. Five representative spectra are shown in Experiment 44, Fig. 44.1. The photograph shows the redshift of the spectral lines, and their corresponding recessional velocities. Hubble found a direct correlation between the distance to a galaxy and its redshift. In 1929 Hubble published this discovery, which today is known as Hubble's law.

The Hubble law written in equation form is

$$\text{velocity } (v) = H \times \text{distance}(d)$$

where H is a constant called the Hubble constant.

Hubble's law provides a method to measure the distance to the most distant galaxies. First, the recessional velocity must be determined. This is accomplished by measuring the redshift of the galaxy's spectral lines. Second, use the Doppler equation to convert wavelength change to recessional velocity. The Doppler equation is written

$$\frac{\Delta\lambda}{\lambda_0} = \frac{v}{c}$$

where $\Delta\lambda$ is the change in wavelength: λ_0 is the original wavelength from the laboratory: v is the recessional velocity, and c is the velocity of light. The Doppler equation is valid only when v is small compared with c. Third, plot a graph of recessional velocity versus distance for several relatively close galaxies. The plotted data will show a straight line.* Use this graph to obtain the distance to the more remote galaxy. Locate the remote galaxy's recessional velocity on the straight line, and read the corresponding distance from the horizontal axis.

New valid data for the age of the universe has been published this year (2004). Data from the Wilkinson Microwave Anisotropy Probe or WMAP, a NASA orbiting laboratory; placed in orbit June 30, 2001. The WMAP research group calculated the universe to be 13.7 billion years old, plus or minus 1 percent. Since the data has validity, the value for the age can provide a new method for calculating the distance to galaxies. When we know the recessional velocity of a remote galaxy, we can use Hubble's law to find the distance.

$$\text{recessional velocity } (v) = H \times \text{distance}(d)$$

$$d = \frac{v}{H}$$

* See Laboratory Experiment 1.

EXPERIMENT 45 NAME (print) ______________________________ DATE ______________
LAST FIRST

LABORATORY SECTION _______ PARTNER(S) ______________________________

CALCULATIONS

1. Show that $H = \frac{1}{t}$. Where $H = \frac{v}{d}$ Hubble's law, and by definition velocity (v) = distance (d)/time (t).

2. If the age of the universe is really 13.7 billion years, calculate the value of the Hubble constant H.

3. Calculate the distance to a galaxy that has a recessional velocity of 50,000 km/sec using the value of H that you calculated above in Question 2.

Questions that expand your thinking concerning the universe.

1. The universe of homogeneous—that is, when observing the large-scale structure of the universe, it appears to have the same composition everywhere. The universe is also isotropic—that is, it looks the same in any direction. These two observations together are known as the cosmological principle. Astronomers believe the principle to be correct. If so, which of the following are true statements?

   (a) The universe has no center.
   (b) The universe has no edge.
   (c) The universe will expand forever.

2. What do you think exists beyond the Horizon Volume?

3. The universe is expanding at an ever-increasing rate. That is—the universe is undergoing an acceleration expansion. How does the expansion affect the speed and size of the universe? How does the accelerating expansion influence the observers' farthest view of the universe?

4. The total universe is everything that is—all mass, energy, space, and time. What is the total universe expanding into? *Hint*: Think scientific evidence or absence thereof.

5. Compose three or more answers to the following statement. The mass and energy that make up the universe originated from—

   (a)

   (b)

   (c)

6. How big is the total universe? You calculated the radius of the Hubble Volume (the observable universe) in Procedure 3. Is it possible to calculate the radius of the total (everything that is) universe? Justify your answer. (*Hint*: Consider the expansion rate.)

Experiment 46

Air Pressure

INTRODUCTION

Pressure is defined as force per unit area:

$$P = \frac{F}{A}$$

where P = pressure
F = force in dynes, newtons, or pounds
A = area in square centimeters, square meters, or square inches

The meteorologist uses the millibar as a unit of pressure. The *millibar* is one-thousandth of a bar. The *bar* is defined as one million dynes per square centimeter. The relationships are expressed as follows:

$$1 \text{ millibar} = \frac{1}{1000} \text{ bar}$$
$$1 \text{ bar} = 10^6 \text{ dynes/cm}^2$$
$$1 \text{ millibar} = 10^3 \text{ dynes/cm}^2$$

Air pressure is due to the weight of air. At Earth's surface (45°N, sea level), the total weight of all the air above one square inch of area is 15.7 lb. This is equivalent to a column of mercury 30 inches high. This pressure is also known as one atmosphere. When measured in millibars, one atmosphere equals 1013 millibars.

As described by Newton's law of universal gravitation, the density of the atmosphere varies inversely with height above Earth's surface. Density is greater at the surface and decreases with increasing elevation. The density of the atmosphere at sea level is about 1.3×10^{-3} g/cm^3 and decreases to about one-half this value at 18,000 ft.

Swiss physicist Daniel Bernoulli (1700–1782) stated that the properties of gases could be explained with the assumption that any gas consists of minute corpuscles in rapid motion that undergo elastic collisions with one another and with the walls of the container holding them. This bombardment of the walls of the container gives rise to what we call pressure. Bernoulli's statement was made more than 100 years before the kinetic theory of gases was propounded.

Bernoulli also stated a principle about fluids (liquids or gases) in motion. Today this is known as *Bernoulli's principle.* In simple terms, it states that where the velocity of a fluid is maximum, the pressure is minimum and where the velocity of a fluid is minimum, the pressure is maximum. For example, in Fig. 46.1 the baseball is moving to the right through the air and spinning counterclockwise. This creates a lower pressure on the upper side of the baseball, because the motion of the baseball through the air plus the spin increases air velocity. A higher pressure is created underneath the baseball because the motion of the baseball through the air minus the drop in pressure due to the decreased air velocity beneath the moving ball that is caused by its spin. With the lower pressure on top and higher pressure underneath, the baseball curves upward.

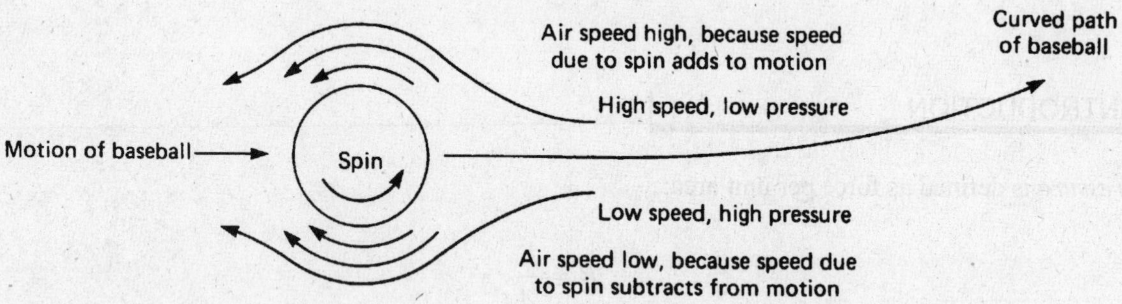

Figure 46.1 Spinning baseball curves because of changes in air pressure.

LEARNING OBJECTIVES

After completing this experiment, you should be able to do the following:

- ▼ Define *pressure,* and state its units of measurement.
- ▼ State the cause of air pressure at Earth's surface and how air pressure varies with altitude.
- ▼ State some of the effects of air at rest.
- ▼ State Bernoulli's principle.
- ▼ State some of the effects produced by air in motion.

APPARATUS

Meter stick, support base and holder to balance meter stick, two toy balloons (round, same size), two rubber bands (small), enamel tray (large white), glass tumbler (round with smooth edge), index card (cover for glass tumbler), two plastic soda straws, straight pin, rubber hose (¼ in diameter, 2 ft in length), vacuum base, glass plate (4 × 6 in), two Ping-Pong balls, two plastic bottles (250 mL), large beaker (1 L), pinch clamp.

EXPERIMENT 46 NAME (print) ______________________ DATE __________
LAST FIRST

LABORATORY SECTION ______ PARTNER(S) ______________________

PROCEDURE

1. Determine the least count of the barometer located in the laboratory. ________

2. Determine the present barometric pressure in

   (a) inches of mercury. ________

   (b) centimeters of mercury. ________

3. Inflate the two balloons to approximately the same size and with rubber bands secure each near the opposite ends of a meter stick. Balance the meter stick on the support base, and then observe the direction the meter stick moves when one of the balloons is punctured with a straight pin. Explain.

4. Place the glass tumbler in the large white enamel tray, and fill the tumbler until it overflows. (Use a large beaker for a supply of water from the water tap.) Position an index card over the open end of the glass tumbler to cover the open end completely. Holding one hand firmly on the index card to keep it tight against the rim of the glass, invert the tumbler; then remove your hand from the card. Observe that the water stays in the tumbler even though the tumbler is upside down. Explain.

5. Close one end of a plastic soda straw with a pinch clamp. You may have to double the end of the straw back slightly before applying the clamp. Puncture the soda straw with a straight pin about 1 in above the clamp. Fill the large enamel tray with water to a depth of about an inch and submerge the straw horizontally in the tray of water. Allow the straw to fill completely with water. Position your finger over the open end of the straw and raise the straw out of the water to a vertical position with the clamped end downward. Intermittently remove and replace your finger on the open end of the straw. Explain what you observe.

6. Place a large beaker about three-fourths full of water in the large enamel tray that contains about 1 in. of water. Lay the rubber hose flat in the tray of water so that it will become filled with water. Holding a finger over each end of the rubber hose so that the water cannot escape, place one end of the rubber hose in the beaker of water and leave the other end in the tray. Remove both fingers from each end of the rubber hose, being careful to keep the end of the hose you have placed in the beaker below the level of the water. Raise the beaker of water and observe what happens. Lift the lower end of the hose above the water level in the tray. This is called a *siphon.* Explain how it operates.

7. Position the vacuum base in the center of the glass plate, and pull the lever forward. Explain why the glass plate is held firmly by the rubber base.

EXPERIMENT 46 NAME (print) ______________________________ DATE ______________
LAST FIRST

LABORATORY SECTION ______ PARTNER(S) ______________________________

8. Hang two Ping-Pong balls on separate threads so that they are suspended about 20 cm from a support rod and are about 2 cm apart. Using a clean soda straw blow a stream of air between the two balls from a position in front of the balls and level with them. What do you observe? Explain.

9. On a glass plate placed at the edge of the laboratory table, position two plastic bottles on their sides so that they are about 2 cm apart with their long axes parallel to one another. Using a clean soda straw, blow a stream of air between the bottles in a direction parallel to the long dimension of the bottle. Explain what you observe.

QUESTIONS

1. If the atmosphere were condensed to the density of mercury, how high above sea level would the upper level of the liquid be? Use the data from this experiment to answer the question.

2. What is the existing local air pressure in millibars? Show your work for obtaining the answer.

_________ mb

3. Is the air pressure in Question 2 considered to be a low or a high pressure when compared to normal atmospheric pressure? Explain.

4. What pressure units is a meteorologist using when he or she says the barometric pressure is 29.68 and falling?

5. Using information gained from this experiment, explain why:

   (a) a ship moving in a narrow canal may collide with the side of the canal.

   (b) a person in a parked automobile along a freeway feels the car being "pulled" toward a semitrailer truck as it passes by.

Experiment 47
Humidity

INTRODUCTION

Humidity refers to the amount of moisture or water vapor in the air. The air in our environment is rarely ever dry. The amount of moisture present varies up to about 4 percent by volume.

Moisture present can be expressed in terms of absolute humidity or relative humidity. *Absolute humidity* is defined as the amount of moisture present in a specified volume of air and is expressed as the actual moisture content (AH). The actual moisture content is measured in grams per cubic meter in the SI system where temperature is recorded in degrees Celsius or in grains per cubic foot in the British system where temperature is given in degrees Fahrenheit. (One grain = 1/7000 lb.)

Actual moisture content of an air sample = absolute humidity (AH)

The *relative humidity* is defined as the ratio of the amount of moisture present in one cubic foot of air (actual moisture content) to the amount of moisture one cubic foot of air can hold (maximum capacity) at the existing air temperature:

$$\text{RH} = \frac{\text{AH}}{\text{MC}}$$

where RH = relative humidity at a given temperature
AH = absolute humidity
MC = maximum moisture capacity at that same temperature

Relative humidity can be expressed either as a fraction or as a percentage. The fraction can be converted to a percentage by multiplying the fraction by 100:

$$\%\ \text{RH} = \frac{\text{AH}}{\text{MC}} \times 100$$

The absolute humidity of an air sample can be determined if the dew point is known. The *dew point* of the air refers to the reading on the thermometer at the time when dew begins to form or the air becomes saturated. If the temperature of the air is gradually lowered, a point (temperature) will be observed at which the air can no longer hold the moisture present and the water vapor condenses to liquid water. Knowing this temperature allows us to determine the amount of moisture present in grams per cubic meter of air. For example: Refer to the Table of Absolute Humidity. If the dew point temperature is 30°C (86°F), the absolute humidity is 30.4 g/m^3 (or 12.9 $\text{grains}/\text{ft}^3$).

LEARNING OBJECTIVES

After completing this experiment, you should be able to do the following:

- ▼ Define *humidity, absolute humidity,* and *relative humidity,* state their units of measurement, and give an example of each.
- ▼ Determine experimentally the absolute and relative humidity of the air.
- ▼ Calculate the relative humidity from given and inferred data.

APPARATUS

Thermometer, small beaker (250 mL), calorimeter cup, ice cubes, small plastic spoon, stirring rod.

Table of Absolute Humidity (maximum amount of moisture in an air sample) as a function of temperature in both degrees Celsius and degrees Fahrenheit in SI units (grams per meter cubed) and in British units (grains per cubic foot)

Absolute Humidity Values

Air Temp.	Air Temp.	Absolute Humidity (g/m³)	Absolute Humidity (grains/ft³)
50ºC	122ºF	83.0	27.4
45ºC	113ºF	65.4	23.9
40ºC	104ºF	51.1	21.2
35ºC	95ºF	39.6	17.1
30ºC	86ºF	30.4	12.9
25ºC	77ºF	23.0	10.1
22ºC *	72ºF	20.2	8.7
20ºC	68ºF	17.3	7.9
15ºC	59ºF	12.8	5.6
10ºC	50ºF	9.4	4.1
5ºC	41ºF	6.8	2.9
0ºC Water freezes	32ºF	4.8	2.0
–5ºC	23ºF	3.4	1.4
–10ºC	14ºF	2.3	0.9
–15ºC	5ºF	1.6	0.6

* 72°F (22°C) is generally considered to be "normal" room temperature

EXPERIMENT 47 NAME (print) ____________________ DATE ____________
LAST FIRST

LABORATORY SECTION ______ PARTNER(S) ____________________

PROCEDURE

1. Fill the glass beaker half full of water. Insert the thermometer; then add small pieces of ice to the water, stirring gently with the stirring rod. The temperature of the water and the beaker will begin to lower. Allow the temperature to decrease until you notice moisture beginning to collect on the outside of the beaker. Note the temperature as this occurs and record it in Data Table 47.1.
2. Remove all ice from the glass beaker, and allow the water to increase in temperature. Stir gently with the thermometer and determine the temperature at which the moisture on the outside of the beaker disappears. Record this temperature in the data table.
3. Repeat Procedures 1 and 2, using the calorimeter cup in place of the glass beaker. Record these temperatures in Data Table 47.1.

DATA TABLE 47.1

Procedure	Dew Point, in degrees Celsius	
1. Glass beaker—cooling		
2. Glass beaker—warming		______ Average dew point
3a. Calorimeter cup—cooling		
3b. Calorimeter cup—warming		

4. Find the absolute humidity of the air, from Data Table 47.1 using the average value for the dew point. Refer to the example in the introduction.

   AH = AC at the dew point temperature = ____________ $grams/m^3$

5. Using your thermometer, measure the temperature of the air in the room.

   Temp. of air = ____________ °C

6. From the Table of Absolute Humidity Values read and record the maximum moisture capacity for the air in the room at its present temperature.

   MC of air in room = ____________ $grams/m^3$

CALCULATIONS

1. Determine the relative humidity of the air in the lab room as a fraction or decimal quantity. Show your work.

2. Calculate the relative humidity of the air in the lab room as a percentage. Show your work.

QUESTIONS

1. Explain the difference between *absolute humidity* and *relative humidity.*

2. The air temperature in the laboratory is lowered 5°C. Assuming the absolute humidity remains constant, would the relative humidity increase or decrease? How much? Show your work.

3. If the relative humidity of the air in the laboratory is 25 percent at its present temperature, what is the absolute humidity?

Experiment 48

Weather Maps (Part 1)

INTRODUCTION

Air masses are large bodies of air that take on the characteristics of the area over which they originate. Air masses originating over Canada are cold and dry, whereas air masses originating over the Gulf of Mexico are warm and humid. An air mass is fairly homogeneous in temperature and pressure across its expanse, but it shows considerable vertical variation. *Isotherms* and *isobars* are used to indicate equal temperature and equal pressure contours on a weather map. In general, the pressure and temperature of an air mass decrease with increasing elevation. Because of unequal heat received by Earth's surface from the Sun, the temperature changes; the density of the air is thereby caused to change, and thus the pressure. Such changes coupled with Earth's rotation bring about movement of the air masses over the surface of Earth.

The boundary between two air masses is called a *front*. If a cold air mass is moving into a warm area, the front is labeled a *cold front*. If warm air is moving into a cold region, the front is called a *warm front*.

The term *weather* refers to the day-to-day variations in the general physical properties of the lower part of the atmosphere, known as the *troposphere;* this region varies from 7 to 11 miles in height. The physical properties that play an active role in our daily lives are temperature, pressure, humidity, precipitation, wind, and extent of cloud formation.

The U.S. National Weather Service is the federal organization that provides national weather information. It is part of the National Oceanic and Atmospheric Administration (NOAA—pronounced "Noah"), which was created within the U.S. Department of Commerce in 1970. Under NOAA, the National Weather Service reports the weather of the United States and its possessions, provides weather forecasts to the general public, issues warnings against tornadoes, hurricanes, floods, and other weather hazards, and records the climate of the United States.

The nerve center of the National Weather Service is the National Meteorological Center (NMC), located just outside Washington, D.C., at Suitland, Maryland. It is the NMC that receives and processes the raw weather data taken at thousands of weather stations. Analyses of data are made twice daily on observations taken over the northern hemisphere at 0000 hours and 1200 hours Greenwich Mean Time (7 A.M. and 7 P.M. EST).

To help analyze the data taken at the various weather stations, maps and charts are prepared that display the weather picture. Figure 48.1 shows some of the important symbols used to represent the weather in these charts. Since these charts present a synopsis of the weather data, they are referred to as synoptic weather charts. There are a great variety of weather maps with various presentations. However, the most common are the daily weather maps issued by the U.S. National Weather Service. A set consists of a surface weather map, a 500-millibar height contour map, a map of the highest and lowest temperatures, and a map showing the precipitation areas and amounts. (See the Daily Weather Maps, weekly series, supplied by the laboratory.)

The Surface Weather map presents station data and analysis for 7 A.M. EST for a particular day. This map shows the major frontal systems and the high- and low-pressure areas at a uniform height. An *isobar* is a line drawn through points of equal pressure. They are represented by solid lines on the weather map. An *isotherm* is a line drawn through points of equal temperature. Prevalent isotherms are indicated by dashed lines. On the Surface Weather map a station model, shown in Fig. 48.2, gives the pertinent weather data at the station's location. The stations shown on the weather maps in this experiment are only a fraction of those included in the National Weather Service's operational weather maps on which the analyses are based.

LEARNING OBJECTIVES

After completing this experiment, you should be able to do the following:

- ▼ Define *air mass, warm front, cold front, isobar, isotherm,* and *wind.*
- ▼ Read weather information from surface weather maps issued by the National Weather Service.
- ▼ Plot station reports similar to those found on Surface Weather maps.

APPARATUS

This *Laboratory Guide* and daily weather maps (weekly series) from the National Weather Service.

Internet Sources for Weather Data:

Today the Internet offers an even more precise and often real time alternative to printed weather maps. There are many local, state and regional sites online that provide very accurate and timely meteorological data. One of the most comprehensive and user-friendly Internet sources is provided by the University of Illinois at *http://ww2010.atmos.uiuc.edu.* If you have Internet access in your laboratory room, we recommend that you visit this site to see what kind of information is currently available from such a source.

The University of Illinois site provides an easy tutorial for learning how to use the weather material presented there. Special interest topics such as "Hurricanes" and "El Nino" are given, plus the "Current Weather" page gives you a good overview of meteorological conditions across the U.S. and around the world. There is also an extensive section of "Online Guides" to provide you with many multimedia instructional modules and ideas for class activities and curriculum projects on this subject.

Another good Internet source for weather information is the Meteorology Education and Training (MetEd) site set up by the Cooperative Program for Operational Meteorology, Education and Training (COMET) at *http://meted.ucar.edu/index.htm.* This is a very comprehensive source, but it requires that you be a registered MetEd user to access many of its modules. There is no cost to register so this is also a viable resource for weather information if you decide to use the Internet as part of your study of meteorology.

One of the major differences between the data found on printed weather maps and the material found on the Internet is that on the printed maps, a large amount of information is contained on each map. On the Internet, separate maps are often provided for each type of data such as surface temperature, humidity/dew points, precipitation, or cloud cover. This means that you have to click on several separate maps to get all of the weather information that you require, but the data you have to work with is usually more detailed and a lot more current.

Experiments 48 and 49 in this laboratory manual can still provide a good way to learn the fundamentals of meteorology and become familiar with the notation and symbols used on the Internet sites. Once you have worked through the basics in these experiments, you will have the background necessary to access the web and get a feel for how modern weather studies are being made.

EXPERIMENT 48 NAME (print) ______________________________ DATE ____________________
LAST FIRST

LABORATORY SECTION ______ PARTNER(S) ____

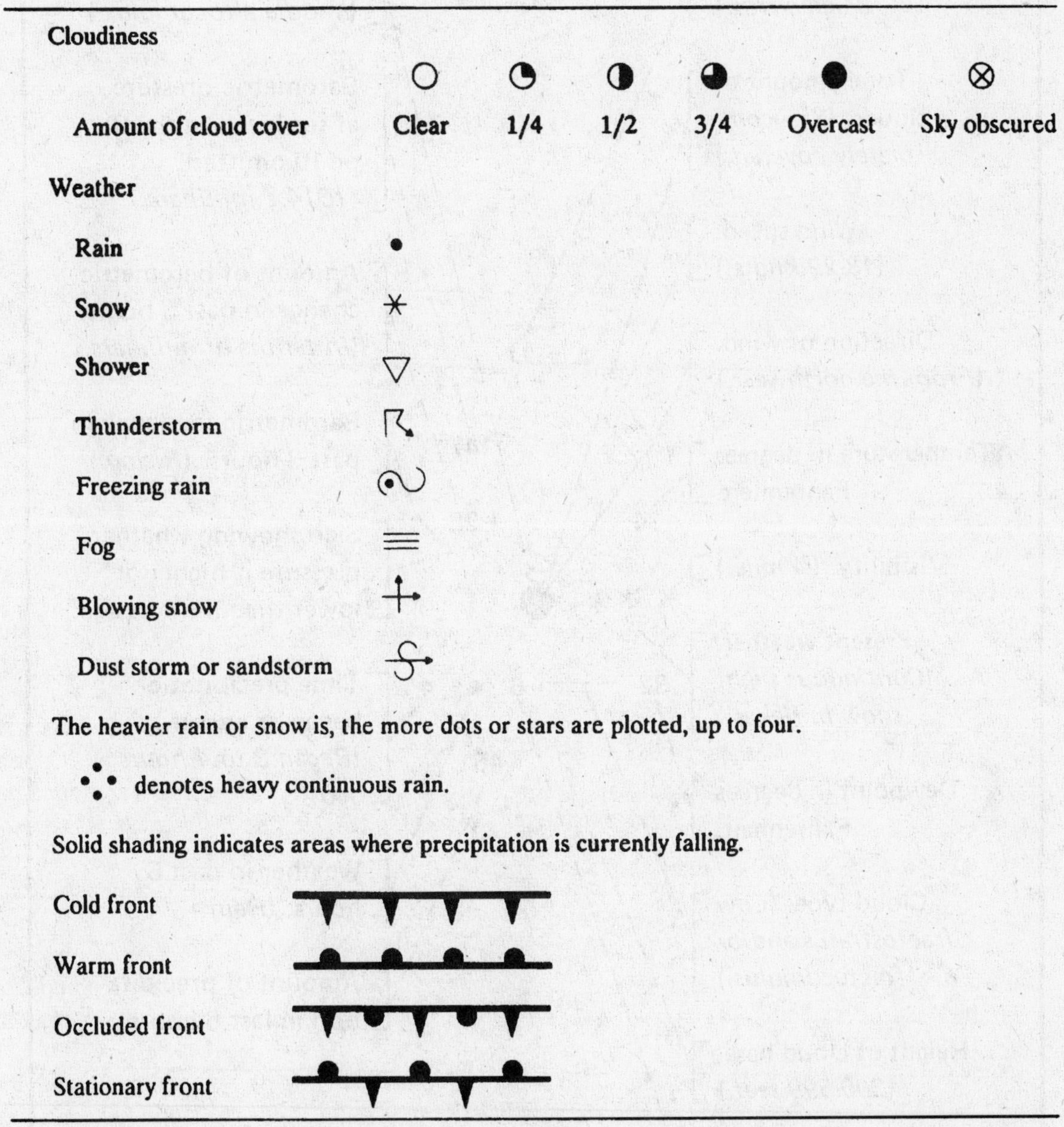

Figure 48.1 Some of the most important symbols used to display weather conditions.

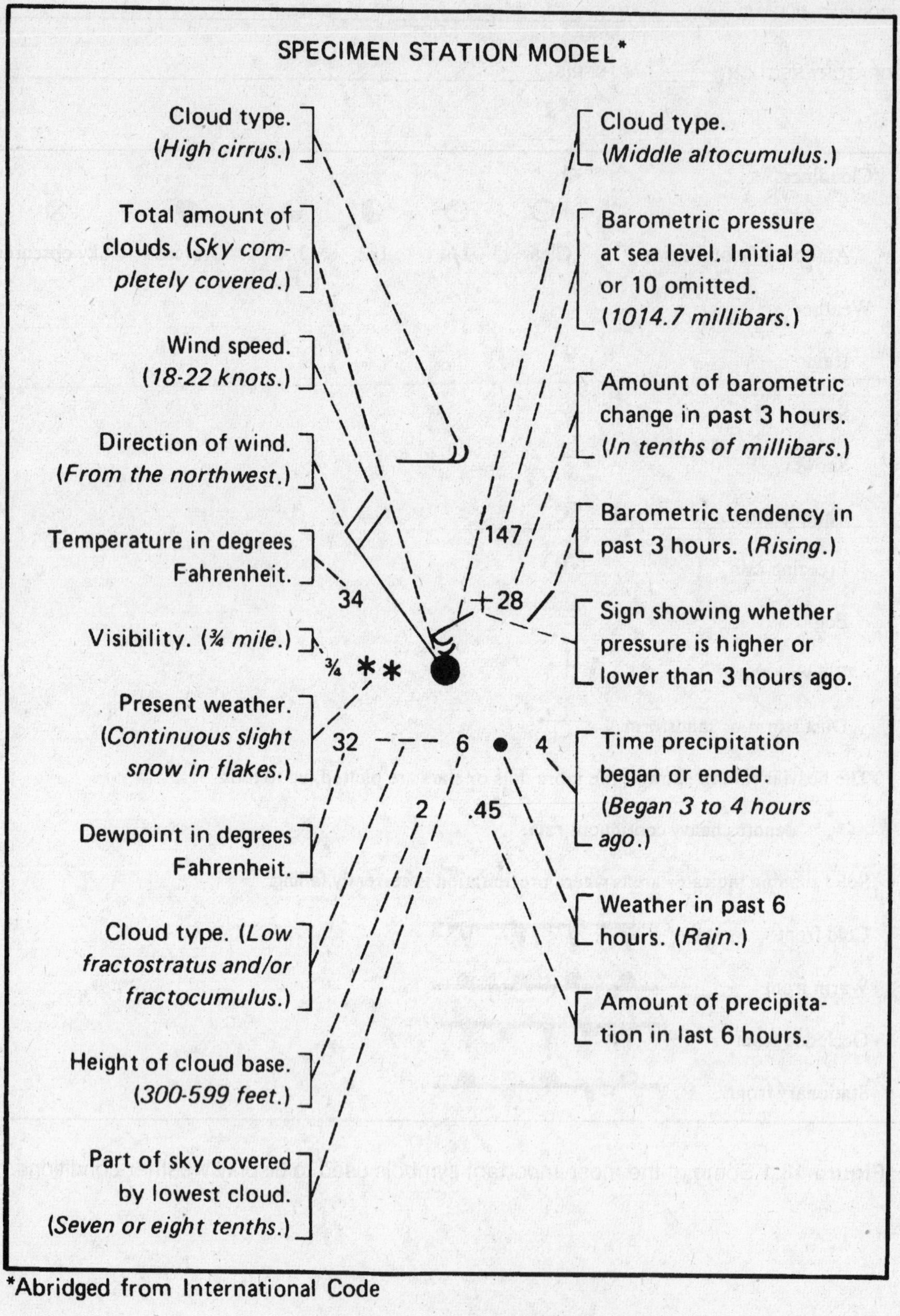

Figure 48.2 Specimen station model.

EXPERIMENT 48 NAME (print) ______________________ DATE ______________
LAST FIRST

LABORATORY SECTION ______ PARTNER(S) ______________________

PROCEDURE

1. Record the information called for in Data Table 48.1 for each city listed. The instructor will indicate which weather map is to be used to complete Data Table 48.1.

DATA TABLE 48.1

	Bismark, ND†	Oklahoma City, OK	Montgomery, AL	Houston, TX	New Orleans, LA
Temperature, in degrees F					
Dew point, in degrees F					
Barometric pressure, in millibars					
Wind direction*					
Wind magnitude, in knots					

* A wind is a horizontal movement of air over the surface of Earth. The direction the wind is coming from is the direction of the wind.

† Use these five cities or others announced by the laboratory instructor. If new cities are to be used, cross out listed names and write in the new city names.

2. Plot the information given below for the three cities (A, B, and C) indicated on the next page.

	City A	City B	City C
Air temperature	62°F	88°F	32°F
Air dew point	60°F	52°F	26°F
Air pressure	1022.6 mb	987.4 mb	1042.8 mb
Wind direction	SW		NW
Wind magnitude	24 knots	Calm	10 knots
Cloud coverage	0.8	Fair (No clouds)	0.5
Present weather	Continuous rain		Intermittent fall of snowflakes, moderate

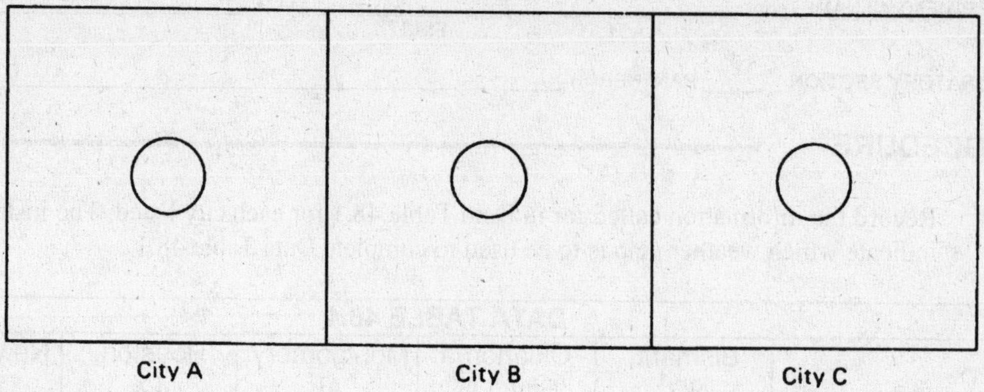

3. Determine the speed of a cold front by observing the front on two consecutive days on the two weather maps in the laboratory manual (Figs. 48.3 and 48.4).

4. Using the weather map supplied by the laboratory instructor, select a cold front and a warm front and record the air temperature that exists about 200 miles on either side of the fronts and indicate whether the temperature rises or falls as the front is crossed at right angles from front to back.

	Temperature ahead	Temperature behind	Rise or fall
Cold front	________°F	________°F	________°F
Warm front	________°F	________°F	________°F

5. Record the barometric pressure that exists about 200 miles on either side of the fronts you selected in Question 4, and indicate whether they rise or fall as the front is crossed at right angles from front to back.

	Pressure ahead	Pressure behind	Rise or fall
Cold front	________ mb	________ mb	________ mb
Warm front	________ mb	________ mb	________ mb

EXPERIMENT 48 NAME (print) ________________________________ DATE ____________________
LAST FIRST

LABORATORY SECTION _______ PARTNER(S) ______________________________________

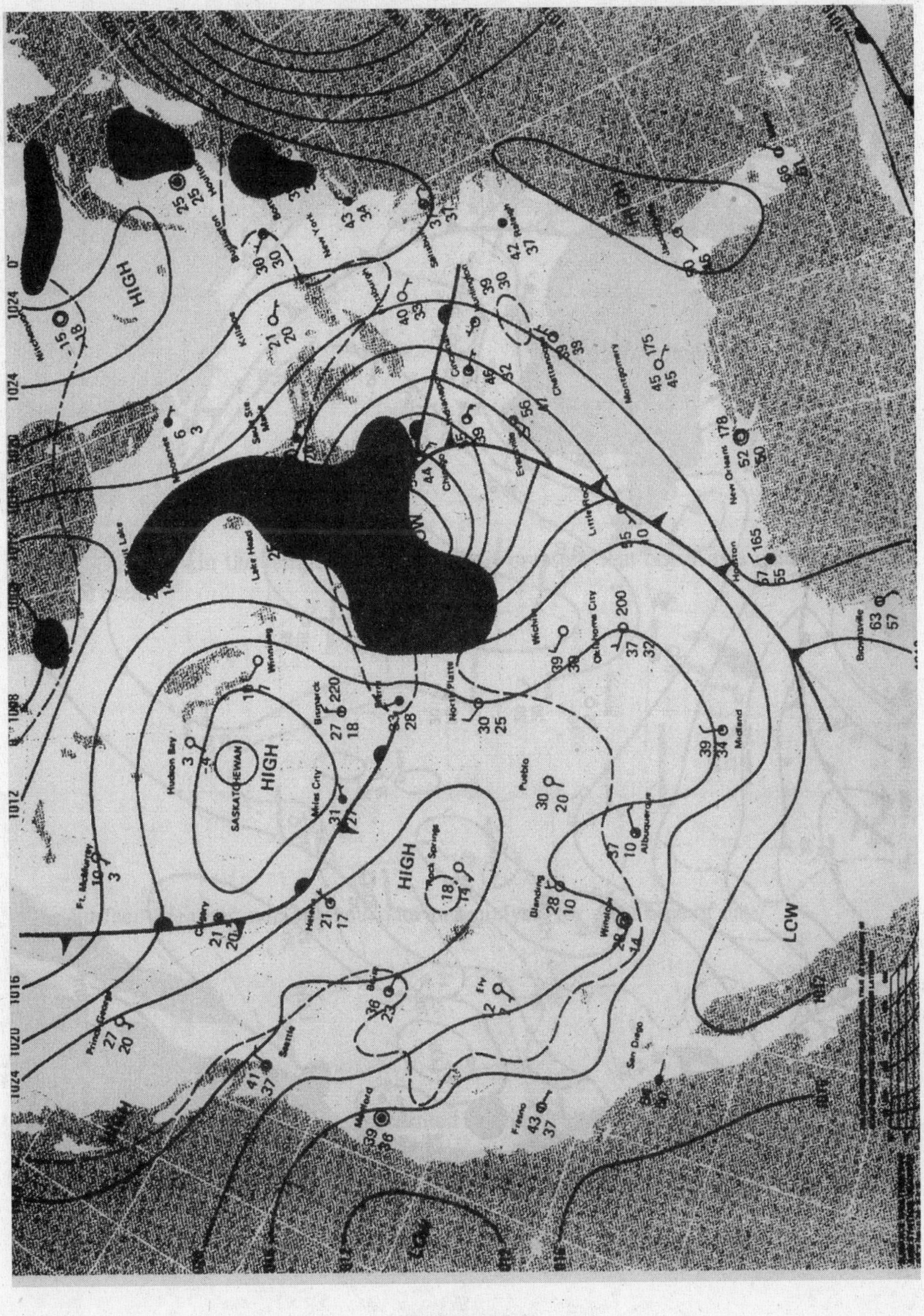

Figure 48.3

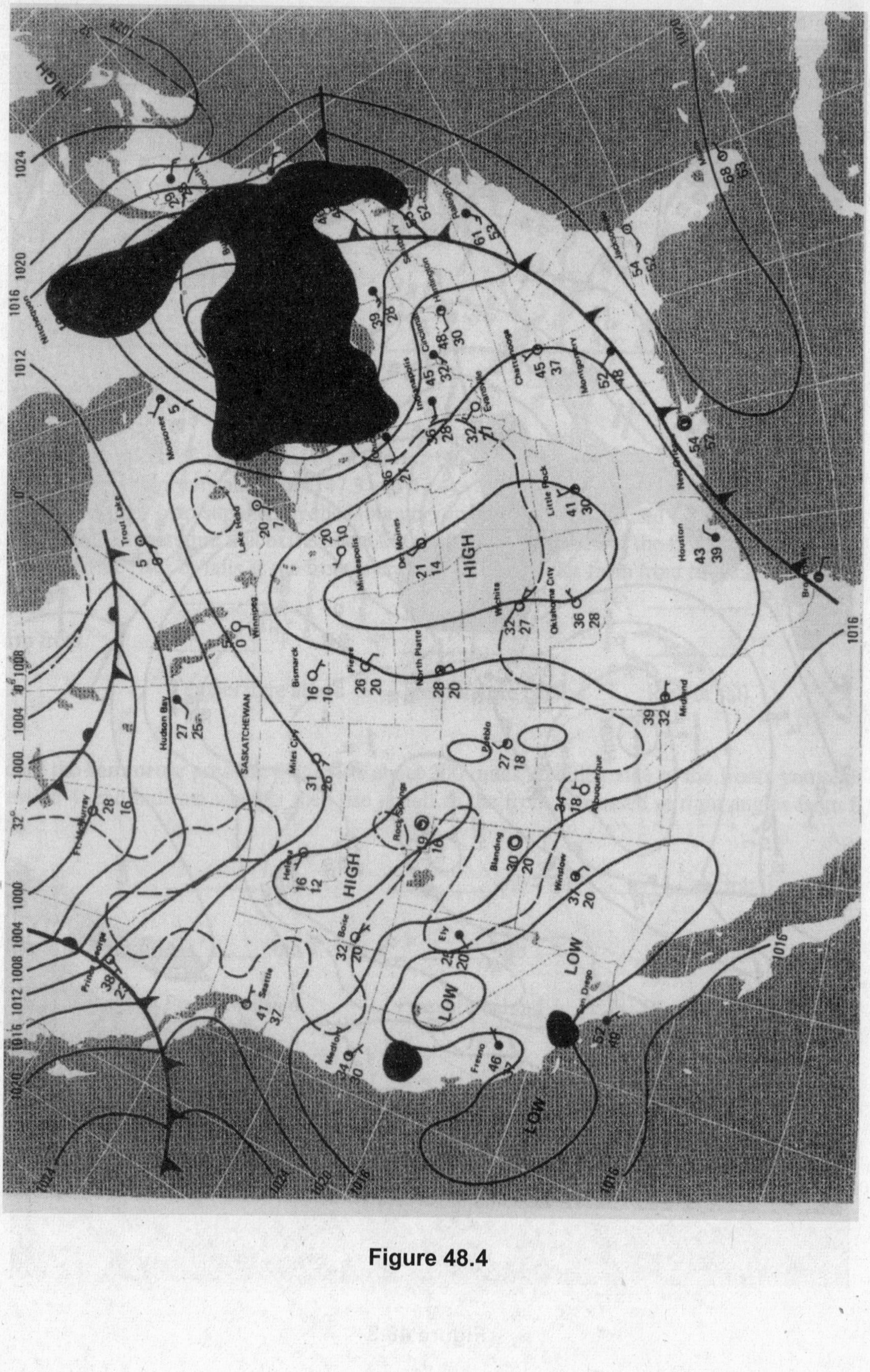

Figure 48.4

EXPERIMENT 48 NAME (print) ______________________________ DATE ______________
LAST FIRST

LABORATORY SECTION _______ PARTNER(S) ____

QUESTIONS

1. List the changes in the temperature and the air pressure that occur as a warm front passes over a surface area.

2. List the changes in the temperature and the air pressure that occur as a cold front passes over a surface area.

3. The Surface Weather map presents data and analysis for what time of day?

4. Define an *isobar.* How are they represented on the weather map?

5. Define an *isotherm*. How are they represented on the weather map?

6. How is the amount of precipitation indicated on the weather map?

7. What is the lowest pressure indicated by an isobar on Fig. 48.3?

8. What is the highest pressure indicated by an isobar on Fig. 48.3?

Experiment 49

Weather Maps (Part 2)

INTRODUCTION

The most outstanding features shown on a daily weather map (see Experiment 48) are the warm and cold fronts.

In addition to the long black lines with their rounded or sharp-pointed projections that symbolize a warm or cold front, there are plain dark lines with open ends, and others that form complete circles. These lines, called *isobars,* are lines of equal pressure, and their central areas indicate regions of high or low pressure.

A study of the weather map will show many facts concerning the weather elements associated with these pressure cells. The first noticeable feature is the way the isobars are plotted on the map. Note the unit of measurement and the difference between any two isobars drawn on the map. It is clearly seen that reporting stations seldom report these particular values of pressure; the isobars are drawn between the known values indicated at the individual stations.

Note how the pressure varies as one approaches the center of the high- or low-pressure cell. Also note the fact that no two isobars ever cross. Why?

LEARNING OBJECTIVES

After completing this experiment, you should be able to do the following:

- Define *high-pressure cell, low-pressure cell,* and *isobars.*
- Identify and explain the physical characteristics of weather associated with high-pressure and low-pressure cells.
- Plot pressure values on a map and draw lines of equal pressures (isobars) on the map.

APPARATUS

This *Laboratory Guide* and a weekly series of the daily weather maps from the National Weather Service.

City	Pressure (millibars)
Denver, Colorado	1016.8
Pueblo, Colorado	1015.2
Chicago, Illinois	1014.1
Moline, Illinois	1017.1
Springfield, Illinois	1016.8
Des Moines, Iowa	1022.4
Sioux City, Iowa	1024.5
Concordia, Kansas	1021.8
Dodge City, Kansas	1018.2
Goodland, Kansas	1019.6
Wichita, Kansas	1018.9
Winnipeg, Manitoba, Canada	1019.8
Escanaba, Michigan	1015.8
Duluth, Minnesota	1020.7
International Falls, Minnesota	1020.6
Minneapolis, Minnesota	1022.0
Kansas City, Missouri	1020.1
Kirksville, Missouri	1020.1
Springfield, Missouri	1019.1
Billings, Montana	1017.9
Miles City, Montana	1019.6
North Platte, Nebraska	1023.4
Omaha, Nebraska	1023.7
Valentine, Nebraska	1025.7
Bismarck, North Dakota	1024.5
Fargo, North Dakota	1024.3
Minot, North Dakota	1020.7
Armstrong, Ontario, Canada	1018.7
Thunder Bay, Ontario, Canada	1018.7
Estevan, Saskatchewan, Canada	1019.4
Huron, South Dakota	1026.1
Pierre, South Dakota	1025.7
Rapid City, South Dakota	1024.5
La Crosse, Wisconsin	1017.3
Milwaukee, Wisconsin	1014.0
Wausau, Wisconsin	1016.8
Casper, Wyoming	1018.1
Cheyenne, Wyoming	1020.2

EXPERIMENT 49 NAME (print) ______________________________ DATE ______________
LAST FIRST

LABORATORY SECTION ______ PARTNER(S) ______________________________

PROCEDURE

1. A high-pressure area is shown over Saskatchewan, Canada on both weather maps in Exp. 48. Record the wind direction at approximately 90° intervals around the high-pressure cell.

   Wind direction north of central area __________

   Wind direction east of central area __________

   Wind direction south of central area __________

   Wind direction west of central area __________

   What is the direction (clockwise or counterclockwise) of the circulation pattern around

   a high-pressure cell? ______________________________

2. A low-pressure area is shown over eastern Iowa on the weather maps in Exp. 48. Record the wind direction at approximately 90° intervals around the low-pressure cell.

   Wind direction north of central area __________

   Wind direction east of central area __________

   Wind direction south of central area __________

   Wind direction west of central area __________

   What is the direction (clockwise or counterclockwise) of the circulation pattern around

   a low-pressure cell? ______________________________

3. Record the maximum pressure indicated on any isobar shown on a weather map in the *Laboratory Guide.*

   ____

4. Record the minimum pressure indicated on any isobar shown on a weather map in the *Laboratory Guide.*

   ____

5. Plot the following pressures, given in millibars, on the map of the United States (Fig. 49.1) and draw in the 1020 and the 1024 isobars. See Weather Service map for location of cities.

LAKE SUPERIOR
MICHIGAN
Escanaba
ILLINOIS
Chicago
Moline
Springfield
WISCONSIN
Wausau
Milwaukee
La Crosse
MISSOURI
Kirksville
Kansas City
Springfield
IOWA
Des Moines
Sioux City
ONTARIO
Armstrong
Thunder Bay
International Falls
MINNESOTA
Minneapolis
MANITOBA
Winnipeg
NORTH DAKOTA
Fargo
Bismarck
Minot
SOUTH DAKOTA
Huron
Pierre
Rapid City
NEBRASKA
Omaha
North Platte
Valentine
KANSAS
Concordia
Wichita
Goodland
Dodge City
SASKATCHEWAN
Estevan
MONTANA
Miles City
Billings
WYOMING
Casper
Cheyenne
COLORADO
Denver
Pueblo

Figure 49.1

EXPERIMENT 49 NAME (print) ______________________________ DATE ______________________
LAST FIRST

LABORATORY SECTION ______ PARTNER(S) ____

QUESTIONS

1. How is the unit of pressure, the bar, defined? (See Experiment 46.)

2. What is the normal range of atmospheric pressure (in millibars) at Earth's surface?

3. What is the difference in pressure indicated by two adjacent isobars on a weather map?

4. What is the direction of the air circulation around a low-pressure cell in the northern hemisphere?

5. What kind of weather is usually associated with a low-pressure cell in North America? Why?

6. What is the direction of the air circulation around a high-pressure cell in the northern hemisphere?

7. What kind of weather is usually associated with a high-pressure cell? Why?

8. A pressure of 1012 mb is indicated on a weather map. Is this pressure high or low? Explain.

Experiment 50

Topographic Maps

INTRODUCTION

A *topographic map* is a scaled-down representation of a surface area of Earth as viewed from a high altitude. The maps are generally drawn by means of contour lines plus color and shading to represent specific areas.

The Global Positioning System (GPS) is a fully functional Global Satellite System composed of dozens of GPS satellites in medium height Earth orbit. This system has been developed by the United States Department of Defense and has not only become the backbone of worldwide navigation, but it is also an extremely important tool for surveying and map-making. Using the simultaneous interaction with 3 or more of these satellites, any specific location can be determined on the Earth's surface with a horizontal accuracy of better than 100 m resolution and a vertical accuracy that is nearly that precise. The highly detailed Geospatial data thus provided can be used for many applications in geology and geography.

Needless to say, GPS has made more accurate contour mapping much easier and more accurate. With this new, improved data acquisition service the maps produced by the United States Geological Survey are now better than ever and although digital data can completely define the surface structure of the Earth in great detail, printed contour maps are still the best way to visualize the surface structure of any given area.

To see the operation parameters and useful applications of the GPS system there are two good web sites that you can access. One is *http://en.wikipedia.org/wiki/Global_Positioning_System** and the other is *http://www.colorado.edu/geography/gcraft/notes/gps/gps.html*. These two sites along with the United States Geological Survey home page at *http://www.usgs.gov/* will allow you to further research up-to-date GPS applications and operations. The United States Geological Survey site can also provide you with current information on where and how to obtain printed quadrangle maps such as the ones used in this experiment for all areas of the United States.

An explanation of some of the features on a quadrangle map.

Map Orientation: When a quadrangle map is positioned for normal viewing, geographic north is at the top of the map. The meridian lines traverse the map from bottom to top, and the parallels traverse the map from left to right. The latitude can be obtained by reading the degree values at the side of the map, and longitude can be obtained by reading the degree values at the bottom or at the top of the map.

* Please note that all website addresses included in this manual were accurate at the time of publication. Houghton Mifflin is not responsible for content changes of any websites referenced by an external link.

Colors and Symbols: Color, shading, and symbols are used on the quadrangle map to indicate specific items of interest. Green is used to indicate forests. Blue indicates a body of water such as a lake, river, or stream. Salmon color indicates a city or town. Black is used for the boundary lines and for artificial structures. Brown is used for the contour lines. Red lines and broken red with white lines indicate primary roads. Secondary roads are drawn as solid faint black or dashed faint black lines. Railroads are indicated as a faint black broken line with black circular dots. Other details are indicated with various shadings of color.

Scale: The reduced scale of the map is given as a ratio or fraction, and is usually given at the bottom of the map. A scale of 1:62,500 means that one inch on the quadrangle map represents 62,500 inches on Earth's surface. Earth's surface is considered to be horizontal.

Contour Lines: Contour lines are drawn in brown and show details of Earth's surface. Any contour line drawn on the map indicates a line of equal elevation. The interval between contour lines is usually 20 ft. This means that there is a 20-ft elevation difference between any two lines. Generally every fifth line is drawn broader and darker than the others. These darker lines are numbered with the correct elevation in feet above mean sea level.

Contour lines that are close together indicate a steep slope; that is, change in elevation takes place rapidly, or the terrain is very steep. See Fig. 50.1 for methods used to draw contour lines to indicate surface profile.

Elevation: The elevation of a particular place on the map is indicated in feet above mean sea level. As mentioned above, the dark brown contour lines give the elevation in places that have been determined with care. These are marked BM, meaning bench mark.

The elevation at the bench mark is also indicated in feet. The bench mark is usually a circular brass plate located at the point indicated on the map.

Streams: When contour lines are drawn on the map, the direction of a slope can be determined by noting the pattern of certain contour lines. When these lines form a series of V-shaped lines, the point of the V indicates upstream, or to a direction point of higher elevation. Thus, the direction of a stream can be determined. Check this by observing a stream (dotted blue line) on the quadrangle map.

LEARNING OBJECTIVES

After completing this experiment, you should be able to do the following:

- ▼ Read and record data from a quadrangle map.
- ▼ Determine the relationships of groundwater to land surface.
- ▼ Draw a contour map when the surface profile is given.
- ▼ Draw the surface profile represented by a contour map.

APPARATUS

U.S. Geological Survey topographic quadrangles (maps): (1) Local area; (2) Interlachen, Florida.

For information on where and how to obtain these maps go to the home page of the U.S. Geological Survey at http://www.usgs.gov/.

EXPERIMENT 50 NAME (print) ______________________________ DATE ______________
LAST FIRST

LABORATORY SECTION ______ PARTNER(S) ______________________________

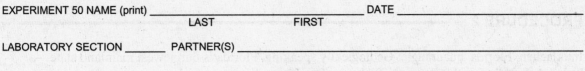

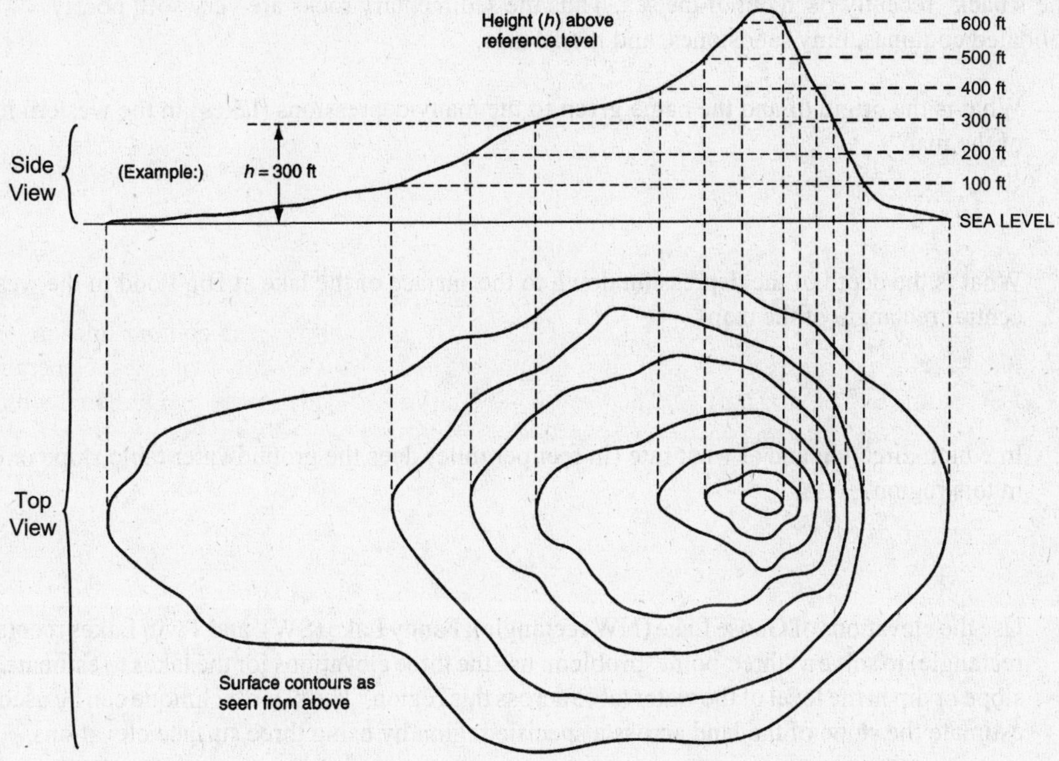

Figure 50.1 Surface profile and topographic map (contour interval = 100 ft)

PROCEDURE 1

Read the introductory material on topographic maps designated by the instructor and answer the following questions for the specific U.S. Geological Survey topographical quadrangle map provided to you. The instructor may change and add questions to fit the quadrangle map in use.

(a) What is the map scale?

(b) What is the total area (in square miles) represented by the quadrangle map?

(c) What is the latitude and longitude of the *City Hall*? OR __________ .
Latitude _______
Longitude _______

(d) What is the location (direction) of the *airport* OR _________ to *city hall* OR _________ ?

PROCEDURE 2

Interlachen, Florida, quadrangle: Geologically speaking, Florida is our newest mainland state—a "whale's back" recently risen out of the sea. Thus the sedimentary rocks are very soft, poorly consolidated coquinas, limy sandstones, and mudstones.

(a) What is the origin of and the name given to the many depressions (lakes) in the western half of the map?

(b) What is the depth of the depression down to the surface of the lake at Big Pond in the west-central rectangle of the map?

(c) In which direction and at what rate (in feet per mile) does the groundwater table slope or dip in this region?

(d) Use the elevations of Goose Lake (NW rectangle), Fanny Lake (SW), and Twin Lakes (center rectangle) to solve a "three-point" problem: use the three elevations for the lakes to estimate the slope or dip in the level of the water table across this region. The same technique can be used to estimate the slope of the land across a specific region by using three surface elevations.

(e) Plot the slope of the ground surface at any scale you wish. Now superimpose on the first slope plane the slope of the water table as determined in the question above. Both profiles should run from the northwestern part of the map to any point within the broad, swampy area in the eastern part of the map.

(f) Do these slope lines explain why the swamps exist where they do? Comment.

(g) What areas on the map have the least slope?

(h) What is the elevation of *Goose Lake* OR __________ ?

What is the elevation of *Big Pond* OR __________ ?

EXPERIMENT 50 NAME (print) ______________________ DATE ____________
LAST FIRST

LABORATORY SECTION ______ PARTNER(S) ______________________

(i) What is the highest elevation shown on the map? __________ft

(j) State the location of the highest elevation as given in Question (i).

(k) What is the direction of flow of ____________ Creek (or River)?

PROCEDURE 3

Draw a contour map to represent the following surface profile.

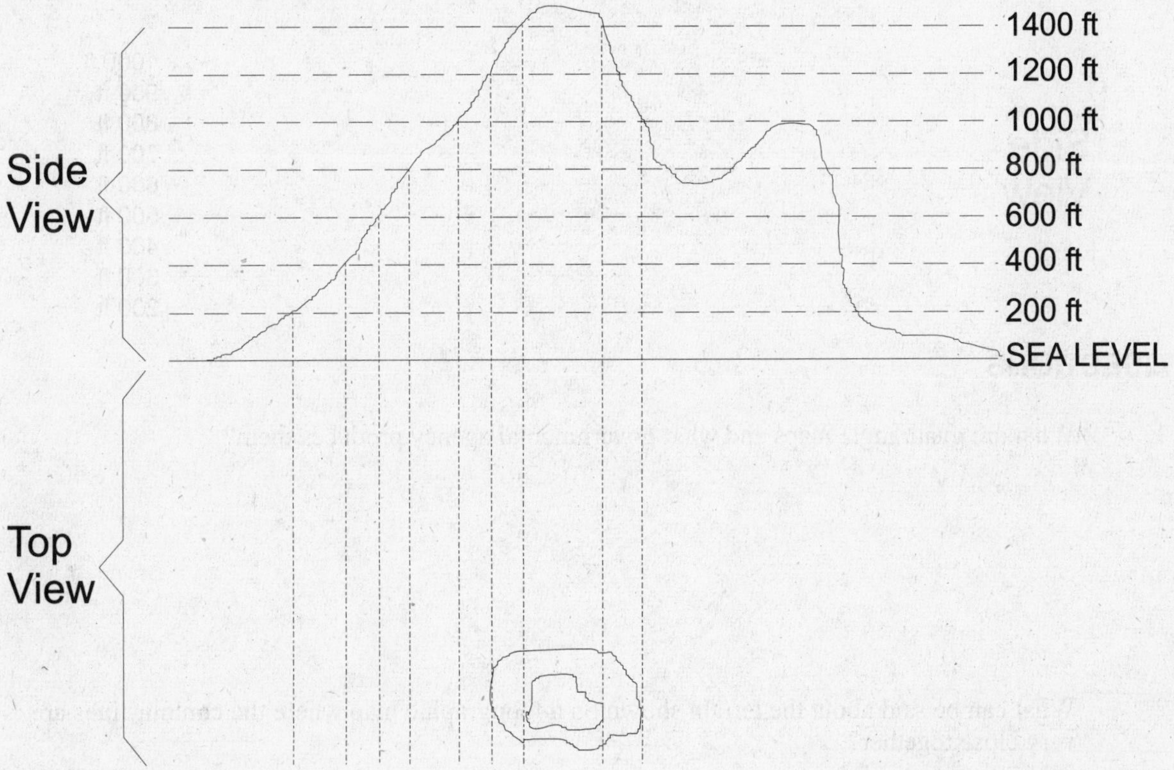

Note: Contour slopes drawn are arbitrary since you do not know the actual topography. Make your contour map follow the proper height contours but there is no right or wrong slope for this drawing.

PROCEDURE 4

Draw the surface profile represented by the following contour map.

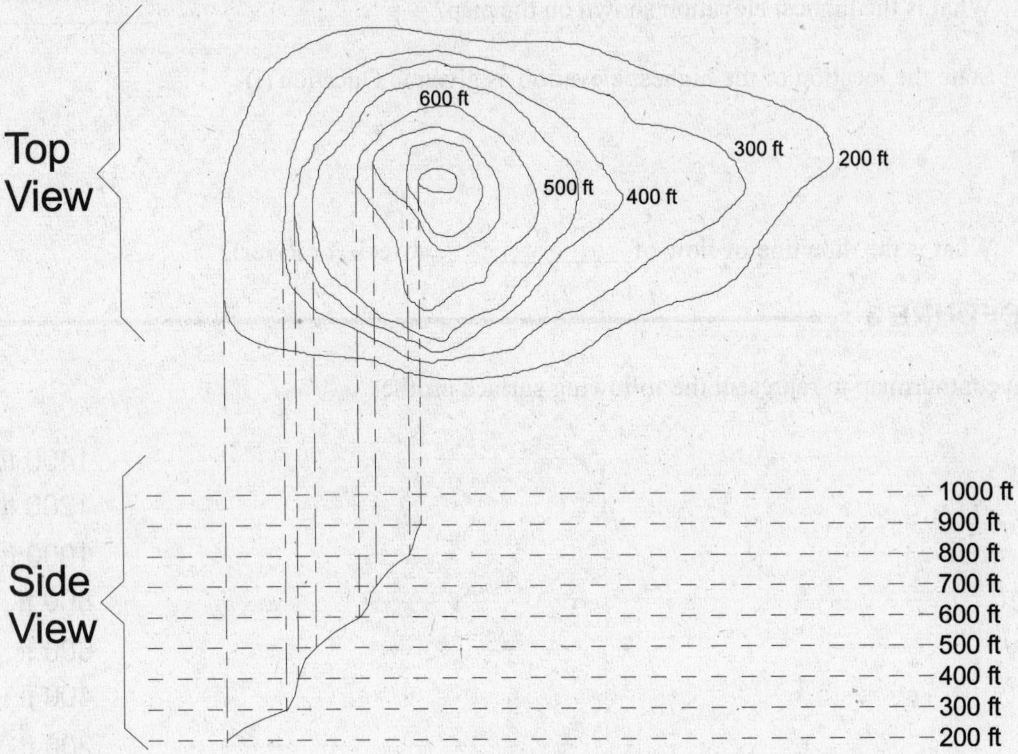

QUESTIONS

1. What are guadrangle maps and what governmental agency produces them?

2. What can be said about the terrain shown on a topographic map where the contour lines are very close together?

Experiment 51

Minerals

INTRODUCTION

A *mineral* is a solid, homogeneous, inorganic substance (compound or element) found occurring naturally in Earth's crust. Minerals possess a fairly definite chemical composition and a distinctive set of physical properties that include hardness, cleavage, color, streak, luster, crystal form, fracture, tenacity, specific gravity, magnetism, fluorescence, and phosphorescence.

Hardness refers to the ability of one mineral to scratch another. The scale (Fig. 51.1) given below is used as a basis for comparing the hardness of some common minerals.

Softest ↑	1. Talc
	2. Gypsum
	Fingernail
	3. Calcite
	Copper coin
	4. Fluorite
	5. Apatite
	Steel knife, plate glass
	6. Feldspar
	Steel file
	7. Quartz
	8. Topaz
	9. Corundum
↓ Hardest	10. Diamond

Figure 51.1 Moh's Hardness Scale

Cleavage refers to the tendency of some minerals to break along definite smooth planes. The mineral may exhibit distinct cleavage along one or more planes, or it may exhibit indistinct cleavage or no cleavage at all. The degree of cleavage that a mineral exhibits is a clue to the identification of the mineral.

Color refers to the selective reflecting of light of one or more specific wavelengths. Although the color of a mineral may be impressive, it is not usually a reliable property for identifying the mineral, since the presence of small amounts of impurities may cause drastic changes in the color of some minerals.

Streak refers to the color of the powder of the mineral. A mineral may exhibit an appearance of several colors but it will always show the same streak color. A mineral rubbed (streaked) across the surface of an unglazed porcelain tile will thereby be powdered and will show its true streak color.

Luster refers to the appearance of the mineral's surface in reflected light. Mineral surfaces appear to have a metallic or nonmetallic luster. A metallic luster has the appearance of polished metal; a nonmetallic appearance may be of varying lusters and the lusters likened to the materials as listed below:

adamantine	appearance of	diamond
greasy	appearance of	oily glass
pearly	appearance of	pearl
resinous	appearance of	yellow stains
silky	appearance of	silk
vitreous	appearance of	glass

Crystalline structure refers to the way the atoms or molecules that make up the mineral are arranged internally. This arrangement is a function of the size and shape of the molecules and the forces that bind them together.

Fracture refers to the way a mineral breaks. The mineral may break into splinters, ragged or rough irregularly surfaced pieces, or shell-shaped forms known as conchoidal fractures.

Tenacity refers to the ability of the mineral to hold together. Some minerals are tough and durable and others are fragile and brittle.

Magnetism refers to the property of possessing a magnetic force field. A mineral possessing magnetism can be detected by a magnetic compass.

Fluorescence refers to the emission of light (to which human eyes are sensitive) by a mineral that is being stimulated by the absorption of ultraviolet or x-ray radiation.

Phosphorescence refers to the emission of light by a mineral after the stimulating source (rays or ultraviolet radiation) has been removed.

LEARNING OBJECTIVES

After completing this experiment, you should be able to do the following:

- ▼ Define a *mineral.*
- ▼ State the methods and techniques for identifying minerals.
- ▼ Identify some common minerals.

APPARATUS

Mineral specimens, copper coin, plate glass, knife blade or nail, steel file, magnetic compass, magnifying glass, ultraviolet lamp.

Reference for Experiments 51 and 52: Michael O'Donoghue, *An Illustrated Guide to Rocks and Minerals* (San Diego: Thunderbay Press, 1996).

KEY TO MINERALS

1. Minerals harder than steel or glass (5.5)*
    - 1.1 Minerals with distinct cleavage or with distinct crystal form
        - 1.11 Minerals with colored streak
            - 1.111 Pyrite
        - 1.12 Minerals with uncolored, white, or pale streak
            - 1.121 Hornblende
            - 1.122 Potash (or potassium), feldspars (variety: orthoclase) (variety: microcline)
            - 1.123 Plagioclase feldspar (variety: albite) (variety: labradorite)
            - 1.124 Rock crystal quartz
    - 1.2 Minerals with indistinct cleavage or with no cleavage
        - 1.21 Minerals with colored streak
            - 1.211 Hematite
            - 1.212 Magnetite
        - 1.22 Minerals with uncolored, white, or pale streak
            - 1.221 Chert
            - 1.222 Flint
            - 1.223 Milky quartz
            - 1.224 Rose quartz
            - 1.225 Jasper
            - 1.226 Olivine
2. Minerals softer than steel or glass (5.5)
    - 2.1 Minerals with distinct cleavage or with distinct crystal form
        - 2.11 Minerals with colored streak
            - 2.111 Galena
            - 2.112 Graphite
        - 2.12 Minerals with uncolored, white, or pale streak
            - 2.121 Biotite
            - 2.122 Calcite
            - 2.123 Gypsum (variety: selenite, satin spar, alabaster)
            - 2.124 Halite
            - 2.125 Muscovite
            - 2.126 Azurite
            - 2.127 Sphalerite
            - 2.128 Fluorite
            - 2.129 Talc
    - 2.2 Minerals with indistinct cleavage or with no cleavage
        - 2.21 Minerals with uncolored, white, or pale streak
            - 2.211 Malachite
            - 2.212 Bauxite

* Hardness scale: 2.5, fingernail; 3.0, copper penny; 5.5, knife blade, window glass; 6.5, steel file.

INDIVIDUAL MINERAL DESCRIPTIONS

1.111 PYRITE
Hardness: 6–6.5
Streak: black to greenish
Luster: metallic
Color: brass yellow
Chem. Comp.: FeS_2
Sp. Grav.: 5.02
Comments: Often cubic in shape with striations on faces; also comes massive; conchoidal fracture; called "fool's gold."

1.121 HORNBLENDE
Hardness: 5–6
Streak: green-gray
Luster: vitreous
Color: greenish black
Chem. Comp.: Ca, Mg, Fe, Al silicate
Sp. Grav.: 3.2
Comments: Long, columnar crystals; visible cleavage in 2 directions; widespread occurrence in igneous and metamorphic rocks, particularly the latter.

1.122 POTASH (or POTASSIUM) FELDSPARS
Variety: ORTHOCLASE
Hardness: 6.2
Streak: white
Luster: vitreous
Color: white to gray
Chem. Comp.: $K(AlSi_3O_8)$
Sp. Grav.: 2.57
Comments: Common rock-forming mineral; monoclinic; two prominent cleavages making angle of 90° with each other.
Variety: MICROCLINE
Hardness: 6
Streak: uncolored, white, or pale
Luster: vitreous
Color: buff, green, pink, gray
Chem. Comp.: $K(AlSi_3O_8)$
Sp. Grav.: 2.54—2.57
Comments: Well-developed cleavage in two directions.

1.123 PLAGIOCLASE FELDSPARS
Variety: ALBITE
Hardness: 6
Streak: uncolored, white, or pale
Luster: vitreous
Color: white or light-colored
Chem. Comp.: $Na(AlSi_3O_8)$
Sp. Grav.: 2.62
Comments: Two-directional cleavage.

Variety: LABRADORITE
Hardness: 6
Streak: uncolored, white, or pale
Luster: vitreous
Color: dark color
Chem. Comp.: $Ca(AlSi_3O_8)$
Sp. Grav.: 2.69
Comments: Bluish sheen produced by striations; occasionally shows iridescence.

1.124 ROCK-CRYSTAL QUARTZ
Hardness: 7
Streak: uncolored, white, or pale
Luster: vitreous
Color: colorless
Chem. Comp.: SiO_2
Sp. Grav.: 2.65
Comments: Six-sided crystals with striations on crystal faces; conchoidal fracture.

1.211 HEMATITE
Hardness: 5.5–6.5
Streak: red-brown, except when in powder, then streak is Indian red
Luster: metallic (when in crystals)
Color: black, dark brown, or red
Chem. Comp.: Fe_2O_3
Sp. Grav.: 5.26
Comments: Has a metallic luster when in crystals; also comes in granular, fibrous, or massive forms.

1.212 MAGNETITE
Hardness: 6
Streak: black
Luster: metallic
Color: black
Chem. Comp.: Fe_3O_4
Sp. Grav.: 5.18
Comments: The variety of magnetite called Lodestone is a natural magnet and will strongly deflect a compass needle.

1.221 CHERT
Hardness: 7
Streak: uncolored, white, or pale
Luster: greasy
Color: variable, but usually light
Chem. Comp.: SiO_2
Sp. Grav.: 2.6
Comments: Opaque on thin edge; compact (dense); conchoidal fracture.

1.222 FLINT
Hardness: 7
Streak: uncolored, white, or pale
Luster: greasy
Color: variable, but usually dark
Chem. Comp.: SiO_2
Sp. Grav.: 2.6
Comments: Compact (dense); conchoidal fracture; translucent on thin edge.

1.223 MILKY QUARTZ
Hardness: 7
Streak: uncolored, white, or pale
Luster: vitreous or greasy
Color: milky
Chem. Comp.: SiO_2
Sp. Grav.: 2.65
Comments: Translucent; no cleavage; milky color due to minute liquid inclusions.

1.224 ROSE QUARTZ
Hardness: 7
Streak: uncolored, white, or pale
Luster: vitreous or greasy
Color: pale pink to rose red
Chem. Comp.: SiO_2
Sp. Grav.: 2.65
Comments: Color due to minute traces of titanium; occasionally coarsely crystalline but usually without crystal form.

1.225 JASPER
Hardness: 7
Streak: pale
Luster: vitreous
Color: red to brown
Chem. Comp.: SiO_2
Sp. Grav.: 2.65
Comments: Red color caused by inclusion of hematite; conchoidal fracture. It is a cryptocrystalline quartz (submicroscopic crystals).

1.226 OLIVINE
Hardness: 6.5–7
Streak: uncolored, white, or pale
Luster: vitreous
Color: olive to gray green
Chem. Comp.: $(Mg, Fe)_2SiO_4$
Sp. Grav.: 3.27–4.37
Comments: Granular appearance in mass; transparent variety known as peridot; fracture is visible in peridot; conchoidal.

2.110 GALENA
Hardness: 2.5
Streak: gray or black
Luster: metallic
Color: lead-gray
Chem. Comp.: PbS
Sp. Grav.: 7.4–7.6
Comments: Opaque, heavy, often in cubes, cubic cleavage in three directions at right angles.

2.112 GRAPHITE
Hardness: 1
Streak: gray or black
Luster: greasy
Color: gray
Chem. Comp.: C
Sp. Grav.: 2
Comments: Has greasy feel; perfect cleavage in one direction.

2.121 BIOTITE (BLACK MICA)
Hardness: 2.5-3
Streak: uncolored, white, or pale
Luster: vitreous
Color: dark green to black
Chem. Comp.: (K, Mg, Fe, Al) silicate
Sp.Grav.: 2.8–3.2
Comments: Basal cleavage prominent (perfect in one direction), flakes in flat sheets; flexible and elastic; transparent to translucent.

2.122 CALCITE
Hardness: 3
Streak: uncolored, white, or pale
Luster: vitreous
Color: sometimes colorless but usually white, though it may be any color
Chem. Comp.: $CaCO_3$
Sp. Grav.: 2.7
Comments: Transparent to translucent, rhombohedral cleavage (three inclined directions *not* at right angles); Iceland Spar variety: clear and double refractive. (Make pencil dot on paper, place calcite over it, and dot appears double.)

2.123 GYPSUM
Hardness: 2
Streak: uncolored, white, or pale
Luster: vitreous (Satin Spar variety-silky luster)
Color: white or light-colored
Chem. Comp.: $CaSO_4 \cdot 2H_2O$
Sp. Grav.: 2.32
Comments:
Variety: Selenite—often clear and transparent; flexible but inelastic sheets; cleavage in three directions, one very good, the others poorer.
Variety: Satin spar—silky luster, fibrous-forming veinlets.
Variety: Alabaster—composed of aggregates of fine crystals.

2.124 HALITE
Hardness: 2.5
Streak: uncolored, white, or pale
Luster: vitreous
Color: colorless to white
Chem. Comp.: NaCl
Sp. Grav.: 2.16
Comments: Salty taste; water soluble; transparent to translucent; cubic cleavage (perfect in three directions at right angles).

2.125 MUSCOVITE (LIGHT-COLORED MICA)
Hardness: 2–2.5
Streak: uncolored, white, or pale
Luster: vitreous
Color: white or light shades of green or yellow
Chem. Comp.: $KAl_2(AlSi_3O_{10})(OH)_2$
Sp. Grav.: 2.76–3.1
Comments: Basal cleavage prominent (perfect in one direction), flakes in flat sheets; flexible and elastic; transparent to translucent.

2.126 AZURITE
Hardness: 3.5–4
Streak: uncolored, white, or pale
Luster: vitreous
Color: azure blue
Chem. Comp.: $Cu_3(CO_3)_2(OH)_2$
Sp. Grav.: 3.77
Comments: Transparent to translucent when in crystal form.

2.127 SPHALERITE
Hardness: 3.5–4
Streak: yellow brown to nearly white
Luster: resinous
Color: black to colorless (colorless rare)
Chem. Comp.: ZnS
Sp. Grav.: 3.9–4.1
Comments: Often shows distinct cleavage planes.

2.128 FLUORITE
Hardness: 4
Streak: uncolored, white, or pale
Luster: vitreous
Color: can be colorless, yellow, blue, green, or violet
Chem. Comp.: CaF_2
Sp. Grav.: 3.18
Comments: Transparent to translucent, cubic crystals, good cleavage in four directions.

2.129 TALC
Hardness: 1
Streak: uncolored, white, or pale
Luster: pearly or greasy
Color: white, gray-green, silver-white
Chem. Comp.: $Mg_2(Si_4O_{10})(OH)_2$
Sp. Grav.: 2.7–2.8
Comments: Very soft; greasy or soapy feel, foliated or compact masses; perfect cleavage in one direction, translucent on thin edge.

2.211 MALACHITE
Hardness: 3.5–4
Streak: uncolored, white, or pale
Luster: vitreous (when in crystals)
Color: bright green
Chem. Comp.: $Cu_2CO_3(OH)_2$
Sp. Grav.: 3.9–4.03
Comments: Silky in fibrous varieties, dull in earthy type.

2.212 BAUXITE
Hardness: 1–3
Streak: uncolored, white, or pale
Luster: earthy
Color: gray to buff or brown
Chem. Comp.: mixture of aluminum hydroxides
Sp. Grav.: 2.0–2.55
Comments: An earthy rock, composed chiefly of gibbsite, boehmite, and diaspore.

EXPERIMENT 51 NAME (print) ______________________________ DATE ______________
LAST FIRST

LABORATORY SECTION ______ PARTNER(S) ______________________________

PROCEDURE

Using the mineral specimens provided, examine and test the samples for as many of the physical properties listed in Data Table 51.1 as you can.

1. To determine the hardness of a sample, first try to scratch a piece of glass with the sample itself. Be careful to choose several areas on the sample to attempt this scratch test because some harder material may be imbedded in the basic material. If the sample scratched glass it is obviously harder than steel or glass. If the sample is very large you can attempt to scratch it with a file or knife blade to make the same determination but THIS SHOULD NOT BE DONE on smaller samples as they can easily be damaged so badly that the next group of students will not be able to use them. Once you know if your sample is harder or softer than steel or glass, you have determined whether it fits into class 1, or class 2, and you have eliminated about one half of the choices in the Key to Minerals. If you know that the sample is softer than steel or glass, you can try to scratch a knife blade, a penny, or your fingernail with the mineral sample to further determine its hardness classification. The note below the Key to Minerals gives you a scale to use to assign some approximate numbers to your hardness determinations.

2. Flat, usually parallel sides on a mineral sample are a good indication that cleavage is present. Look carefully at the sample and try to decide if a distinct cleavage is present or if the mineral shows only indistinct cleavage or no cleavage at all. This breaks your choices up into subclasses .1 or .2 and begins to narrow your choices further.

3. Next draw the mineral sample across the white porcelain streak plate with light to medium pressure. You should observe a dark or colored streak or perhaps an uncolored, white or pale streak. This determination allows you to eliminate many more choices from those possible in the Key to Minerals and thus breaks your choices into subclasses .01 or .02.

The above three procedures help you narrow down the choices for each mineral sample but you are not quite finished yet. Now turn to the detailed Mineral Description table and look at the specific properties of the individual choices given to you in the laboratory manual. These are not all on the minerals that you might find on Earth, but they represent the most plentiful and important ones and should cover any sample given to you in this lab to identify.

Look for luster, color, magnetic properties using a compass, and any other identifying characteristics of each sample. Then compare these characteristics with the information given in the Mineral Descriptions. If you are still in doubt there is one more thing you can try. If you measure the mass of a sample and can determine its volume by emerging it in a partially filled graduated cylinder or by some other method, see experiment 28 in this manual for a more complete explanation of how to do this, you can determine the density (specific gravity*) of the mineral. Specific gravity values are given for each mineral in the Mineral Descriptions and this can be used to lock down your identification with a good deal more confidence.

* Specific gravity is defined as the mass of a sample divided by the mass of an identical volume of pure water. Since the density of pure water in 1.00 g/cm^3, the numerical value for specific gravity is the same as that of the density of a material but because units cancel out it is a unitless quantity that relies on the density of pure water for its actual physical interpretation. Example: if the density of a mineral turns out to be 5.18 g/cm^3, its specific gravity is 5.18 and if the sample is black with a metallic luster, a black streak, and it will NOT scratch glass, the specimen is undoubtedly MAGNETITE. Check the Mineral Descriptions and see if you agree.

DATA TABLE 51.1

Specimen Number	Hard-ness	Cleav-age	Streak	Luster	Color	Magne-tism	Specific Gravity	Mineral Name
1								
2								
3								
4								
5								
6								
7								
8								
9								
10								
11								
12								
13								
14								
15								
16								
17								
18								
19								
20								

Experiment 52

Rocks

INTRODUCTION

A *rock* is an independent geologic unit of Earth's crust. Most rocks are composed of a single mineral or combination of two or more minerals, but some are composed of organic materials (for example, bituminous coal). Rocks are placed in three main classes according to their origin:

1. *Igneous rocks,* formed from the cooling and solidification of molten volcanic material called *magma.*
2. *Sedimentary rocks,* formed by the accumulation of sediments caused by weathering.
3. *Metamorphic rocks,* formed by a change in the structure of igneous and sedimentary rocks through the application of high pressure and temperature.

Rocks are identified on the basis of texture (coarse grain, fine grain, or glassy) and color, which is a function of the mineral content. The identification of rocks in this experiment will be done with charts that give the description of the most common rocks.

Remember the key to crystal size lies primarily in the time that is available while the rock cools. The slower the cooling the larger the crystals will usually be. Since slower cooling generally takes place deep under ground, large crystal structure usually means the magma cooled into rock deep beneath the ground under conditions of very high pressure. Slow cooling also means that separation of minerals is more pronounced because they have a longer time to consolidate because of density differences or varying crystallization temperatures. Surface cooling means rapid formation of the rock and little or no crystal formation occurs during the formation of these surface rocks.

LEARNING OBJECTIVES

After completing this experiment, you should be able to do the following:

- Define the term *rock,* and state the three classifications of rocks.
- Identify some common rocks.

APPARATUS

Numbered rock specimens and magnifying glass.

IGNEOUS ROCKS

1.	Light-colored rocks		
	1.1	Coarse grain, visible to the eye; grains even-sized	
		1.11 Quartz present	*Granite*
		1.12 No quartz present	*Syenite*
	1.2	Fine grained	*Rhyolite*
	1.3	Glassy; no graining visible; glassy luster visible under magnifying lens	
		1.31 Smooth texture; color light; light in weight; vesicular	*Pumice*
2.	Dark-colored rocks		
	2.1	Coarse grain, visible to the eye; grains even-sized	
		2.11 Hornblende present	*Diorite*
		2.12 Olivine present	*Olivine gabbro*
	2.2	Fine-grained; grains even-sized	
		2.21 Nonvesicular	*Basalt*
		2.22 Vesicular; glassy under lens	*Scoria*
	2.3	Glassy; no graining visible; smooth texture, nonvesicular; conchoidal fracture	*Obsidian*

METAMORPHIC ROCKS

1.	Coarse foliation; banded, even-grained, grains fine to coarse; contains feldspars, quartz, and biotite mica		*Gneiss*
2.	Fine foliation, even- or uneven-grained, grains medium to coarse		
	2.1	Essential minerals mica and quartz	*Mica schist*
	2.2	Essential minerals mica and quartz, accessory mineral garnet	*Garnetiferous mica schist*
	2.3	Green color, silky texture	*Chlorite schist*
3.	Slaty cleavage; grains even in size and very fine-grained		*Slate*
4.	Massive; little or no foliation; even-grained, fine to coarse grains		
	4.1	Essential mineral calcite or dolomite	*Marble*
	4.2	Essential mineral quartz	*Quartzite*

EXPERIMENT 52 NAME (print) ______________________________ DATE ______________
LAST FIRST

LABORATORY SECTION ______ PARTNER(S) ______________________________

SEDIMENTARY ROCKS

1. Typically coarse, *angular* fragments cemented in finer materials, vary from pebble size upward — *Breccia*

2. Typically coarse, rounded fragments cemented in finer materials, vary from pebble size upward — *Conglomerate*

3. Sand grains, cemented together by quartz, calcite, or iron oxides; rounded quartz grains most abundant constituent. Granular; medium to fine grains (size of sugar grains)
   - 3.1 Iron oxide cement
     - 3.11 Red color — *Red sandstone*
     - 3.12 Brown; contains abundant mica, usually muscovite — *Micaceous sandstone*

4. Clay and silt particles; grains smaller than sand and barely visible. Well compacted; clayey odor.
   - 4.1 Red, brown, or yellow color — *Ferruginous shale*

5. Altered plant remains; black, massive, unlaminated; grains indistinguishable; smooth, dull glassy luster; does not soil hands and paper — *Cannel coal*

6. Variable texture, crystalline, granular, coarse or fine; effervesces in hydrochloric acid
   - 6.1 Very fine-grained; compact, dense; conchoidal fracture — *Lithographic limestone*
   - 6.2 Composed largely of shells — *Shell limestone*
   - 6.3 Composed of small masses about the size of a pinhead or smaller — *Oolitic limestone*

PROCEDURE 1

Complete the information for the introduction to igneous, sedimentary, and metamorphic rocks in Data Table 52.1.

DATA TABLE 52.1 Introduction to Igneous, Sedimentary, and Metamorphic Rocks

Examine the numbered specimens and decide on their most distinctive features, keeping in mind that some of the features may also be visible in other types or groups and that some may not appear in other specimens of the same type or group. The following questions should be answered concerning properties of the rock name and rock group given.

1. Rock name **Obsidian** Rock group **Igneous**

a. Is this rock intrusive or extrusive by formation? (Circle one.)

b. What evidence is there to substantiate your answer?

____

c. What do you suppose the general mineral composition of the rock might be?

____

2. Rock Name **Granite** Rock group **Igneous**

a. Compared to obsidian, did this rock cool rapidly or did it cool at a slower rate? (Circle one.)

b. What evidence do you find in this rock to substantiate your answer?

____

c. What are two essential minerals that compose this specimen?

(1) ________________ (2) ________________

d. What is the black accessory mineral found in the specimen? ________

3. Rock name **Micaceous sandstone** Rock group **Sedimentary**

a. What is the composition of the lighter colored grains?

____

b. Are they equal-granular (well sorted)? ________________

c. What rock sources might these grains be derived from?

(1) ________________ (2) ________________

EXPERIMENT 52 NAME (print) ______________________ DATE ______________________
LAST FIRST

LABORATORY SECTION ______ PARTNER(S) ______________________

DATA TABLE 52.1—continued

4. Rock Name **Chlorite schist** Rock group **Metamorphic**

   a. This rock has a foliated texture. How is this texture formed?

   ____

5. Rock name **Marble** Rock group **Metamorphic**

   a. What is the relative hardness of this rock sample? ______________________

   b. What is the chemical composition of this rock sample? ______________________

   c. This rock has undergone metamorphism. What do you think the original rock might have been?

   ____

6. Rock name **Lithographic limestone** Rock group **Sedimentary**

   a. What is the relative hardness of this rock sample? ______________________

   b. What is the chemical composition of this rock sample? ______________________

   c. What may be a good interpretation for the environment of the deposition for this sample (fluvial, marine, aeolian, continental)? (Circle one.)

7. Rock name **Scoria** Rock group **Igneous**

   a. Is this rock intrusively or extrusively formed? (Circle one.)

   b. What evidence is there to substantiate your answer?

   ____

8. Rock name **Sandstone, red** Rock group **Sedimentary**

   a. What causes the red coloring in this sample? ______________________

   b. What is the mineral composition of the grains in this sample? ______________________

DATA TABLE 52.1—continued

9. Rock name **Pumice** Rock group **Igneous**

a. How does this sample differ from sample No. 7? ____________________

____

b. Is this sample intrusively or extrusively formed? (Circle one.)

10. Rock name Shell **limestone** Rock group **Sedimentary**

a. What types of fossils are found in this sample? What part of the ocean did they probably come from (beach, reef, deep ocean)? (Circle one.)

____

b. What is the chemical composition of this rock sample? ____________________

____

11. Rock name **Quartzite** Rock group **Metamorphic**

a. This is a granular metamorphic rock consisting essentially of quartz. What do you think the original rock was before metamorphism changed it?

____

12. Rock name **Slate, gray** Rock group **Metamorphic**

a. This is a foliated rock. What general rock type could it have been before the effects of metamorphism?

____

b. What is the economic use of this rock? ____________________

13. Rock name **Basalt** Rock group **Igneous**

a. What is the probable composition of this rock? ____________________

b. Is it intrusively or extrusively formed? (Circle one.)

EXPERIMENT 52 NAME (print) ______________________________ DATE ______________
LAST FIRST

LABORATORY SECTION ______ PARTNER(S) ______________________________

DATA TABLE 52.1—continued

14. Rock name **Gneiss** Rock group **Metamorphic**

a. Explain the parallel alignment of the individual layers.

____

b. Which has undergone a higher degree of metamorphism, rock no. 12 or rock no. 14? (Circle one.)

15. Rock name **Cannel coal** Rock group **Sedimentary**

a. This is a variety of bituminous coal of uniform and compact fine-grained texture with general absence of banded structure; composed predominantly of plant spores. Can you see this structure with a magnifying glass? ______________

PROCEDURE 2

Identify each of the specimens provided, and record in Data Table 52.2.

DATA TABLE 52.2

Specimen Number	Rock Name	Texture	Composition or Mineral Content
1			
2			
3			
4			
5			
6			
7			
8			
9			
10			
11			
12			
13			
14			
15			
16			
17			
18			
19			
20			

Experiment 53

Rock-Forming Minerals

INTRODUCTION

Of the more than one hundred chemical elements, only eight (oxygen, silicon, aluminum, magnesium, iron, calcium, sodium, and potassium) combine in different proportions and manner to make up most of the common minerals. Of the two thousand or so recognized species of minerals, only a score comprise by far the greater amount of the common rocks observed on Earth's surface. These few prevalent minerals can be termed the *rock-forming minerals.*

Most igneous rocks consist of *crystals* of two or more of the following minerals: olivine, pyroxene, amphibole, plagioclase feldspar, potassium feldspar (orthoclase), biotite, muscovite, and quartz.

Some sedimentary rocks, notably the evaporites, also consist of crystals, but very often of only a single mineral. For example, rock gypsum is made up of the mineral gypsum, and rock salt is largely the mineral halite; some limestones are almost pure calcite. On the other hand, most sedimentary rocks contain more or less rounded particles or grains of several minerals cemented together. Although a few sandstones may have over 90 percent quartz, most sandstones are generally mixtures of clay minerals and several of the other minerals named above; that is, the rock is the expected residue of weathered igneous rocks.

Most metamorphic rocks also contain several of the rock-forming minerals, but in addition they usually have suites of less common minerals that were created by pressures, heat, and chemical solutions acting upon and altering the original materials during the metamorphic processes.

LEARNING OBJECTIVES

After completing this experiment, you should be able to do the following:

- ▼ Name the principal rock-forming minerals.
- ▼ From supplied minerals, determine the bonding, or consolidation mechanism of igneous rocks, as contrasted with that of sedimentary rocks.

APPARATUS

Mineral specimens, plate glass, knife blade, steel file, magnifying glass, key to mineral identification, microscope, thin section slides of granite and sandstone.

Caution Safety Tips

Be careful when using sharp or pointed instruments during rock testing procedures.

EXPERIMENT 53 NAME (print) ______________________ DATE ______________
LAST FIRST

LABORATORY SECTION ______ PARTNER(S) ______________________

PROCEDURE

1. From the minerals in the study tray, identify by number the igneous rock-forming minerals listed below. The instructor will have a set of labeled minerals so that you may later check your identifications.

DATA TABLE 53.1

Specimen Number	Mineral	Identifying Remarks
	Orthoclase feldspar	
	Plagioclase feldspar	
	Olivine	
	Quartz	
	Amphibole pyroxene	
	Biotite	
	Muscovite	
	Pyrite	
	Magnetite	

2. The instructor will provide a specimen of coarsely crystalline granite.

(a) List the minerals you can identify in the granite.

(b) Which component minerals show cleavage?

Which important component mineral has no cleavage?

(c) Examine the granite specimen with the magnifying glass or under the microscope. Are there any rounded grains?

Which minerals show the best crystal shapes?

Which single mineral shows the greatest range in size and the poorest geometric shape?

Igneous rocks that form from magma slowly cool and crystallize. Which mineral appears to have crystallized early?

Which mineral seems to have formed quite late?

What is your reasoning for the answer to the above question?

3. (a) How are granites and other igneous rocks bonded or held together? As an aid in answering this question, examine under the microscope the thin section provided by the instructor.

(b) As a contrast, examine the thin section of sedimentary rock (in this case a "clean" sandstone) under the microscope. How do the mineral grains in this rock differ in shape from those in any other igneous rock? Of what mineral is this rock almost wholly composed? (See the hand specimen also.)

(c) How are sedimentary rocks, in general, consolidated or held together? (Ask the instructor for a hint, if necessary.)

Experiment 54

Igneous Rocks and Crystallization

INTRODUCTION

Most igneous rocks are composed of two or more minerals that are silicates. Experiments done in the laboratory show that the crystallization of silicate materials from molten magma occurs within a temperature range of 1200 to 600°C. The minerals that crystallize out at the higher temperatures show well-defined crystal forms. This is to be expected, because they have greater freedom to grow within the molten melt. Minerals forming at the lower temperatures will develop with restricted freedom and the crystals thus formed will have less well-formed faces.

It is not a difficult task to let a mineral cool and note the crystallization, but it should be remembered that the actual conditions within a cooling magma are complex, and the resulting minerals formed by crystallization depend on many factors.

LEARNING OBJECTIVES

After completing this experiment, you should be able to do the following:

- Describe crystallization of a warm melt, which is analogous to the crystallization of igneous rocks from a magma.
- Identify a few igneous rocks and ascertain their textures.

APPARATUS

Igneous rock specimens, glass slides and/or evaporation dishes, forceps, liquid dropper or wire loops, thymol crystals, saturated solutions of several salts. Note: Thymol may irritate the eyes or skin; handle its solution with the dropper and its crystals with the forceps.

Safety glasses or goggles are recommended. The chemical thymol can irritate skin and must not get into eyes. Use care when handling all chemicals and wash hands thoroughly after experiment.

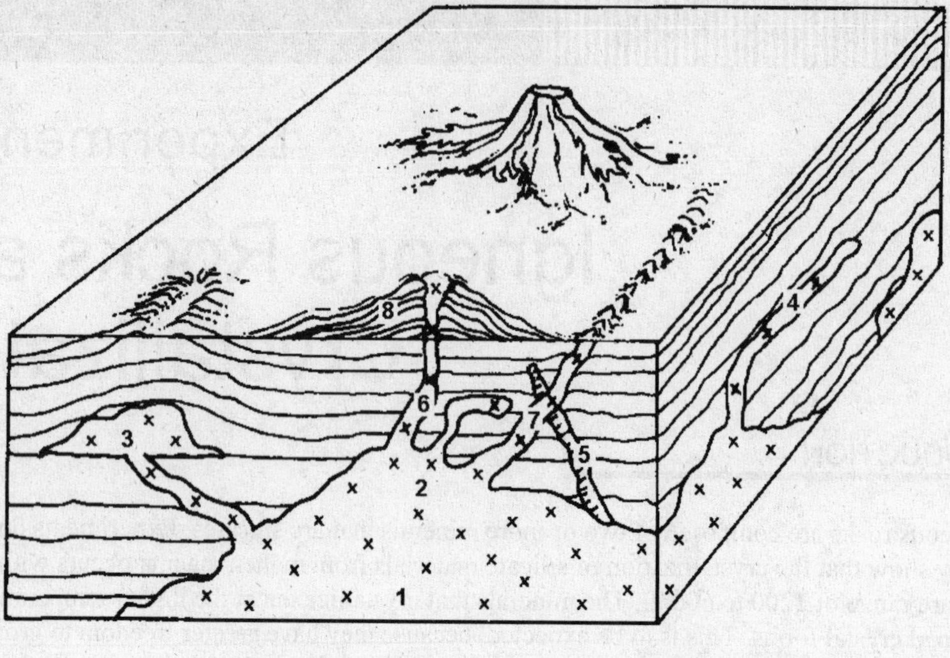

Figure 54.1 Diagram of igneous features.

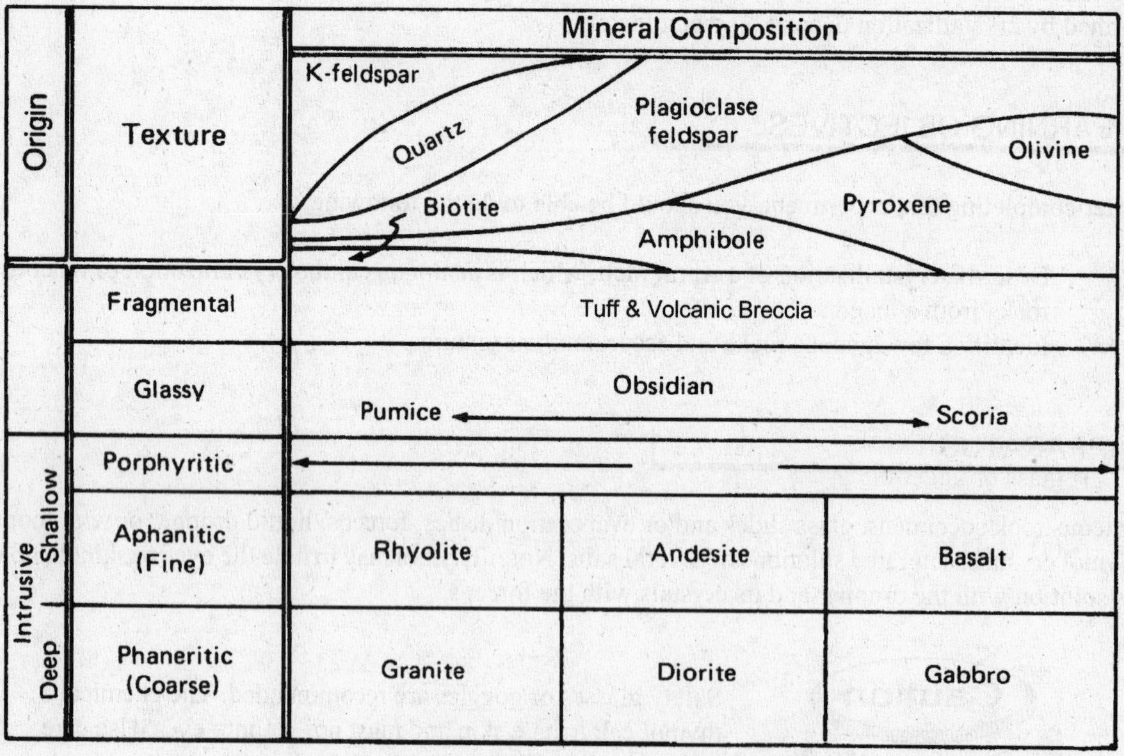

Figure 54.2 An aid in rock identification

EXPERIMENT 54 NAME (print) ______________________________ DATE ______________
LAST FIRST

LABORATORY SECTION ______ PARTNER(S) ______________________________

PROCEDURE

1. Name each of the igneous features in Fig. 54.1 and answer the following questions.

1. ____________ 2. ____________ 3. ____________ 4. ____________

5. ____________ 6. ____________ 7. ____________ 8. ____________

(a) Name any two features that usually have crystalline texture.

(b) Suggest one feature that might show porphyritic texture.

(c) Where would you expect to find rocks showing glassy texture?

(d) Which igneous body is older, 5 or 7? Why ?

2. On the front table are vials of saturated solutions of sodium chloride, sodium nitrate, potassium aluminum sulfate (alum), and copper acetate. Place two or three drops of several of the solutions on a glass slide or in an evaporating dish. Label each specimen, allow to evaporate slowly in sunlight or over a hot plate set on very low heat.

3. From the melted thymol that the instructor provides, place a large drop near the end of a glass slide and allow it to spread to the size of a dime. Using forceps, drop two or three crystals of thymol at scattered points into the melted thymol and observe quickly with the magnifying glass or under the microscope.

(a) Do the crystals grow by internal expansion or by external accretion?

(b) Do the crystals show growth lines?

(c) What is the effect of limited space upon the shapes of some crystals?

4. Place several crystals of thymol near the end of a glass slide and crush them to powder with another glass slide or other smooth object. Using a dropper or wire with a tiny loop, place a drop of melted thymol on the powdered thymol. Examine the crystallization as before.

   (a) How does the crystallization in this experiment differ from that in the above?

   (b) Explain the difference in terms of the spontaneity of crystallization. The same result can be produced by placing melted thymol on a glass slide and then immediately cooling it on an ice cube for some seconds.

   (c) What effect, then, does the rate of cooling have on crystal size?

5. Although all igneous rocks crystallize from a cooling silicate melt (a magma), some sedimentary rocks, the evaporites, crystallize by the evaporation of cool saline solutions. Examine the specimens you prepared previously in Procedure 1. Note that, for any particular crystalline substance, the angle between corresponding crystal faces is always the same.

   (a) Does each compound have a unique crystal shape?

   (b) Do you think that some minerals may be identified by crystal shape alone?

6. Refer to Fig. 54.2 and identify the igneous rocks provided in the specimen tray. In the spaces in Data Table 54.1, briefly describe the texture of each rock.

DATA TABLE 54.1

Rock	Number	Texture
Granite		
Basalt		
Obsidian		
Andesite		
Scoria		

Experiment 55

Sedimentary Rocks

INTRODUCTION

Four processes are involved in the creation of sedimentary rocks: (1) physical and chemical weathering of the parent rock, (2) transportation of the weathered products by gravity, running water, wind, or ice, (3) deposition, with some sorting according to sizes of grains, in sedimentary basins such as lakes, stream channels, and especially the continental shelves of the oceans, and (4) compaction and cementation of the sediment into solid rock.

Sedimentary rocks are composed largely of one or more of only four materials: quartz, clay, rock fragments, and shells composed of calcium carbonate (calcite). Most sedimentary rocks are termed *detrital;* that is, they consist of "broken" particles weathered from preexisting rocks. The second most important group of sedimentary rocks are of biochemical and chemical origin, and some, like rock salt and gypsum, are crystalline in texture, having been formed by the evaporation of saline lake and ocean waters. Rarely, organic matter alone forms sedimentary rock, such as coal. The two groups of sedimentary rocks are outlined in Table of Sedimentary Rocks.

LEARNING OBJECTIVES

After completing this experiment, you should be able to do the following:

- ▼ Describe several properties of sedimentary rocks.
- ▼ Identify some common sedimentary rocks.

APPARATUS

Coarse sand, mixture of sands, a glass jar, a graduated cylinder, a millimeter scale, tray of rock specimens, dilute hydrochloric acid, dropper.

Caution

Safety Tips

Safety glasses or goggles are recommended. Be careful when handling or pouring acid. The HCl used here is dilute, but any contaminated skin surface should be thoroughly flushed with clear water. Do not get in your eyes.

Table of Sedimentary Rocks

DETRITAL ROCKS		
Texture	Composition	Rock Name
Coarse-grained (over 2 mm)	Rounded fragments of any rock type, usually quartz, quartzite or chert	Conglomerate
	Angular fragments of any rock type	Breccia
Medium-grained (1/16 mm to 2 mm)	Over 90% quartz	Arkose
	Quartz with 25% feldspar	Graywacke
	Quartz, clay, and rock chips	Siltstone
Fine-grained (1/256 mm to 1/16 mm)	Quartz and clay	Siltstone
Very fine-grained (less than 1/256 mm)	Clay and quartz	Shale
CHEMICAL ROCKS		
Medium-to-coarse crystalline		Crystalline limestone
		Lithographic limestone
Very fine crystalline		
Composed of tiny pellets with concentric internal structure		Oolitic limestone
Fragmented shells loosely cemented	Calcite ($CaCO_3$)	Coquina
		Fossiliferous limestone
Abundant shells in dense calcareous matrix		
Microscopic shells and clay		Chalk
Banded calcite, often spongy		Travertine
Crystalline replacement of limestones	Dolomite ($CaMg(CO_3)_2$)	Dolomite
Microcrystalline	Chalcedony (SiO_2)	Chert
Fine-to-coarse crystalline	Gypsum $CaSO_4 \cdot 2H_2O$	Gypsum
Medium to-coarse crystalline	Halite (NaCl)	Rock salt
Dense, banded	Altered plant remains	Coal

EXPERIMENT 55 NAME (print) ______________________________ DATE ______________
LAST FIRST

LABORATORY SECTION ______ PARTNER(S) ______________________________

PROCEDURE

1. Measure out on a piece of paper enough of the coarse sand at the front desk to make a column about 1 in high in the glass jar or graduated cylinder. Drop or quickly pour all of the sand into the jar or cylinder without agitating the vessel. Measure carefully the height of the top of the sand above the tabletop.

   (a) Height of sand above the table __________ mm

   Now tamp the base of the jar or cylinder lightly half a dozen times on the palm of your hand and remeasure the height of the sand top.

   (b) Height of sand after tamping __________ mm

   All newly deposited sediments eventually compact and become denser because of the weight of other sediments being deposited on top.

   (c) By what proportion or percentage did the sample compact?
   Percentage = 100 × decrease/original.

   Ignore the thickness of the bottom of the container. Leave the sample in the container to be used in the next procedure.

2. Calculate in cubic centimeters the overall volume of the cylindrical sample used above. If the sample is in a graduated cylinder, volume can be read directly; one millimeter equals approximately one cubic centimeter.

   Volume $= 3.14(R^2 H)$, where R is radius and H is height of column after tamping.

   (a) Bulk volume of sample, V_b __________ mL

   Noting carefully the amount of water used, very slowly add water to the sample so that it comes just to the top of the sand.

   (b) Volume of water added, V_w __________ mL

   (c) Calculate the porosity or void space of the sand sample:

   Percent porosity $= \dfrac{V_w}{V_b} \times 100$ __________ %

3. Obtain a few spoonfuls of the mixed sands at the front desk. Fill a jar or cylinder somewhat more than half-full of water. Add the mixed sand and check to see if different beds or layers of sediment have formed on the bottom. If not, shake and invert the jar and then allow the sediment to resettle to the bottom.

   (a) Observe the two different layers. What properties of the sediments tell you that there are indeed separate layers present?

   (b) If sand and silt were being transported by a stream, which would be moved along the bottom and which carried by suspension?

   (c) If sand and silt were being delivered simultaneously to the sea, which would be deposited first?

4. The tray on the laboratory table contains several sedimentary rocks. Use Table of Sedimentary Rocks and the information on sedimentary rocks to identify the rocks listed in Data Table 55.1.

DATA TABLE 55.1

Number	Rock Name	Texture	Mineral Composition
	Quartz sandstone		
	Fossiliferous limestone		
	Shale		
	Gypsum or rock salt		
	Arkose (sandstone)		
	Coquina		
	Oolitic limestone		
	Conglomerate		

Appendix I

Measurement and Significant Figures

A *measurement* is a comparison of an unknown quantity with a precisely specified quantity called a standard unit. Measurements should be made and recorded as accurately as possible. The term *accuracy* refers to how close the measurement comes to the exact value. *Accuracy* depends on the careful use of measuring instruments and the ability of the individual taking the measurements. Errors will be made in every measurement, but the magnitude of the errors can be kept small when the observer is conscientious and uses an instrument with great care.

Precision refers to the degree of reproducibility of a measurement; that is, to the maximum possible error of the measurement. Precision may be expressed as a plus or minus correction. All measurements are approximate. It is impossible to know the "exact" length, mass, or amount of anything. The observer and the measuring instrument place a limit on the accuracy of all measurements.

Several readings that are grouped closely together have a high precision even though their average value may not be very close to the accepted value. Such a group of readings would have good precision but poor accuracy. The reverse can also be true. A measurement consisting of a sloppy set of readings that are not well grouped but whose average value just happens to be quite close to the accepted value, could be said to have high accuracy but low precision.

Instruments for taking measurements are constructed with a calibrated scale for obtaining numerical values concerning the property being measured. The smallest division on the calibrated scale that can be read by the observer, without guessing, is known as the *least count* of the instrument. For example: A meter stick is divided into 100 equal divisions and marked on the stick as centimeters. Each centimeter is further divided into 10 equal divisions. Thus, the least count of the meter stick is one tenth of one centimeter, or one-thousandth of one meter.

When the meter stick is used to take a measurement of an unknown quantity (for example, length), the observer can always obtain a value within one-thousandth (0.001) of a meter of the exact value of the unknown length, without guessing. But an additional step can increase the accuracy by one doubtful digit, because the observer can estimate a fractional part of the smallest division on the meter stick. For example, if one end of the meter stick is placed at one end of the object to be measured, and the other end of the object falls between two of the smallest divisions, the observer estimates this additional value and adds it to the known scale reading. This estimated digit is significant and is the last digit recorded when taking a measurement.

When a measurement is made, the number recorded should contain all known digits of the measurement plus one doubtful digit. The number of digits recorded for a measurement is referred to as the number of significant figures in the measurement. By definition, a significant figure is a known digit or a doubtful digit that has been determined and recorded when the measurement was taken. Remember, a proper measurement always contains all known digits that were taken utilizing the entire last count of the measuring device that was used, plus one doubtful digit that had been estimated by the person taking the reading.

Recorded numbers may contain the digit zero (0), which may or may not be significant. When a zero digit is used to locate the decimal point, it is not significant. For example, the numbers 0.048, 0.0032, and 0.00057 each have two significant digits. When a zero appears between two nonzero digits in a number, it is significant. For example, 2.04 has three significant digits; 8.002 has four significant digits. Zeros appearing at the end of a number may or may not be significant. For example, in the number 5480 if the digit eight is a doubtful digit, then the zero is not significant. If the eight is a known digit and the zero is a doubtful digit, then the zero is significant.

The powers-of-10 notation can be used to remove the ambiguity concerning a number like 5480. When using the powers-of-10 notation, we first write all the significant digits of the number. This is followed by 10 to the correct power to locate the decimal point. For example, for three significant digits we write 5.48×10^3 for four significant digits we write 5.480×10^3.

The following procedure is usually used to round off significant figures to fewer digits. If the last significant digit on the right is less than 5, drop it and insert zero instead. If the last significant digit on the right is 5 or greater, drop it and increase the preceding digit by one.

EXAMPLES

Round off the following numbers to two significant digits.

247 Because the last digit on the right is greater than 5, drop it and increase the preceding digit by one, for the result 250.

243 Because the last digit on the right is less than 5, drop it and insert zero instead, for the result 240.

245 Because the last digit on the right is 5 or greater, drop it and increase the preceding digit by one, for the result 250.

As a general rule, the number of significant digits of the product or the division of two or more measurements should be no greater than that of the measurement with the least number of significant digits. For example, suppose the area of a table is to be determined. This is accomplished by measuring the length and width of the table, then multiplying one value by the other. In this procedure, it is inaccurate and meaningless to calculate and give an answer indicating greater accuracy than justified by the original data. For example: the length of a table is measured with a meter stick as 1.8245 m (5 significant figures) and the width as 0.3672 m (4 significant figures). The area $A = 1.8245\ \text{m} \times 0.3674\ \text{m} = 0.6703213\ \text{m}^2$ as shown on a calculator. This six-figure number is not justified as the correct area of the table as given by the two measurements. The correct value for the area is $0.6703\ \text{m}^2$. This value has four significant digits corresponding to the least number of significant digits in the two numbers making up the original data.

When adding or subtracting significant figures use the following two rules:

Rule 1. A known digit plus or minus a doubtful digit will give a doubtful digit.
Rule 2. Only one doubtful digit is allowed in a significant figure.

For example:

2.34	the 4 is doubtful
+ 16.5	the 5 is doubtful
18.84	the 8 and the 4 are doubtful

From Rule 1, the known digit 3 plus the doubtful digit 5 equals a doubtful digit 8. Therefore, the correct answer is 18.8. This is a significant figure with only one doubtful number, which satisfies Rule 2.

Appendix II

Conversion of Units

The solution to many problems in physical science requires changing units of measurements from one system to another. The following method is simple and easy once it is understood. An increased understanding of this method can be gained by solving several practice problems.

For example: What is the distance in feet traveled by an automobile in 10 s, if the average speed of the car is 60 mi/h? Because we want to know the distance traveled in feet, we must convert miles per hour to feet per second. This can be done as follows:

1. Write down the term you want to convert and set it equal to itself. At the same time, make a mental or written note of the terms to which you wish to convert.

$$\frac{60\text{ mi}}{\text{h}} = \frac{60\text{ mi}}{\text{h}}$$

2. Multiply the right side of the equation by 1 or 1/1, which is the same thing. It does not change its actual value.

$$\frac{60\text{ mi}}{\text{h}} = \frac{60\text{ mi}}{\text{h}} \times \frac{1}{1}$$

Where do we get these 1/1 factors? We must find an appropriate conversion factor and make it into a fraction that is equal to 1/1. As a simple example, let us look at the conversion factor "1 yd = 3 ft". This conversion factor will equal 1/1 if we set it up as 1 yd/3 ft or as 3 ft/1 yd. Which one do we use? When converting yards to feet we must cancel the yd in the original value so we use 3 ft/1 yd. If we were converting feet to yards we would use the other fraction.

3. Find the appropriate conversion factor and arrange it as a fraction so that the unit in the numerator or denominator of the 1/1 will cancel out the unwanted unit in the original term. In this case, you can use the conversion factor: 1 mile = 5280 ft. If you do not know the conversion factor off hand, you can find the appropriate values in the table of conversion factors that has been conveniently placed on the inside back cover of this manual. Place the term "1 mile" in the denominator and at the same time place an equal term, "5280 feet" above this newly introduced denominator in order to keep the value equal to 1/1. Then you can cancel like units and calculate the final numerical value for the original quality that will now be in the new units that you want.

$$\frac{60\text{ mi}}{\text{h}} = \frac{60\,\cancel{\text{mi}}}{\text{h}} \times \frac{5280\text{ ft}}{1\,\cancel{\text{mi}}}$$

4. The next step is to eliminate the unit hour in the denominator. This can be accomplished by multiplying again by 1/1 and placing the unit hour in the numerator of the 1/1 term. At the same time, an equal unit should be placed in the denominator of the 1/1 term in order to keep the value equal to 1/1. Because 60 min is equal to 1 h, we can use it and then cancel both "hour" terms.

$$\frac{60\text{ mi}}{\text{h}} = \frac{60\cancel{\text{mi}}}{\cancel{\text{h}}} \times \frac{5280\text{ ft}}{1\cancel{\text{mi}}} \times \frac{1\cancel{\text{h}}}{60\text{ min}}$$

5. At this point, we have feet per minute, but we want feet per second; therefore we must convert the minutes to seconds. Again we multiply by 1 or 1/1 and add the appropriate units for conversion. Because 60 sec is equal to 1 min, we obtain

$$\frac{60\text{ mi}}{\text{h}} = \frac{60\cancel{\text{mi}}}{\cancel{\text{h}}} \times \frac{5280\text{ ft}}{1\cancel{\text{mi}}} \times \frac{1\cancel{\text{h}}}{60\cancel{\text{min}}} \times \frac{1\cancel{\text{min}}}{60\text{ s}}.$$

6. After all like units are canceled, the numbers are canceled insofar as is possible; we arrive at

$$\frac{60\text{ mi}}{\text{h}} = \frac{528\text{ ft}}{6\text{ s}}$$

or, after dividing,

$$\frac{60\text{ mi}}{\text{h}} = \frac{88\text{ ft}}{\text{s}}.$$

The problem can now be solved using 88 ft/s for 60 mi/h. Knowing that

$$d = vt$$

and substituting

$$d = \frac{88\text{ ft}}{\text{s}} \times 10\text{ s}$$

we obtain

$$d = 880\text{ ft}.$$

Appendix III
Experimental Error

The process of taking any measurement always involves some uncertainty. Such uncertainty is usually called experimental error. Two methods are used to calculate the amount of error: (1) When a standard or accepted value of the physical quantity is known, percent error is calculated. (2) When no accepted value exists, the percent difference must be calculated.

METHOD 1

$$\text{Percentage error} = \frac{\text{Absolute difference}^{*}}{\text{accepted value}} \times 100$$

$$= \frac{|E_V - A_V|}{A_V} \times 100$$

where E_v is the experimental value and A_v is the accepted value (also known as the standard value).

* Absolute difference means that the smaller value is subtracted from the larger value.

METHOD 2

$$\text{Percentage difference} = \frac{\text{largest experimental value} - \text{smallest experimental value}}{\text{Average of all experimental values}} \times 100$$

To find the average:

$$\text{Average} = \frac{E_1 + E_2 + E_3 + \ldots E_n}{n}$$

where E_1, E_2, E_3, and E_n represent the individual experimental values and n represents the number of experimental values being averaged. The percent difference equation can also be written as:

$$\text{Percentage difference} = \frac{(E_L - E_s)}{\text{Average}} \times 100$$

where E_L is the largest experimental value and E_s is the smallest experimental value.

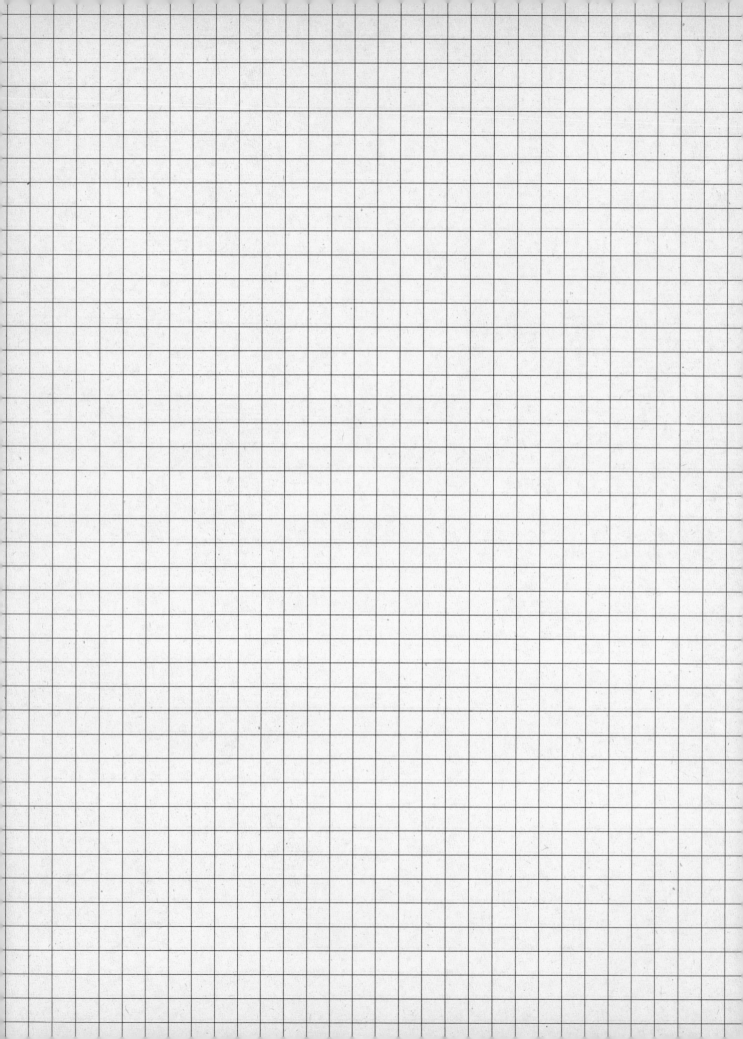

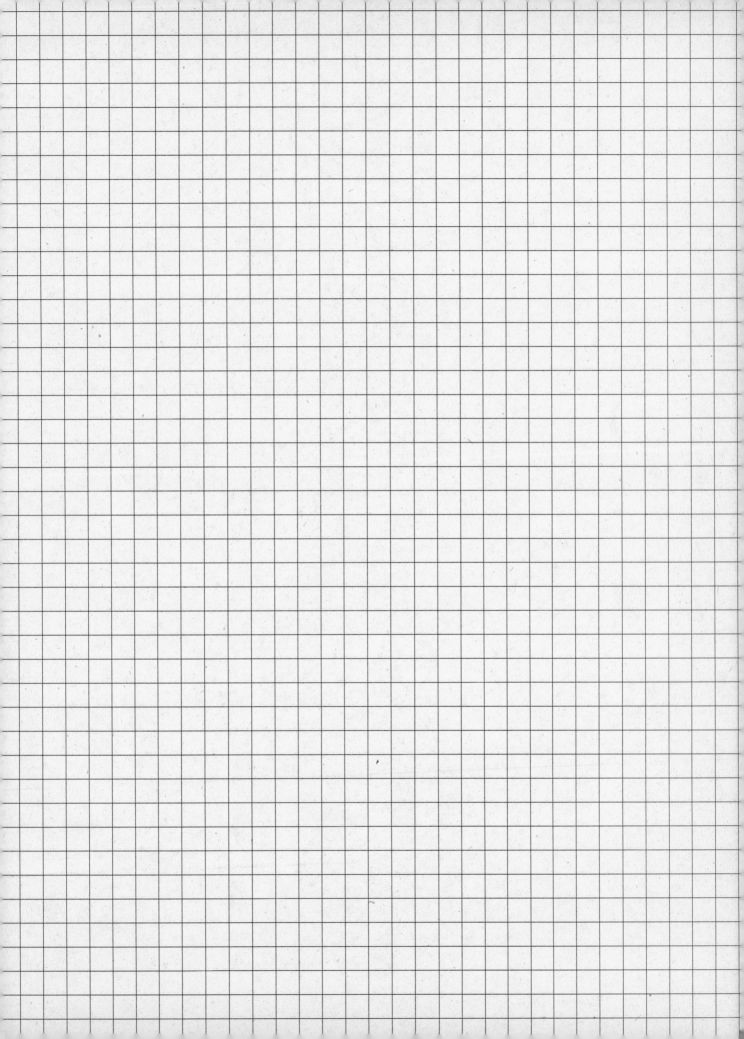

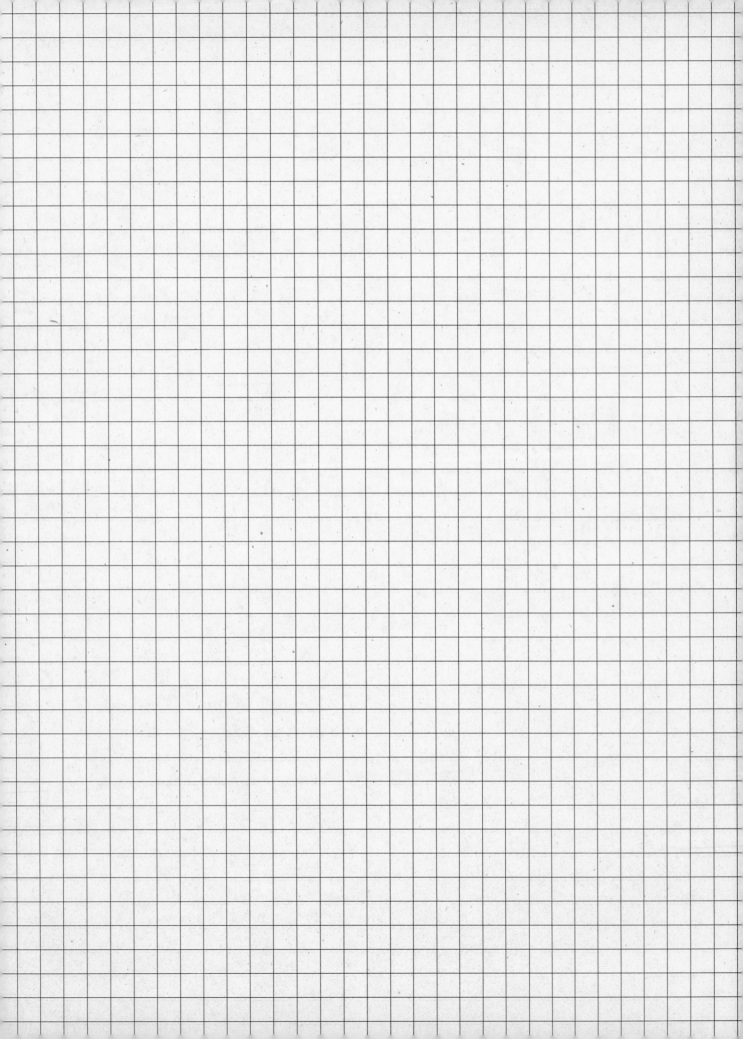

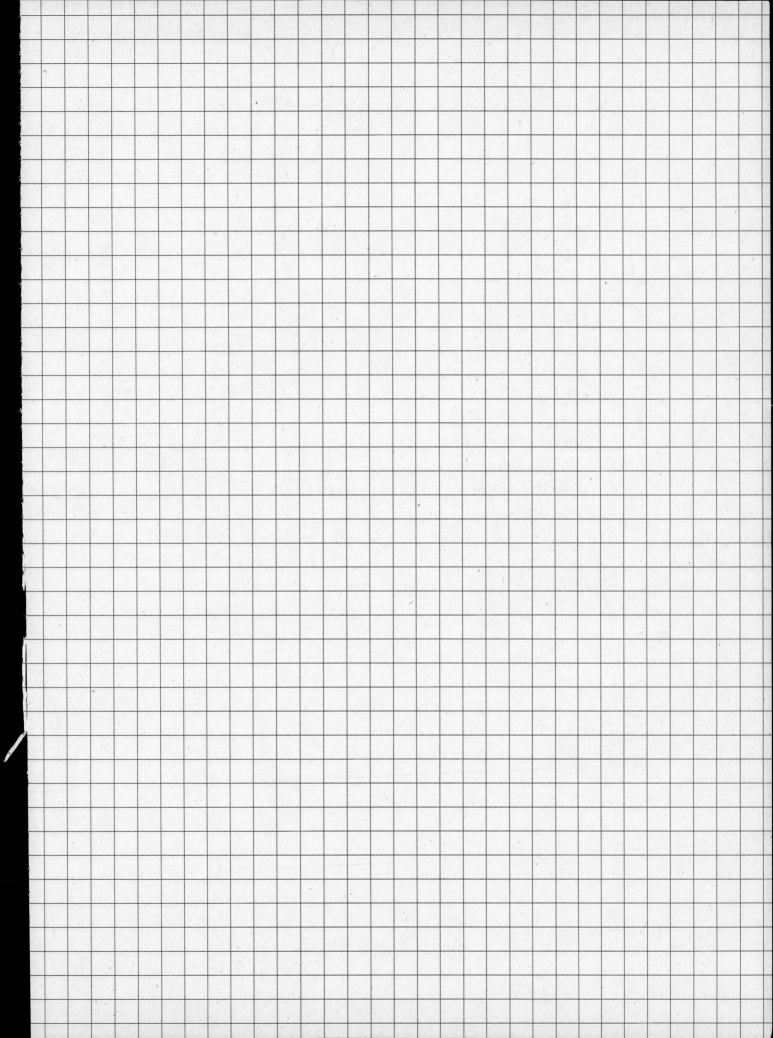

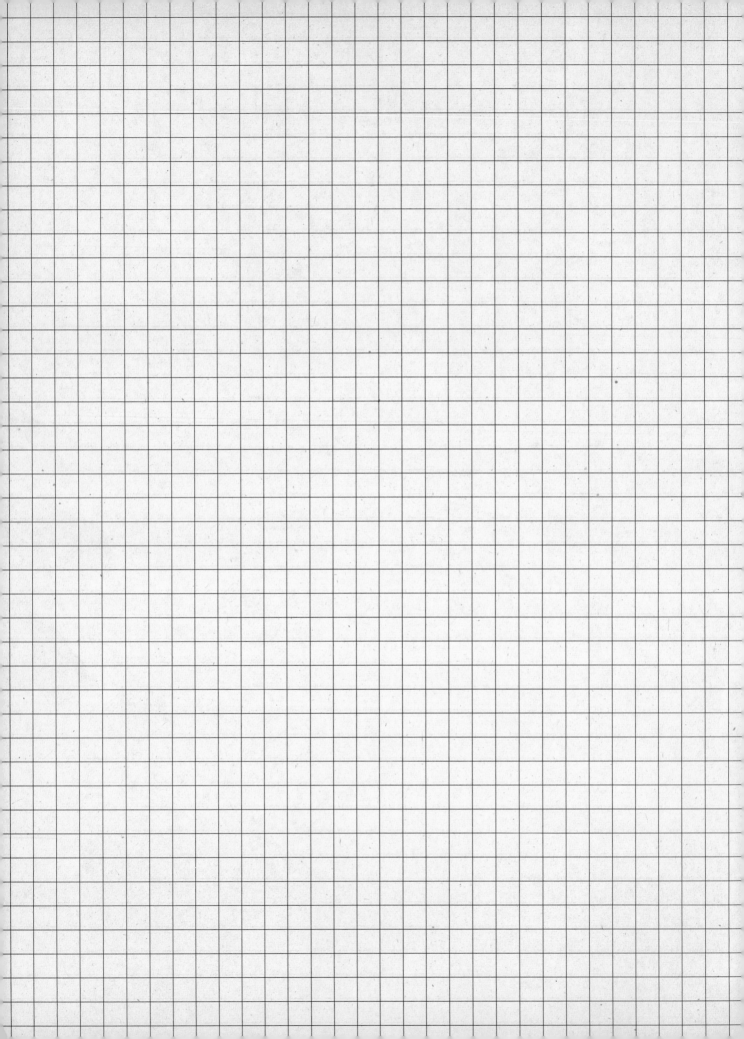